New Perspectives on Aristotle's *De caelo*

Philosophia Antiqua

A Series of Studies on Ancient Philosophy

VOLUME 117

New Perspectives on Aristotle's *De caelo*

Edited by

Alan C. Bowen and Christian Wildberg

BRILL

LEIDEN • BOSTON
2009

This book is printed on acid-free paper.

Library of Congress Cataloging-in-Publication Data

New perspectives on Aristotle's De caelo / edited by Alan C. Bowen and Christian Wildberg.
p. cm. – (Philosophia antiqua ; v. 117)
ISBN 978-90-04-17376-7 (hardback : alk. paper)
1. Aristotle. De caelo. 2. Astronomy, Greek. 3. Cosmology. I. Bowen, Alan C. II. Wildberg, Christian.

QB41.A73N49 2009
520–dc22

2009029137

ISSN: 0079-1687
ISBN: 978 90 04 17376 7

PRINTED IN THE NETHERLANDS

CONTENTS

Indices

PREFACE

This volume is the first collection of scholarly articles in any modern language devoted to Aristotle's *De caelo*. It grew out of series of workshops held at Princeton, Cambridge, and Paris in the late 1990's. The project was one of international collaboration; and in our opinion, the level of insight and sophistication reached could not have been attained by any single scholar. The many advantages of this collaborative effort are, we believe, palpable in this collection; and we hope that this volume will prove an effective tool for teaching and scholarship. Since Aristotle's *De caelo* had a major influence on cosmological thinking until the time of Galileo and Kepler and helped to shape the way in which Western civilization imagined its natural environment and place at the center of the universe, familiarity with the main doctrines of the *De caelo* is a prerequisite for an understanding of much of the thought and culture of Antiquity and the Middle Ages.

Alan C. Bowen
Christian Wildberg
Princeton, NJ
March 2009

ACKNOWLEDGEMENTS

We are very pleased to acknowledge most gratefully Veronica Shi for her assistance and Andrea Falcon for his generous advice in compiling the extensive bibliography that supplements this volume. The entries on the Arabic tradition have been most kindly supplied by Gerhard Endress, who is currently editing the numerous Arabic versions of the *De caelo*. We also thank Henryk Jaronowski and Thu Ta for their help in compiling the indices.

INTRODUCTION

The ancient cosmological myths and theories of Western civilization from Hesiod and the *Bible* to Lucretius and Plotinus profoundly influenced the way in which large parts of humanity have thought and continue to think about themselves, their past, and their future, as well as their place in the universe. For scribes and scholars, scientists and philosophers throughout the ages, two towering cosmological works were of preeminent importance and influence, Plato's "dialogue" *Timaeus* and Aristotle's response to it, the physical tractate *De caelo* (*On the Heavens*). For centuries after their composition in classical Athens, these two works initiated a lively and increasingly sophisticated ancient and medieval cosmological debate. Their profound intellectual influence extended well into the 17th century.

Whereas Plato's *Timaeus* starts out as a dialogue, it quickly mutates into an elaborate cosmological narrative that presents in deceptively traditional form a *mélange* of Platonic speculation with the latest discoveries of mathematics. The narrator, however, presents his account not as a matter of simple fact but as a likely story (εἰκὼς μῦθος, 29c–d). Precisely what this means is a matter of contention. It may mean, for example, that Plato's cosmology fails to count as knowledge because it proffers only approximations, preliminary explanations that wait to be surpassed and improved by the theories of future experts; or it may mean that, according to Plato, *any* cosmology is going to be epistemologically deficient because the objects of the physical world that it deals with are necessarily but fleeting likenesses of certain other items which alone are the objects of epistemic certainty. In any case, one of Aristotle's goals in his *De caelo* (which cannot have been written very long after the *Timaeus* as it bears the hallmarks of an early Aristotelian work) is to formulate a critical response to the "unscientific" character of his teacher's natural philosophy. Aristotle is explicit about his desire to give a rigorous and scientific theory of celestial phenomena broadly conceived—phenomena that are so patently obvious yet at the same time too remote for direct empirical study. However, with the help of observation, speculative common sense, well-chosen analogies, and mathematical reasoning, he strives to offer something that amounts to, or at least promises, the certainty expected of a comprehensive scientific account, and is not just a story of any sort.

Whereas Aristotle's *Physics* offered an in-depth analysis of central notions of any physical inquiry, such as matter and form, causation, motion, time and space, the *De caelo* is born out of a subsequent ambition to understand the universe in its entirety. In fact, the range covered by the inquiry is captured nicely in the ambiguity of its title in Greek, Περὶ οὐρανοῦ. For some parts of the treatise deal with the nature and movement of the sphere of the fixed stars that encompasses the geocentric universe; other parts deal with the planetary movements and other phenomena in the region between the sphere of the fixed stars and the Moon; and other parts again concern the universe as a whole (including the Earth) and the natural motion of all the elementary bodies in it. The same ambiguity is found in the Latin "*caelum*"; hence, the Latin version of the title of Aristotle's treatise, *De caelo*, is felicitous. The standard English translation of the title as *On the Heavens* is, however, only partially successful since our "heavens" does not include the sublunary world, as "οὐρανός" and "*caelum*" may.

A number of works have been published in the last decades on the *Timaeus*, but there are only a few probing studies of, or commentaries on, the *De caelo* itself.[1] It was not, however, our intention in conceiving this collection to offer a comprehensive running commentary on that work. To be sure, the *De caelo* does contain fascinating material bearing on questions of Aristotle's scientific methodology as well as highly influential and perplexing trains of thought the study of which can contribute much to our understanding of Aristotle's philosophy of nature. But there is also much that is plainly straightforward and of less philosophical interest as to warrant comprehensive commentary. So, rather than covering every aspect of the treatise, we thought that it would be useful to provide a collection essays on the *De caelo* that address challenging issues in the *De caelo* by acquainting the reader with some of the latest and most exiting aspects of current scholarship on Aristotle's natural philosophy. Thus, the present collection is designed to provide useful in-depth discussion of some important ideas, or of difficult passages and chapters in the *De caelo*, and thereby to deepen the reader's understanding and critical appreciation of Aristotle's cosmology.

---

[1] We should point out that there is a good translation into English with notes by the late Cambridge scholar W.K.C. Guthrie, published in Harvard's Loeb collection in 1939, and a more recent annotated text and translation of books 1 and 2 by Stuart Leggatt (1995).

Accordingly, in the initial paper of this collection, Thomas Johansen reflects on the fact that the *De caelo* stands as a critical rejoinder to Plato's *Timaeus*. To understand the nature of this response, Johansen undertakes a careful comparison of the two treatises, paying attention first to the scope and subject matter of the two works, and then to the seeming absence of the world soul in the cosmology of the *De caelo*. By examining the role that these two works assign to soul in explaining why the heavens move as they do, Johansen brings to light far-reaching differences regarding the relation of soul and body, the nature of motion and the sense in which motions are said to be natural, as well as the place of teleology in the Platonic and Aristotelian expositions.

Next, Sarah Broadie carefully examines Aristotle's "general" attack on the thesis in Plato's *Timaeus* that the cosmos came into being and will never cease to be. This general or dialectical attack proceeds on two related fronts. In the first, Aristotle argues that what always is (or what always is-not) cannot ever not-be (or be). In the second, he maintains that, if something is γενητόν (*scil.* has come into being, can come into being, or must come into being), it is also φθαρτόν; and that, if something is φθαρτόν (*scil.* has ceased to be, can cease to be, or must cease to be), it is also γενητόν. Though, as she shows, Aristotle's arguments ultimately fail, Broadie's explanation of why they fail brings to light an important moment in the understanding of key modal concepts.

In the third paper, Rob Bolton asks a fundamental methodological question that has puzzled many readers of the *De caelo*, "How can Aristotle on some occasions dismiss arguments that posit and apply explanatory principles without regard for the perceptual phenomena and yet on others offer arguments of just that sort?" According to Bolton, Aristotle is not inconsistent. Rather, in the *De caelo* (as elsewhere in his writings), Aristotle distinguishes scientific argumentation that draws on perceptual phenomena and proceeds to conclusions from well-established causal principles, and dialectical argumentation which does not require detailed, specific knowledge of the subject at hand but draws on common, general ideas that are credible to us and reaches conclusions by logical inference. Yet, as Bolton explains, Aristotle does recognize the limitations and uncertainties of dialectical argument in physical theory and regards it as inferior to properly scientific argumentation. But he still holds that dialectical argument is indeed warranted so long as the relevant phenomena of the heavens are inaccessible to us or when fundamental assumptions of physical theory are at issue.

Next, Jim Hankinson analyzes one of Aristotle's most notorious contribution to cosmological theory, his various arguments for the existence of a fifth elementary body, aether, the nature of which is to move in a circle. Hankinson's aim is to show that there is a way of construing these arguments which does for the most part avoid confusing two critical senses of contrariety lurking in the distinction of the natural and the unnatural: one in which *x* is contrary to *y* as *upwards* is to *Tuesday*, and another in which *x* is contrary to *y* as *upwards* is to *downwards*. In the course of his analysis, Hankinson focuses on the problem posed by Aristotle's assertion in the *Meteorologica* that, when fire reaches its natural place, it is drawn around in a circle by the aether immediately above it. As he goes through the arguments, Hankinson asks whether this motion of the fire-sphere is natural, unnatural, or preternatural for Aristotle.

In the fifth article, Mohan Matthen turns to Aristotle's doctrine of natural place. Specifically, he sets out to determine what Aristotle means when he attributes existence and power to the natural places to which the four elementary bodies (earth, air, fire, and water) move. The problem is that these places are but the innermost, motionless boundaries of what contains the elementary bodies when they are said to be in their nature places, and that place is not a cause. Taking earth as his example, Matthen argues that the claim that earth is what is at rest at the center of the Totality (τὸ πᾶν) entails the claim that earth moves to the center if unimpeded. Given this derivation of the nature of earth's motion from its form (which is static), Matthen then demonstrates that the place at which the earth comes to rest exists in the sense that the innermost surface containing the earth belongs to the Totality, itself a proper substance. He proposes, moreover, that this place to which earth moves by nature has power which is manifested in the (eternal) stability and impassivity of the Earth.

Mary Louise Gill likewise focuses on Matthen's question about the locomotion of the sublunary simple bodies or elements. After a brief exposition of how Aristotle understands the phenomenon of change in general, she turns to how he defines these elements as such and introduces their differences in lightness and heaviness. Then, Gill shows how Aristotle distinguishes the motions of these simple bodies from enforced motions and from the motions of self-movers (e.g., animals). She argues that, in his view, these four internally undifferentiated bodies have only a material nature to move up or down (but not to rest)—thus disagreeing importantly with Matthen—and that the form and final cause of their locomotion (viz. the place to which they will move if unimpeded) is external to them. Gill concludes by shedding light on Aristotle's claim that the four elements, though not substances in the final analysis, are most like them.

In the seventh paper, Pierre Pellegrin examines Aristotle's contention that the universe is spherical in shape and the extent to which this contention is based on perceptual information. But first, as preparation, he states his opposition to the idea that the inquiry undertaken in the *De caelo* is astronomical rather than physical, by noting that the physical theory of the *De caelo* draws on contemporary astronomical theory to confirm or support its conclusions, and then by commenting on some problems that arise when one considers the question of the relation between physical and astronomical theory more generally in the *De caelo* and in the Aristotelian corpus. He also attacks the idea that in the *De caelo* Aristotle downplays the role of observation in the physical theorist's study of the heavens and valorizes reasoning from the study of the sublunary domain. To the contrary, Pellegrin holds, in the physical theory of the heavens, arguments involving direct perception are preferred: since, however, such arguments are not always feasible, the physical theorist is obliged to draw on analogies with items and principles evident in the sublunary world and even to offer dialectical arguments, all as a second best. When he comes to the sequence of arguments for the sphericity of the universe, Pellegrin discerns two theses at issue—that the exterior envelope of the universe is spherical and that the universe is spherical by virtue of its being comprised of a series of spherical shells. He analyzes the interconnections between these two theses as they are presented in each of the arguments, and assesses their cognitive worth by looking at the extent to which each thesis is established using perceptual evidence.

Most readers of the *De caelo* will be puzzled by Aristotle's contention that the heavens have a left and a right. In the eighth paper, Jim Lennox looks carefully at Aristotle's argument for this claim and explains both how and why it draws on the study of the motions of animals in *De incessu animalium*. Lennox concludes by explaining that Aristotle engages in this seemingly peculiar sort of argumentation because of the supreme value he puts on knowledge of the heavens, and because he wishes to counter claims made in the *Timaeus* and by Pythagoreans, and to show at the least how one should approach such questions.

Whereas Lennox focuses on Aristotle's use of his study of the motion of animals to illuminate the directionality of the motions of the heavens, Mariska Leunissen considers Aristotle's adaptation of teleological principles found in his biological works to the cosmology of the *De caelo*.

She argues that this cosmology is intended to be a proper physical science, which entails that it include teleological arguments. She observes that, unlike those in the biological works, the teleological arguments in the cosmology do not mention final causes—the evidentiary basis is too slim to permit that—but rely solely on such principles as the principle that nature does nothing in vain or that nature always chooses the best of a number of possibilities. The upshot, she maintains, is that the teleological arguments found in the cosmology, though fully scientific and rooted in perceptual data, are nevertheless qualified as reasonable (rather than offered as true *simpliciter)* precisely because the final causes are unknown, and thus that these arguments bear all the linguistic markers which Bolton and Johansen take to signify philosophical or dialectical argumentation. She distinguishes two main types of teleological argument in the *De caelo*—those independent arguments in which the *presence* of some feature is explained and those arguments in which the *absence* of some feature is explained—and shows in each case how it draws on teleological principles evident in the biological works and yet differs from the sort of argument found in those works.

The concluding paper by Gad Freudenthal takes the reader in a new direction. To this point, each of the papers has aimed to supply a philosophically sound reading of some difficult passages or issues in Aristotle's *De caelo*. Freudenthal likewise starts with a difficult problem in this treatise, the tension in Aristotle's account between his distinction in the matter and nature of the superlunary and sublunary domains, and his contention that the Sun is an efficient cause of change, including generation and corruption, in the sublunary world. But what concerns Freudenthal is not whether we or he can develop a sensible resolution of this tension, but how and why later medieval thinkers elaborated Aristotle's account of the Sun's causal efficacy to include a providential system in which *all* the heavenly motions share this function, and to do this in way that risks lending support to astrology. His main focus is on Alexander of Aphrodisias and how this great defender of Aristotle adapted and generalized key notions in ways not originally intended to address the question of why the sublunary elements persist. As Freudenthal shows, Alexander argued that this is the best of all possible worlds and that the persistence of species and the presence of soul in the sublunary world is the guaranteed outcome of a divine power (otherwise called "nature") that flows downwards to the sublunary world from the heavenly bodies and is configured by their motions, where these motions are caused in turn by God, that is, the Unmoved Mover *qua* Active Intellect. Freudenthal concludes by showing how Averroes built upon Alexander's argument by further adapting select astrological tenets in order to extend the providential role of the heavenly bodies and their motions in the governance of the cosmos.

To enhance the utility of this volume, we have supplemented it with an extensive bibliography that not only includes the works cited in the papers, but also lists the manuscripts of the *De caelo* along with those of the primary medieval commentaries on this text, as well as scholarly monographs and articles pertinent to the study of Aristotle's cosmology today. Since others may wish to supplement it, we have decided to maintain it online (www.ircps.org/dir/bowen.htm) and to invite our readers to send us updates and corrections as they see fit.

FROM PLATO'S *TIMAEUS* TO ARISTOTLE'S *DE CAELO*: THE CASE OF THE MISSING WORLD SOUL

Thomas K. Johansen*

1. *Introduction*

No work by any other philosopher is cited more often in the *De caelo* than Plato's *Timaeus*.[1] The preponderance of references to his teacher's dialogue on cosmology should come as no surprise: as the foundational work in teleological cosmology, the *Timaeus* is the natural reference point for the kind of natural philosophy Aristotle seeks to develop. Yet, as often in Greek philosophy, debt is the mother of polemic and Aristotle's overt treatment of Plato in the *De caelo* is predominantly negative. I shall not here attempt full coverage of all aspects of the *De caelo*'s engagement with Platonic cosmology.[2] Such a project would require a more comprehensive and detailed reading of the *De caelo* than a single chapter allows. Instead, I shall characterize the relationship between the cosmologies of the *De caelo* and the *Timaeus* by first comparing the general scopes and subject matters of the two works, and, then, by focusing on a particular point of comparison, namely, the apparent absence in the *De caelo* of the world soul, which was one of the most important features of Plato's cosmology.

* Brasenose College, Oxford.

[1] Explicit references include: *De caelo* 280a30, 293b32, 300a1, 300b17, 306b19, 308b4. However, the explicit references to the *Timaeus* are only the tip of the iceberg: often the reference to the *Timaeus* is simply assumed: see, for example, 299a6ff. (on the geometrical construction of solids in *Tim.* 53c4–56c7) and 305b31–306b2 (on the mutual transformation of bodies by their resolution into planes discussed in *Tim.* 56c8–57c6).

[2] While one might sympathize with the ambition of Claghorn (1954) to overcome the polarization of Aristotle and the *Timaeus*, too many of his assimilations fail to represent either philosopher. For comparisons between the *Timaeus* and the *De caelo*, the reader is better advised to consult Solmsen (1960).

2. *The scope of the* De caelo

The topic of the *De caelo* is the heavens or in Greek οὐρανός. There is no need to think that the Greek title Περὶ οὐρανοῦ was used by Aristotle himself.[3] Nonetheless, the title reflects Aristotle's own description of the subject matter as "heaven" in *De caelo* 1.9, 278b11. In characteristic fashion, however, he points to three different senses of the word:

> First, however, we must explain what we mean by "heaven" and in how many ways we use the word in order to make clearer the object of our inquiry. In one sense, then, we call "heaven" the substance of the extreme circumference of the whole, or that natural body whose place is at the extreme circumference. We recognize habitually a special right to the name "heaven" in the extremity or upper region, which we take to be the seat of all that is divine. In another sense, we use this name for the body continuous with the extreme circumference, which contains the Moon, the Sun, and some of the stars; these we say are "in the heavens". In yet another sense we give the name to all body included within the extreme circumference, since we habitually call the whole or totality "the heavens". The word, then, is used in three senses.[4] (*De caelo* 278b9–21)

Is there one of these three senses in which Aristotle primarily meant to discuss heaven in the *De caelo*? The question exercised the ancient commentators considerably in their attempt to identify the unitary theme (σκοπός) of the work. The disagreement that arose between Simplicius and Alexander on this question has immediate bearing on the relationship of the *De caelo* to the *Timaeus*.[5] According to Simplicius, Alexander took the subject matter of the *De caelo* to be the οὐρανός in its third sense, that is to say, the whole world or cosmos. But if so, Simplicius suggests, its subject matter would be the same as that of the *Timaeus*, where Plato (27b2–3) refers to "the whole heaven, or the world (κόσμος) or whatever else it might care to be called." Yet, as Simplicius points out,

> Aristotle clearly does not explain the world in this treatise as Plato did in the *Timaeus*, where he treated both of the principles of natural objects, matter and form, motion and time, and of the general composition of the world, and gave a particular account both of the heavenly bodies and

[3] Cf. Allan's note:

> *Titulus Περὶ Οὐρανοῦ nusquam apud ipsum reperitur; titulum de Caelo et Mundo ab Arabicis exemplaribus translatum crediderim; nusquam enim in codicibus Graecis kai κόσμου adiectum videmus.* (Allan 1955, iii).

[4] Unless otherwise indicated, all translations of Aristotle are from Barnes (1985).

[5] See the introduction to Hankinson (2002).

> of those below the moon, in the latter case occupying himself both with atmospheric phenomena and with the minerals, plants, and animals on the earth up to and including the composition of man and his parts.
> (trans. Hankinson 2002, 21)

The final words here suggest that Simplicius is thinking of *Tim.* 27a5–6, where Critias tells us that Timaeus' account will deal with the creation of the heavens (οὐρανός) "down to and including man." To obtain Aristotle's views on the range of issues touched on in the *Timaeus* we would have to read not only the *De caelo* but also, and at least, the *De generatione et corruptione* (for the composition and interaction of simple bodies), the *De anima* (for the nature of soul and its parts), the *Meteorologica* (on geological and meteorological matters), the *Physica* (on the nature of "nature", time, place/space, motion), the *De partibus animalium, De generatione animalium* and *Historia animalium* as well as the *Parva naturalia* (on the composition and function of animal organs, sexual reproduction, and so forth). The *Timaeus* is a work about the οὐρανός in a much more comprehensive sense than in the *De caelo*. In terms of its scope, then, the *De caelo* is not cosmology in the sense of the *Timaeus*.

However, we may also reflect that the border lines between Aristotle's works, as we have them, are much less clearly defined than those between most of Plato's dialogues. So, whether through the efforts of Aristotle or a later editor, the *De generatione et corruptione* reads at least like a continuation of the end of the *De caelo*, whilst the *Meteorologica*, in turn, seems to pick up from the end of the *De generatione et corruptione*. The *De caelo* appears to be part of a series of interrelated treatises within the area of natural philosophy. Rather than saying that Aristotle has narrowed the terms of cosmology in relation to Plato it may, therefore, be more correct to say that the *Timaeus* telescopes into one dialogue a range of discussions that Aristotle develops in much in greater detail throughout a series of contiguous treatises, one of which is the *De caelo*. To get the full picture of Aristotle's response to the *Timaeus* we need to read much more widely than the *De caelo*.

However, if the *De caelo* does not cover the same range of topics as the *Timaeus*, the reason may not be that Aristotle does not want to study heaven in the third sense of "the all," but rather that he has a different sense of "the all" in mind from the Platonic cosmos. For notice how Aristotle presents the third sense of οὐρανός: "we give the name to *all body* included within the extreme circumference, since we habitually call the whole or totality 'the heavens'" (my emphasis). So if we study the heavens in this sense of "the all" (τὸ πᾶν), we will study the universe *qua* all *body* (τὸ πᾶν σῶμα). That Aristotle wants to study the heavens in this sense of the "all" seems to be confirmed by the opening of the *De caelo*. Chapter 1 of book 1 of the *De caelo* introduces the general subject matter of natural philosophy as "bodies and magnitudes and their properties and movements and the principles of this sort of substance" (268a2–4). Chapter 2 opens by postponing the question of whether the universe (τὸ πᾶν) is infinite in magnitude or not (a question picked up in 1.5). Instead, Aristotle makes a start on the question about the different species of its parts. Aristotle explains that we understand such bodies and magnitudes to be natural in so far as they possess a principle of change: "for nature, we say, is their principle of change" (268b16). That is why we take all natural bodies as such to be capable of change of place, in particular (268b14–16). Aristotle goes on to distinguish three kinds of simple locomotion—up, down and circular, understood respectively as motion away from the center, motion towards the center, and motion around the center (268b16–26). The distinction between these three kinds of simple motion allows Aristotle to distinguish three kinds of body according to whether it is in their nature to move upwards (fire and air), downwards (earth and water) or around (αἰθήρ) (268b26–269a6). Fire and air will later be differentiated by the extent to which they naturally move upwards (fire moves towards the very extreme universe, air to a place adjacent to that), just as earth and water will be told apart by the extent to which they move downwards (312b20–313a13).

We are thus given to understand in the first two chapters that the subject of the *De caelo* will be the universe understood as the totality of bodies or magnitudes and their characteristic motions and affections. This characterization of the subject matter seems to contrast in two striking ways with the opening of Timaeus' account of the cosmos. The first is that Aristotle makes no mention of the universe's overall purposefulness, let alone of a purposeful creator. Timaeus, in contrast, posits as the first principle of the cosmos a creator who wanted to make the universe as good as possible (29e1–30a3). The universe for Timaeus was organized not only with a view to making each individual living being as good as possible but also the whole as good as possible (30c4–31a1). In the *Timaeus*, in other words, we have cosmic teleology.

The second difference is that Aristotle does not mention soul. He describes his subject as body or magnitude and its motions. To be sure, Aristotle mentions (*De caelo* 1.1, 268a3), as part of the study of nature, the principles (ἀρχαί) of bodies, which could include soul. However, at least as far as book 1 of the *De caelo* goes, these principles seem to be internal to the simple bodies as bodies and do not seem to involve soul. Timaeus, in contrast, linked the goodness of the cosmos to its possession of soul. God wanted to make the world as good as possible and, therefore, he gave it mind (νοῦς) and, therefore, also soul (ψυχή) (30b1–6). The two points, the greatest possible goodness of the cosmos and its possession of intelligence and soul, are here closely related. The soul is the master and ruler of the body (34c). Thus, the motions of the world soul control the orderly motions of the stars and the planets and ensure that the universe is in the Greek sense of the word a κόσμος, an ordered whole (40a4–b2). For Timaeus, soul is fundamental to how the cosmos works. Cosmology is based on psychology.

3. *The* De caelo *on soul and teleology*

The omission of the world soul from the first book of the *De caelo*, therefore, suggests a striking departure both from Platonic cosmology and from Platonic teleology. We should of course not infer from this omission that Aristotle thereby means to depart from *any* kind of teleology. Part of what it means to say that the motion of each of the five bodies to their proper place is natural is that each moves towards that place *in order to* occupy it. The theory of natural motions is, therefore, inherently teleological. Yet there is no indication in *De caelo* 1 that there is any role for a soul or intellect to coordinate the motions of bodies into a well-ordered whole or κόσμος in Plato's sense. The *De caelo* seems from the opening to be resolutely a treatise about the various kinds of simple bodies and the motions that they naturally, that is, given the kind of body they are, give rise to. Psychology would seem to be alien to this project.

Aristotle's cosmology has so far emerged as a more restricted project than the Platonic precedent in two ways. First, as Simplicius pointed out, the *De caelo* does not pursue cosmology in the broad way that includes, amongst other matters, the details of how living beings are put together and work. But second, even granting that the *De caelo* studies the heavens in the sense of "the all" (τὸ πᾶν), it appears to focus on the all *qua* bodily. Related to this second limitation, the *De caelo* seems to bypass the study of soul and the accompanying cosmic teleology which were fundamental to Timaeus' cosmology.

One possible explanation of the difference between Aristotelian and Platonic cosmology is that in Aristotle nature takes over from soul as an internal principle of motion and rest. Socrates in the *Phaedrus* introduced soul as "the origin and principle (ἀρχή) of motion for other things that are moved as well" (245c8–9). There is a famous crux relating to the compatibility of the *Phaedrus* and the *Timaeus* on the question whether *all* motion (including the pre-cosmic variety) is caused by soul (cf. Skemp 1967, 112–115). However, it seems clear that soul also in the *Timaeus* remains a principle (ἀρχή) with the power to move itself and bodies (34b10–35a1, 36e2–4). So, Aristotle's talk of nature as an inner principle of change and motion, and his ascription of such natures to inanimate bodies, might seem to be a direct move against Plato. One possible answer to the question, "What has happened to Plato's world soul in the *De caelo*?," is, then, "Aristotle does not need it anymore; the 'natures' of the simple bodies do the job of self-motion."

However, this comparison between the roles of soul in Plato and nature in Aristotle may be misleading. On the one side of the comparison, Aristotle is elsewhere leery of ascribing self-motion in a strict sense to anything other than animate beings (*Phys.* 8.4, 254b12–255a20). The simple bodies do not properly speaking have self-motion since they do not have the power to stop themselves and can only move in one way (255a7–11). So, if the *De caelo* suggested that the simple bodies have unqualified self-motion, this suggestion would be in a contrast not only with the *Timaeus* but also with *Physics* 8. In fact, as M.L. Gill shows in this volume (2009, esp. 150–158), the position of the *De caelo* may plausibly be taken as the same as that of *Physics* 8: Aristotle never quite ascribes proper self-motion to the simple bodies. On the side of the *Timaeus*, meanwhile, nothing in this text indicates that the simple bodies need soul in order to move.[6] The simple bodies were originally created by the divine demiurge and in that sense owe their motions to something external. But once created they move by necessity in space (χώρα) given their geometrical natures.[7] To be sure, the world soul may be indirectly involved in their *continued* motion. The rotation of the cosmos, which is governed by the world soul (cf. 34b, 36d–e), shakes up the bodies inside, thereby preventing them from staying in their own proper regions (58a–b). However, the simple bodies by themselves move towards their

[6] For the suggestion that the simple bodies in the pre-cosmos are moved by an irrational soul, cf. Skemp (1967) 106 (following Plutarch). For an alternative answer, see Johansen (2004) 96.

[7] For a full account, see Johansen (2004) 95–99, 120–127.

proper places without psychic guidance. Indeed, likenesses of the simple bodies moved towards certain regions (53a) even before the creation of the cosmos, and so also before the existence of a world soul. If we enter the necessary qualifications and take into account the *Physics*, it is not clear that the simple bodies for Aristotle are self-movers in a significantly stronger sense than in the *Timaeus*.

The part of the *Timaeus* (47e–69a) that sets out the nature of the simple bodies and their characteristic motions is sometimes—somewhat inaccurately—referred to as the "Works of Necessity."[8] This part of the dialogue is sandwiched between a part which primarily deals with the works of reason, and the final part of the dialogue, which, building on the "Works of Necessity," develops an account of the functions of human beings and other animals. The works of necessity are presented as the building materials and auxiliary causes for the divine craftsmen (68e). The focus in the middle part of the dialogue is on the physical constituents of the universe and on the necessary processes and affections that they give rise to. The demiurge does construct the natures of the simple bodies and ensures their proportionality with each other (53b6, 56c3–7), but from then on Timaeus describes the interactions of the bodies as necessary given their natures. The approach to cosmology is here, as one might put it, bottom-up, whereas in the rest of the *Timaeus* it is top-down. From 56c onwards, no use is made in the "Works of Necessity" of any teleological cause, be it god or world soul, which would organize bodily motions towards the good.

It is tempting to insist that the relevant comparison is between the *De caelo* and the "Works of Necessity" specifically. One might then argue that the limitations of the *De caelo* reflect Aristotle's narrower concern in this work, shared by Timaeus in the "Works of Necessity," with the material building blocks of the universe. This is not to say, of course, that there are no significant differences between the ways in which Plato and Aristotle account for the simple bodies. Aristotle's most fundamental disagreement with Plato concerns the composition of the simples. Whereas Aristotle analyzes them in terms of qualities such as hot and cold, wet and dry (cf. *De caelo* 3.4, 302b30), Plato understands their composition geometrically. Aristotle argues that Plato's view faces insuperable difficulties: the geometrical theory cannot, for

[8] After Cornford (1937). The reference is somewhat inaccurate in that this part also contains the description of god's ordering of the simple bodies.

example, account for how the simple bodies change into one another (3.7, 305b27 ff.) or for how the different kinds of body come to have different weights (308b4–28).

These are important differences between the *De caelo* and the *Timaeus*, which Aristotle exploits to full polemical effect. However, the disagreement does not concern the role of soul in the motion of the four simple bodies. Both Aristotle and Plato take it that these bodies move to their proper places of their own accord. To be sure, Plato does not explain these motions, as does Aristotle, in terms of the concept of "nature." Yet, as the winnowing basket analogy implies (*Tim.* 52e5–53a7), the motions of the simple bodies to their proper places does express the peculiar character of each kind of body. Certainly, like the winnowing basket, the receptacle aids the distribution of the simples, but what ultimately determines which kind of body ends up where are clearly the bodies' own properties, just as the grains in the winnowing basket are distributed in accordance with their weight. As part of Aristotle's ongoing polemic in the *De caelo* against the view of the *Timaeus* that the world was created,[9] he remarks that if the simple bodies moved to their proper places even before the creation of the world, then their motions must also then have been natural (i.e., not constrained). But if there were bodies with natures even then, there must have been a cosmos even then (3.2, 300b17). Aristotle thus suggests that Plato explained the motions of the simple bodies as if they had natures of the sort that, according to Plato's own theory, should only exist *after* the creation of the world. In other words, Aristotle's complaint is that Plato implies that the motion of the simple bodies is caused by their proper natures even before the simple bodies were supposed to have such natures. We can see in this objection how close, according to Aristotle, Plato is to helping himself to something like Aristotle's own notion of "nature."

The claim that Aristotle's concept of nature replaces the Platonic soul as a self-mover is thus misleading as far as the four simple bodies go. Soul plays no part in explaining how the four simple bodies move to their proper places in the central part of the *Timaeus* However, the real disagreement about the role of soul in explaining motion seems to arise when we turn from the sublunary world to the outer heavens. Again there are certain initial similarities. The shape of the cosmos as a whole is spherical according to the *Timaeus* (33b), while Aristotle chooses the

[9] See, in particular, *De caelo* 1.10–12.

same shape for his cosmos for parallel reasons (2.4, 286b10–287b21).[10] Both Plato and Aristotle use the sphericity of the universe in support of the claim that its motion is circular (*Tim.* 34a1–2, *De caelo* 2.6). Both characterize the circular motion of the heavens and, in particular, that of the outer heavens as "unceasing"[11] and "uniform."[12]

Notwithstanding their parallel descriptions of the outer heavens, Plato and Aristotle seem to disagree about the cause of their motions. Where Plato takes the heavens to move round because of the world soul, Aristotle takes the revolution to be brought about by a fifth, or as he sometimes calls it, a "first body" (e.g., 284a30), namely, αἰθήρ. The outer heavens are composed of this body, which naturally moves round, just as fire by its nature moves up, and earth moves down (*De caelo* 1.3). Since, as we saw, none of the other simple bodies require soul in order to move, it seems that the introduction of the "first" body may be meant to dispense with the need for a world soul to account for the revolutions of the heavens.

It is in this vein that Friedrich Solmsen writes:

> The first body takes over many functions of the Platonic world soul, but it can also fill the celestial regions in a material sense (which the world soul cannot do), being at once the stuff of which the heavenly bodies consist and the cause of their revolutions. Plato erred in identifying the origin and agent of circular movement. The difference between this movement and rectilinear ones is not that the former is typical of soul and the latter movement is performed by bodies, but that the body engaged in the circular movement is of a superior order, immune to the changes and mutual transformations to which the others are liable.
>
> (Solmsen 1960, 291)

In one way, the first body clearly represents an anti-Platonic stance. Plato takes soul to be both moving and incorporeal. For Plato, the motions of the world soul are not to be understood as the actualization of some bodily potential. Rather, to go by the account of the soul's composition in the *Timaeus* (35a–37c), the world soul seems to move and move the way it does because it is a mixture of being, sameness, and difference, which is mathematically structured. We are told, moreover, that the world soul was created before the world body (34b–c). In the *Timaeus*, there seems, therefore, to be a time in the creation of the world when it is both the

[10] Cf. P. Pellegrin (2009), in this volume.

[11] ἄπαυστον in *Tim.* 36e4, *De caelo* 279b1: cf. συνεχής in *De caelo* 287a24.

[12] κατὰ ταὐτά, in *Tim.* 36c2–3, cf. *Laws* 898a8 ff.; ὁμαλής in *De caelo* 287a24, cf. *Phys.* 228b15 ff., 265b11 ff.; cf. Solmsen (1960) 238.

case that the soul exists and revolves *and* that there is no body attached to these revolutions. Aristotle finds such a view unintelligible. "There is no motion," he says, "without a natural body" (*De caelo* 279a15–16). Motion (κίνησις) is the actuality of the potential *qua* potential (*Phys.* 201b5–6) and potentiality implies matter. Once we have identified a peculiar kind of motion of the heavens it is therefore necessary, at least if this motion is natural, to find some matter whose potential is actualized in this motion. For Aristotle, αἰθήρ is this matter.

However, the fact that for Aristotle the heavenly revolutions manifest the natural potential of αἰθήρ does not by itself mean that these motions are not also a manifestation of soul. From what we have seen so far, it may come as a surprise that Aristotle does speak of the outer heavens as animate and alive:

> Since we have already determined that functions of this kind belong to things which possess a principle of movement, and that the heavens are animate and possess a principle of movement, clearly the heavens must also exhibit above and below. (*De caelo* 2.2, 285a27–30)

Again at *De caelo* 2.3, 286a7–12, Aristotle implies that the heavens are animate. He takes it that the heavens are divine and as such enjoy everlasting *life* (ζωή). It is for that reason, he suggests, that the heavens are made of a body which is by nature such as always to move in a circle.

But now we want to know what role soul plays in relation to αἰθήρ when the heavens move round. We are of course familiar from the *De anima* (cf., e.g., 2.2, 413a20 ff.) with the idea of the soul as a principle of movement of living beings, an idea that Aristotle seems to be referring to in the passage just quoted (*De caelo* 2.2, 285a27–30). Yet given that the αἰθήρ of the heavens itself has the capacity to move round, it is unclear what difference the addition of the world soul makes to its movements. One possibility is that soul actualizes or triggers αἰθήρ's natural potential for circular motion. It is fairly clear that Aristotle in the *De caelo* does not think of soul in relation to αἰθήρ in this way. It is the nature of αἰθήρ to move round just as it is the nature of fire to move up if unimpeded (cf. *Phys.* 8.4, 255b5–12). In the sublunary world there are possible obstacles to the upward motion of fire but there are no superlunary obstacles to the revolutionary tendencies of αἰθήρ.[13]

[13] As Judson (1994) 158 observes, as far as the *De caelo* is concerned, "[Aristotle] takes αἰθήρ's natural capacity for motion not to be a 'dormant' at all, but to be a sufficient explanation of the sphere's rotation."

Another, more promising way of thinking of soul's agency is as a regulatory cause. Αἰθήρ moves round of its own accord but soul determines the manner of its revolutions.[14] We might think of the relationship between soul and αἰθήρ by analogy with the relationship between soul and the fire within us in nutrition. So in *De an.* 2.4, Aristotle explains that:

> Fire is in a sense the helping cause (συναίτιον), but it is certainly not responsible without qualification. Rather the soul [is responsible]; for the growth of fire is unlimited (εἰς ἄπειρον), as long as there is something that can be burned. But all things constructed naturally have a limit (πέρας) and proper proportion (λόγος) of magnitude and growth. And these things are the responsibility of soul, but not of fire, and of the formula (*logos*) rather than the matter. (416a13–18: trans. Gill 1994, 21)

Fire contributes as a helping cause, as a συναίτιον, to nutrition by burning the food. We could even say that fire is responsible for the fact that the food is burned. Yet it is not fire as such that is responsible for the way in which the food is burnt, since fire would continue consuming the food as long as any was available. But there is a limit or proportion (λόγος) to the burning of the food *in nutrition*, and it is soul which imposes this limit. The relationship between fire and soul in nutrition suggests the following analogy with the relationship between αἰθήρ and soul in locomotion. It is αἰθήρ that is responsible for the fact that the heavens move round, but it is soul that is responsible for the way in which they move round. More particularly, on this analogy soul imposes a limit or proportion on the revolutions.

The kind of order imposed by soul on αἰθήρ will of course be rather different from that imposed by soul on fire in nutrition since we are now dealing with a different kind of change, locomotion rather than growth and diminution. However, the passage from *De caelo* 2.2 quoted above gives a good example of how the analogy with nutrition would work. Aristotle mentions the heavens' soul here because he wants to explain the direction of the heavenly revolutions. "Left and right," "up and down," "front and back" are for Aristotle said only with qualification. We can

---

[14] I have been inspired in the following by Judson (1994)160:

> The argument of II.2 may be that, over and above its circularity, the motion of the outermost sphere has a "structure" or form that cannot be explained by reference to the αἰθήρ but that mirrors, and can be explained by, the form of the being whose matter the αἰθήρ is. More precisely, the circular motion possesses a particular direction (which Aristotle analyzes in terms of right/left, above/below).

distinguish at least two ways in which we talk about directions. We talk about right and left, back and front, and up and down in relation to ourselves as observers. However, in animate beings, Aristotle claims, we also talk about left and right, up and down according to functions (ἔργα). Again *De an.* 2.4 provides a good example.[15] The roots of a plant seem to grow down in relation to a normal upright-standing observer, as well as in relation to the universe as a whole. However, in terms of nutrition, up is where the food enters, down where the waste exits. So, functionally speaking, for plants the roots are up. Similarly, Aristotle argues in *De caelo* 2.2 that we can distinguish up and down and right from left in locomotion as follows:

> up is the part from which (ὅθεν) movement originates, the right the part from which (ἀφ' οὗ) it proceeds, and the front the direction in which it is directed (ἐφ' ὅ). (285a23–24)

In functional terms, right is, therefore, where the heavenly motion proceeds from. Aristotle distinguishes between a primary and a secondary heavenly motion. He takes the primary motion to be that of the outer heavens, the stars. The outer heavens revolve from the right to the left and back to the right (ἐπὶ δεξιά),[16] and from the top to the bottom. Inside the primary motion, there is also a secondary motion going from left to right (and back again) to which the planets are subject.

For Aristotle, it is because the heavens have soul that we can differentiate directionality by function.[17] Such a functional account of the heavenly motions may itself be taken to be teleological: the heavenly motions of the outer planets are organized by soul *so as* to move from right to left, from up to down. However, there is a further teleological aspect to the account of the heavenly motions as animate which introduces a more

---

[15] As Lennox (2009) shows in this volume, the *De incessu animalium* is the primary text for the functional differentiation of directions in animals.

[16] When Aristotle talks of the motion of the stars as beginning from the right and rotating to the right (ἐπὶ δεξιά), we should probably think of this motion as running counter-clockwise; cf. Moraux (1965) xciii, n. 5 on the ambiguity of the expression.

[17] *De caelo* 284b30–34:

> One should not seek the up and down and the right and left and the front and back in all body, but only in those that possess in themselves a principle of movement because they are animate; for in no inanimate body do we see the part from which the beginning of the movement derives. (trans. Leggatt 1995, 117)

See further the chapter by Lennox in this volume for a careful discussion of this claim and its implication for the explanatory relationship between cosmology and zoology.

explicit value judgment. Aristotle takes the right, up, and front to be *better* than their opposites, left, down, and back. *De caelo* 2.2, 284b14 refers us, in particular, to the *De incessu animalium*,[18] which says that the starting point of locomotion is honorable, and that "the upper is more honorable than the lower, the front than the back, and the right than the left" (706b12–14). Enjoying as they do the best life, it is not surprising, therefore, that the outer heavens display the more honorable motions from right to left and from up to down, whilst the somewhat inferior planets are also subject to the inferior motion from left to right.[19]

Plato would agree that rotation to the right is better. At 36c6 Timaeus says that the single most rational motion of the Same, which governs the heavenly motions, moves "to the right (ἐπὶ δεξιά)."[20] Moreover, Timaeus, like Aristotle, makes the revolutions of the seven planets subject to a further motion in the opposite direction. The less rational circle of the Different, which moves inside and is governed by the circle of the Same, is thus divided into seven circles running in different directions at different speeds (36d2–7). Beyond this it is hard to make clear comparisons between the *De caelo* and the *Timaeus*, for example, on the relative speeds and courses of the heavenly bodies. Aristotle refers such matters to astronomy (*De caelo* 2.10, 291a31), while Timaeus postpones discussion of the details of the planetary motions to a later occasion (38e). However, the distinction between the superior motion of the stars "to the right" and the inferior contrary motion of the planets seems common to both.

So far we have seen Aristotle make use of soul when he wants to explain the direction of the heavenly revolutions, a phenomenon which was not accounted for simply by the fact that the heavenly bodies are composed of αἰθήρ. Because the heavens have a soul, we can apply to them the functional notion of directionality that we use to explain animal motion. And when we talk about "up and down," and "left and right" in functional terms, we are also led to think of one of each of these pairs as better and the other as worse. The deployment of soul and that of teleology go hand in hand.

---

[18] See Lennox (2009) in this volume for a full discussion of the *De caelo*'s debt to this work.

[19] As Bolton (2009) 56f. shows, this argument has the status of being no more than reasonable (εὔλογον), as opposed to a stricter form of argument based on perception, no doubt because an argument about a subject matter so remote from us could be no more than εὔλογον, cf. Bolton (2009) 58–64.

[20] Note, however, that according to Moraux (1965) xcv, n. 5, the motion ἐπὶ δεξιά is taken to run clockwise in the *Timaeus*.

Aristotle again makes explicit reference to soul in his explanation of the relative complexity of the motions of the various heavenly bodies. Soul is brought in to deal with the ἀπορία of "why we find the greatest number of movements in the intermediate bodies, and not rather, in each successive body a variety of movement proportionate to its distance from the primary motion" (*De caelo* 2.12, 291b29–32). The primary motion is here that of the outer heavens, that is, of the stars. The outer heavens move with a single uniform motion, from right to left to right, as we have seen. We might expect those planets that are closest to the outer heavens to move in a manner most similar to these; that is to say, with the fewest number of motions.[21] Similarly, with the Earth at the center of the universe, we might expect the motions nearest the Earth, and furthest from the fixed stars, to be the most complex. Yet, Aristotle claims, the movements of the Sun and the Moon are fewer than those of some of the planets closer than these to the outer heavens (291b35–292a1).

The ἀπορία is solved, Aristotle argues, by thinking of the heavenly bodies as animate:

> We think of the stars as mere bodies, and as units with a serial order indeed but entirely inanimate; but we should rather conceive of them as enjoying action (πρᾶξις) and life (ζωή).[22] (292a18–21)

It is tempting to think that Aristotle by "we" here means to reflect not just on the thinking that generally leads people to the ἀπορία, but also on the wider tendency in the *De caelo* (which I noted at the outset of this chapter) to treat the universe as a mere body. This approach, we are now told, leaves an explanatory deficit in the account of the heavenly

[21] Aristotle is thinking here of the number of interconnecting moving spheres needed in order to produce the observed motion of a planet; cf. Leggatt (1995) 246:

> A planet's apparent movement is explained as the resultant of several movements in the following manner: the planet is fixed to a rotating sphere, the poles of which are affixed to a second, encompassing and rotating sphere; this second sphere is in turn attached by its poles to a further sphere, and so on.

[22] Ἀλλ' ἡμεῖς ὡς περὶ σωμάτων αὐτῶν μόνον, καὶ μονάδων τάξιν μὲν ἐχόντων, ἀψύχων δὲ πάμπαν, διανοούμεθα· δεῖ δ' ὡς μετεχόντων ὑπολαμβάνειν πράξεως καὶ ζωῆς. Leggatt translates ὡς in 292a20 as "as though", with the implication that the stars do not in fact have soul. However, linguistically, the parallel with the first occurrence of ὡς (a18) suggests that we should understand ὡς rather in the sense of "as" or "*qua*." We have been thinking of the stars only as bodies so far, now we should think of them *qua* ensouled. The attribution of soul to the stars is thus made no more counterfactual by the second use of ὡς than the attribution of body to the stars is by the first use. J.L. Stocks' translation is correct (Barnes 1985, 481), as is that of Lennox (2009) in this volume.

motions. We need the concept of soul to explain the relative complexity of the various heavenly motions. Each living being seeks by its actions to achieve its proper good. However, different kinds of living being pursue this aim in different ways. Higher animals typically display more complex actions than lower animals, with the exception of the very highest life-form, which enjoys the good with no effort, like a man who enjoys good health without need of exercise. At the bottom of the hierarchy are plants, which "perhaps have little action and of one kind only" (292b7–8). Aristotle suggests that we should think of the planets' motions in the same way.[23] The most complete motion is that of the fixed stars, which is simple and effortless. The planet farthest from the outer heavens and closest to the Earth has, like plants, the fewest actions, whereas those closer to the outer heavens have the most complex motions, like man and the higher animals. For, Aristotle explains, "while it is clearly best for any being to attain the real end (τέλος), yet, if that cannot be, the nearer it is to the best the better will be its state" (292b17–19)

Again, in this argument, we see the close connection for Aristotle between the ascription of soul to the heavenly bodies and the teleological explanation of their motions (cf. Lloyd 1996, 171). The complexity of their motions is explained by the different actions required for them to best attain their ends. The possibility and appropriateness of the planets' engaging in actions with these degrees of complexity are, in turn, explained by the fact that they have souls.

For Aristotle, then, we need to say that the heavenly bodies have a soul and not just a body to explain several things about how they move. The fact that they are made of αἰθήρ explains that they move round, but it does not explain *how* they move round, their direction or their courses. To explain how they move round in the proper teleological way, that is to say, in the way that shows why *these* motions are the best, we need to think of the heavenly bodies as beings with a soul, which because of their souls live different kinds of lives.

It is clear that for Aristotle the soul operates on the inherent ability of αἰθήρ to move in a circle. The αἰθήρ of the heavenly bodies revolves independently of soul, yet soul can be said to impose a certain form and structure on its revolutions. Soul can to that extent be thought of as a formal cause of the heavenly motions with αἰθήρ as the material cause.

---

23 The general teleological nature of the argument may contribute to its being no more than εὔλογον, cf. Bolton (2009) 68 n. 23

In a similar way we saw that whilst fire would burn foodstuff whether or not it was part of a living being, soul would in nutrition impose a limit and proportion to the way the fire burned. Soul and αἰθήϱ might then with justification be thought of as related as form to matter in the revolutions of the heavens.

In the *Timaeus*, meanwhile, the soul is created separately from the body, inserted into the body from without, and moves it in a way that seems not to express any inherent potentiality of body.[24] For Timaeus there is no special body with the natural potentiality to move round and, therefore, no special first body, such as αἰθήϱ, for the heavenly bodies to be made out of. Rather, the heavenly bodies, or at least their majority, as Timaeus puts it, are made out of fire,[25] whose natural motion is rectilinear,[26] not circular. When the soul is put in charge (36e3), we might therefore think that it imposes its circular motions on the body of the universe.

It is for this reason, I take it, that Aristotle gets the impression (*De caelo* 2.1, 284a27) from Plato that the continuous circular motion of the outer heavens is enforced on the heavenly bodies by soul:

> Yet nor is it reasonable that it [the outer heavens] is preserved by soul constraining (ἀναγκούσης) it to remain everlasting; for such a life of the soul cannot be free from pain or blessed; since [the soul] is in fact accompanied by force in its movement, if the soul causes the first body to move when the latter is of a nature to move otherwise, and cause it to move continuously, then this soul must be without rest and deprived of all relief of mind (ἔμφϱονος). (284a27–33: trans. Leggatt 2002, 115)

Aristotle seems to have in mind two passages in the *Timaeus* in particular: (a) 34b8–9, where Timaeus describes the life of the universe as that of a happy god (εὐδαίμονα θεόν); and (b) 36e2–5, where Timaeus says that, when the world soul had been added to the world body,

[24] Notwithstanding *Tim.* 34a1–4: here circular motion is said to be proper to the world body because of its spherical shape, not because of its material constitution.

[25] Cf. 40a2–4: τοῦ μὲν θείου τὴν πλείστην ἰδέαν ἐκ πυϱὸς ἀπηϱγάζετο, ὅπως ὅτι λαμπϱότατον ἰδεῖν τε κάλλιστον εἴη. 45b4–5 mentions light as a particularly gentle sort of fire. However, even if we imagined that the Sun were made out of this kind of fire, this would still not count as a fifth kind of body, like αἰθήϱ, but rather as a species of fire. The claim at *Tim.* 54c4–6 that there was a fifth kind of geometrical body on which the demiurge modeled the universe seems irrelevant here, since the fifth body is not mentioned as a material constituent in the universe.

[26] Cf. *Tim.* 43b–c, where the four bodies bombard the circular soul with the six rectilinear motions.

> the soul was woven right through from the center to the outermost heavens, which it enveloped from the outside and, revolving on itself, provided a divine principle of unending and rational life (ἔμφρονος βίου) for all time.

Aristotle's objection is that unless the body that the soul moves has itself a natural circular motion, soul will have to enforce this motion on the body. Within the schema that Aristotle adopts in *De caelo* 1.2, any motion that is not natural to a body is unnatural to it (cf. *De caelo* 1.2, 269a30–b13). But Plato does not present the circular motions as natural to the matter of the heavenly bodies. Aristotle therefore infers that the circular motion must be unnatural to the heavenly bodies. Soul can then only move the heavenly bodies round by forcing them to do so, an effort that is inconsistent with its enjoying a happy and quiet life.[27] In sum, Aristotle's objection is to the *specific* idea that the soul moves the heavenly bodies by force, an idea which seems to follow from Plato's making fire, in particular, the matter of the heavenly bodies.[28] Aristotle is not objecting to the *general* idea that there is a soul which moves the heavens, as some scholars have claimed.[29] For the idea that

[27] It may be that Plato sought to anticipate this kind of objection at 34a5, where, even before the introduction of soul into the account, Timaeus gives a circular motion to the body of the universe as appropriate to its shape. However, this is presented as the result of demiurge "taking away" (ἀφεῖλεν) the six rectilinear motions (34a3–5), which one might take to be its natural motions given its composition of fire, earth, water and air. And Timaeus is talking here about the body of the universe as a whole rather than that of the heavenly bodies, which is a rather different matter.

[28] *De caelo* 2.7, 289a17 may contain a reference to the *Tim.* 40a2–4.

[29] Cf. in particular Solmsen (1960) 244, n. 73:

> *De Caelo* II (284a27ff.) includes a scathing rejection of the notion that a world soul should be responsible for the eternal duration of the world. To my mind the rejection is unqualified and unambiguous, and it should not be argued that Aristotle merely protests against a soul working by *ananke*. Rather what he appears to say is that soul, if credited with such a function, would have to operate by *ananke*. Elsewhere the same book (2.285a29) asserts ὁ οὐρανὸς ἔμψυχος, a significant statement, no doubt, and yet, in view of its uniqueness, its significance should not be overworked. It should not be interpreted as saying that the Cosmos has a soul but merely that it is "alive" (so Guthrie) or "animate", which is anything but trivial. To be sure, Aristotle does not otherwise know of entities that are alive but have no soul, yet we may safely leave the responsibility for this contradiction to himself.

I think we are better advised, if possible, not to ascribe the contradiction to Aristotle. The question may come down (again) to what we mean by οὐρανός in the expression ὁ οὐρανὸς ἔμψυχος. If we mean the outer and inner heavens, the οὐρανός is clearly ensouled, if we mean the entire cosmos, it is not. For more on the distinction see below. I tend to agree with Judson (1994) 159, n. 19, that the reference at 285a29 is to the heavens, though I am not sure that reference is to the *outermost* heavenly sphere exclusively.

the heavens are ensouled is, as we have seen, crucial to Aristotle's own teleological explanation in book 2 of the direction and complexity of the heavenly motions.

4. *Conclusion*

I have argued in this chapter that, despite an emphasis on the bodily constituents of the universe in the *De caelo*, a heavenly soul emerges in the *De caelo* at crucial points where Aristotle wants to offer a certain kind of teleological explanation of the directions and courses taken by the heavenly motions. Aristotle's use of the heavenly soul here looks like a continuation of those passages in the *Timaeus* where Plato accounts for the various revolutions by means of distinctions within the world soul. For Timaeus, we recall, the heavenly motions of the world soul were the prime examples of good order in the universe. It is conspicuous that Aristotle brings in the soul precisely in those contexts where he wants to show that the heavens move in the ways they do because these are the best ways. Despite their different views of the composition of the heavenly bodies, it seems that both Aristotle and Plato would have us believe that the excellence of the heavenly motions ultimately requires us to think of them as animate.

I have been concerned to show that both Plato and Aristotle use notions of soul to give teleological accounts of the heavenly motions. Yet, given this shared background, we can also identify some important differences. We refer to the soul of the *Timaeus* as the "world soul" or the "cosmic soul." This description goes together with the claim that the cosmos as a whole is a single animal (30d1–31a1, 92c5–9). The soul of the heavens is, then, the soul of the entire universe. The *De caelo*, in contrast, never treats the universe as a single ensouled being. In the *De caelo*, the outer heavens, the circle that defines the motions of the stars is ensouled, leading as it does a blissful and eternal life. We are also led to believe that the planets each have a soul in so far as they, like humans, animals and plants, engage in the various kinds of action (πρᾶξις) typical of them. However, there is no indication that these various living beings share a soul that can be described as the world soul, let alone that that they together constitute a single animal. There is, therefore, in the *De caelo* no notion either of a soul that might coordinate all of the activities in the cosmos. That is not to say that a picture of a world order does not emerge from the *De caelo*; it is only to say that there is no world soul that is responsible for any such order. The absence of a soul for the entire world seems to go hand-in-hand here with the absence of any explicit cosmic teleology in the *De caelo*.

Secondly, the *De caelo*, unlike the *Timaeus*, makes little or no use of the world soul to suggest how we should live our lives. Both works present the life of the outer heavens as a prime case of rationality and happiness. However, whereas the *Timaeus* emphatically tells us to imitate this paradigm so that we may become happier and more virtuous (39b–c, 42b–e, 47a–c), the *De caelo* stays clear of any such overtly moral messages. We can certainly discern a *scala naturae* in passages such as *De caelo* 2.12, which implies that the simplicity of a stellar life is superior to the complex endeavors of humans, but Aristotle does not invite us in the *De caelo* to imitate the divine, as he does for example in *Eth. Nic.* 10.[30] Indeed, the *De caelo* makes the celestial life seem distant and hardly attainable for human beings. The complexity of the human practical life contrasts with the effortless simplicity of the outer heavens, which are changeless, eternal, and able to enjoy the best and most independent of lives (*De caelo* 1.9, 279a20–23). The heavens are made of a unique material, "something beyond the bodies that are about us on this Earth, different and separate from them," the "superior glory" of whose nature "is proportionate to its distance from this world of ours" (269b14–17). For Timaeus, in contrast, the human soul has the same ingredients (if slightly less pure) and same mathematical structure as the world soul. Like the world soul, the human soul contains a circle of the Same and a circle of the Different, which move in the same regular way (41d4–7), or at least they did so until we were embodied (43c7–e4). Our souls spent their early days riding round the universe on a star (41d9–e3); and even now, each time we succeed at astronomy, we reproduce in our souls the motions of the heavens. For Timaeus our lives can, in more senses than one, be heavenly again. Astronomy in the *Timaeus* offers a divine template for the ethical life, a notion which Aristotle picks up in his ethics but steers well clear of in the *De caelo*. Here, as on other subject matters, Aristotle divides what Plato's cosmology united.

[30] On Aristotle's debt to the *Timaeus* in the *Ethica Nicomachea*, see Sedley (1997).

Bibliography

Allan, D.J. (1955) ed., *Aristotelis de caelo* (Oxford).
Barnes, J. (1985) ed., *The Complete Works of Aristotle: The Revised Oxford Translation* (Princeton).
Bolton, R. (2009) "Two Standards for Inquiry in Aristotle's *De Caelo*," in A.C. Bowen and C. Wildberg (edd.), *New Perspectives on Aristotle's* De caelo (Leiden) 51–82.
Claghorn G.S. (1954) *Aristotle's criticism of Plato's Timaeus* (The Hague).
Cornford, F.M. (1937) *Plato's Cosmology* (London).
Gill, M.L. (1994) "Aristotle on Self-Motion," in Gill and Lennox (1994) 15–34.
Gill, M.L. and Lennox, J.G. (1994) edd., *Self-Motion from Aristotle to Newton* (Princeton).
Hankinson, R.J. (2002) *Simplicius: On Aristotle's On the Heavens 1.1–4* (Ithaca).
Johansen, T.K. (2004) *Plato's Natural Philosophy* (Cambridge).
Judson, L. (1994) "Heavenly Motion and the Unmoved Mover," in Gill and Lennox (1994) 155–171.
Leggatt, S. (1995) *Aristotle: On the Heavens I and II* (*Warminster*).
Lennox, J.G. (2009) "*De caelo* 2.2 and Its Debt to *De incessu animalium*," in A.C. Bowen and C. Wildberg (edd.), *New Perspectives on Aristotle's* De caelo (Leiden) 187–214.
Lloyd, G.E.R. (1996) *Aristotelian Explorations* (Cambridge).
Moraux, P. (1965) *Aristote, Du ciel* (Paris).
Pellegrin, P. (2009) "The Argument for the Sphericity of the Universe in Aristotle's *De caelo*: Astronomy and Physics," in A.C. Bowen and C. Wildberg (edd.) *New Perspectives on Aristotle's* De caelo (Leiden) 163–185.
Sedley, D. (1997) " 'Becoming like God' in the *Timaeus* and Aristotle," in T. Calvo and L. Brisson, (edd.), *Interpreting the Timaeus-Critias* (Sankt Augustin) 327–339; reprinted in revised form as "The Ideal of Godlikeness" in G. Fine (ed.), *Plato. Oxford Readings in Philosophy* (Oxford, 2000) 791–810.
Skemp, J.B. (1967) *The Theory of Motion in Plato's Later Dialogues* (Amsterdam).
Solmsen, F. (1960) *Aristotle's System of the Physical World: A Comparison With His Predecessors* (Ithaca).

THE POSSIBILITIES OF BEING AND NOT-BEING IN *DE CAELO* 1.11–12

Sarah Broadie*

In *De caelo* 1.10 (279b4–280a35) Aristotle argues on "physical" grounds against those who hold that the world has a beginning but no end. His main target is, of course, the *Timaeus* cosmology understood literally. Then, at the end of the chapter, Aristotle undertakes to show through "general" considerations applying to any sort of thing (καθόλου … περὶ ἅπαντος σκεψαμένοις) whether something can be *genêton* yet not *phtharton*[1] or the reverse (280a32–35; cf. 283b17–18).[2] After laying the ground in chapter 11, he sets about the promised demonstration in chapter 12. It consists in an elaborate argument meant to show the mutual entailment of *genêton* and *phtharton*, and of their contradictories.

However, in the stretch of text we know as *De caelo* 1.12, the question of the mutual entailment of *genêton* and *phtharton* is not broached straightaway. First comes a discussion, which I shall simply call "L_1" (L for *logos*), that starts from a question about the actual being and actual not-being of a thing that has the possibility of being and also the possibility of not-being (281a28–30). Aristotle's answer to this question is that those paired possibilities are each realized at different times (282a4–13). Aristotle reaches this conclusion, which I take to round off the discussion L_1, by means of a famous argument intended to show that that which always is cannot ever not-be, and that which always is-not cannot ever be (281a33–282a4). It seems clear that the discussion L_1 is intended to support the main thesis, carried forward from earlier chapters, that *genêton* ↔ *phtharton*. However, it is less clear exactly how L_1 is supposed to help establish this. Moreover, L_1 is followed by, and indeed seems to turn into, a further stretch of argumentation, L_2, in support of the main thesis. L_2 begins at about 282a14 and continues to 283a3; it elaborates on various

* The University of St Andrews.

1 I leave these terms and their contradictories untranslated, so as not to prejudge whether they mean 'is …', 'can be …', or 'must be generated/destroyed'. But by the end of *De caelo* 1.11–12 the four attributes have emerged as necessary or even essential.

2 The contrast of "physical" and "general" methods in Aristotle is discussed by Robert Bolton (2009) in the present volume.

squares of opposition. It is a rather unfortunate argument for two reasons. First, it involves a *non sequitur*, arising, perhaps, from a false analogy (as I shall suggest in the last section). Secondly, it turns out that Aristotle has adopted an L_1-related starting point ("one of the givens," 283a4–11) which (a) in any case immediately yields the conclusion that *genêton* ↔ *phtharton*, and consequently (b) renders L_2 redundant (if we accept that starting point). All this raises the suspicion that L_1 and L_2 were not first developed together as parts of a single coherent argument.

I shall first mainly discuss L_1 (sections 1–5 below), and shall then examine L_2 (sections 6–15).

(1) The backbone of L_1 is the demonstration that what always is/is-not cannot ever not-be/be. This argument has attracted charges of fallacy and confusion;[3] but according to the account presented here, it is sound on its own terms given certain assumptions.

Our examination begins in *De caelo* 1. 11, where Aristotle clarifies the terms to be used in arguing that *genêton* ↔ *phtharton*. First, he discusses a number of senses of *agenêton* and *genêton*, *aphtharton* and *phtharton* (280b1–281a1). The upshot is that the most central sense (τὸ κυριώτατα λεγόμενον) for *aphtharton* is: '*cannot* perish, or at one time be and at another not-be', and similarly for *agenêton* (281a1–6; cf. 280b31–33). So the central senses of *phtharton* and *genêton* are "*can* be and at another (later) time not-be," and "*can* not-be and at another (later) time be," as is stated at 281b27–29. The next step is a discussion of "can" and "cannot" as used in the above definitions (281a1–26). Whereas the account of *agenêton*, *aphtharton*, and so forth was quite complicated and attentive to a variety of uses, some philosophical and some more ordinary, Aristotle's

[3] E.g., Williams (1965) 100–102 and Judson (1983) 228–233, who diagnose the fallacy differently. The interpretation given here does not depart much from what I gave in Waterlow [S. Broadie] (1982) chapter 4 (clearly summarised by van Rijen 1989, 82–87). (I should be satisfied to have shown that my approach to the backbone argument of L_1 enjoys at least as much textual support as any that charges fallacy; it remains for charity, or some other influence, to tip the scales of judgement.) Waterlow (1982) did not address much of L_2. The book was mainly a response to the influential pioneering work of Hintikka on time and modality in Aristotle, and Hintikka did not focus on Aristotle's specifically cosmological interest in proving that *genêton* ↔ *phtharton*, nor on much of the manoeuvring in L_2. See Hintikka (1973), especially chapter 5, based partly on Hintikka (1957). Judson (1983), by contrast, does justice to the cosmological concern. Van Rijen (1989, 91–94) gives proper weight to the initial question at 281a28–30; thus, he rightly expounds 281a28–b25 as if the main issue is the logical conditions for combining the possibility (in van Rijen, the "natural capacity") of being *F* with that of not-being *F*.

account of "can" here is strikingly simplistic and narrow. He writes as if this auxiliary only governs verbs "φ" such that it makes sense to treat φ-ing quantitatively. "If something *can* move (κινηθῆναι), or *can* lift a weight, we always state it with reference to the maximum, for example '*can* lift a 100 talents or walk a 100 stades'" (281a7–9). Of course, if one of these maximally specified "can φ" expressions applies to some subject, it follows that the subject can φ to a lesser extent too. If one can-maximally walk a 100 stades, it follows that one can walk 50 (281a12–14). However, the "can" in the latter context is obviously not "can-maximally," but has a different and logically weaker sense. The question then arises which of these senses of "can" (no others are considered here) is central, and Aristotle says: the maximal one. "Let what *can* in the central sense (τὸ κυρίως δυνατόν) be defined in accordance with the utmost (κατὰ τῆς ὑπεροχῆς τὸ τέλος)"[4] (281a9–15). This sense now becomes the basis for the argument that is ultimately meant to show that *phtharton* and *genêton* (and likewise *aphtharton* and *agenêton*) necessarily imply one another.

However, there is immediately the awkwardness that the verbs "perish" and "begin" do not in any obvious way govern amounts of anything, so that the allegedly central sense of "can" has no direct application to them. Consequently, Aristotle moves to *what* ceases or starts when something perishes or begins: namely, being and not-being.[5] He explains that this means being/not-being *F*, where '*F*' is a predicate in any one of several categories (281a30–33). We must bear in mind, therefore, that when in the ensuing discussion he speaks of being/not-being, this always implies a subject which is/is-not thus or so. Thus, if it is a case of the non-being of a substance, then the logical situation is that some matter as subject is-not *F*, where "*F*" denotes a substantial form. This fits the target in the *Timaeus*, since, as is well known, Plato's cosmos did not begin to be *ex nihilo* but was formed from pre-existing matter. But now in terms of what medium or dimension are predicative being and non-being to be quantified? The answer is: duration. Speaking strictly, instead of just saying that *x* can be/not-be (i.e., be/not-be *F*), we should specify the temporal maximum for which *x* can be or not-be. To say that *x* can be/not-be for a certain duration, and that *y* can be/not-be for a quantitatively different duration, is to ascribe to *x* and *y* generically different possibilities or capacities for being/not-being.

[4] For the meaning, see Verdenius (1969) 274.

[5] These are the prior terms, i.e., they occur in the analyses of "begins" and "perishes"; see 282a1–4.

(2) Now (281a28–b2), suppose there are things which can be and can not-be. Then they can be for some maximal time, and they can not-be for some maximal time. Examples like maximally lifting 100 talents readily suggest finite temporal maxima; but Aristotle assumes that "for an *unlimited* time" gives a kind of maximum too. The rationale for this is that for a time to be unlimited is (he says) for it to be *greater than* any time specified (281a33–34).[6] So on this account, "for an unlimited time" entails "for the amount of time that is absolutely greatest." And this last phrase is tantamount to "always," a point which Aristotle will soon make explicit (e.g., at 281b25). For "always" indicates more time, and therefore a greater time, than any expression implying "not always," as for instance "only from the birth of Plato onwards." Now suppose further that among the things which can be and can not-be, some can be for an unlimited time and can also not-be for an unlimited time (281a33–b1).[7] Then there would have to be, so to speak, two unlimited times, one in which the object can maximally be and the other in which it can maximally not-be. But this is impossible. (Presumably it is impossible because the two unlimited times added together would constitute a time larger than either, with the result that neither was unlimited after all in the sense just explained. In other words, it is assumed that if there were these two times, they would both be parts of the same history.)

It follows that whatever can be for unlimited time necessarily lacks the possibility of not-being for unlimited time. However, Aristotle is in a position to draw the stronger conclusion that whatever can be for unlimited time lacks the capacity to not-be for any time at all: i.e., what can be always, cannot not-be, and what can not-be always, cannot be. For the above stretch of argument proceeds on the basis of a notion of *can* or *possible* according to which for something to be possible is for it to be possible (e.g.) *now*, and something that does not obtain

[6] In treating unlimited time as a maximum, he treats it as a sort of whole. Similarly, in many passages about the heavens and their motion, he uses "always" (ἀεί) and "everlasting" (ἀΐδιος) to indicate not merely chronological on-and-on-ness, or at-every-pointness, in both directions, but a sort of natural span—as it were, the life-span of a being that lasts for ever; see section 4 below.

[7] Aristotle assumes symmetrical maxima for *can be* and *can not-be*. For a given subject, being and not-being are either both for all time (which is impossible), or both for some-but-not-all time. The idea of temporal maxima for not-being, which is necessary for the argument, seems artificial in the light of some illustrations, not others. Weightlifting and walking suggest maxima only for possibilities of positives, as does the notion of a natural span of life (αἰών); but nature has many alternating spans (e.g., night, day), of which one can figure as negative to the other.

now is nonetheless possible now if and only if no impossibility results from supposing it to obtain at another time. In slightly different words: something that does not obtain is nonetheless a possibility even during the time when it does not obtain, provided that nothing impossible follows from supposing another time in or during which it does obtain (281b2–18). But if something is exercising a possibility of being for unlimited time, there is no other time, limited or unlimited, in which it can exercise a possibility of not-being either with or without temporal limitation. Hence if x is exercising a possibility of being for unlimited time or always, x has no possibility of not-being for any length of time at all. Moreover, since there is no "other time" than the one unlimited time, if the x in question can not-be for any time at all, this must be because it can not-be for (some or all of) the time during which it is exercising its capacity to be. In other words, it can not-be only if it can not-be while at the same time being—which is impossible. Similarly, if something is exercising its possibility of not-being for unlimited time, it cannot be for any time, except on condition that it can be even while in fact not-being—which is impossible. Consequently, whatever has the possibility of being for unlimited time or always, *is* for unlimited time or always, because it cannot not-be ever; and likewise *mutatis mutandis* for whatever has the possibility of not-being always (281b18–25; 29–34).

The next three sections will bring out or fill in various assumptions and ideas that are required for this argument to work.

(3) The argument depends on three ideas unfamiliar in modern treatments of possibility.[8] First, there is the construal of the possibility of a condition S as a possibility belonging to x at, or for, a time (281b15–17).[9] Secondly, there is the idea that, given that S does not belong to x at or for a given time t, the possibility at or during t of x's having S is decided by reference to the non-impossibility of its having S at or for a time other than t (281b17–18). Thus during t, while he is sitting, Callias has the possibility

[8] They are explored in more detail in Waterlow (1982) chs. 2–4. Whether or not any of them hold for Aristotelian modality in general, they are needed to make sense of possibility in *De caelo* 1.12.

[9] Treating the possibility of S as belonging at or for a time of course does not entail holding that the same possibility cannot belong at/for different times. However, Judson, if I understand correctly, thinks otherwise. This assumption of his is an important pillar of his contention that interpreting *De caelo* 1.12 in terms of possibility belonging at/for a time fails to make good sense (given the other conceptual elements) of 'What can be/not-be always, cannot not-be/be'. See Judson (1983) 252, point (1).

of standing if and only if nothing impossible follows from his being on his feet at another time. These two elements of Aristotle's approach in *De caelo* 1.12 stand in contrast to the usual modern understanding. According to the latter, the intuition (however it is achieved) that even if *S* does not obtain in *x* at *t* it is not impossible that it obtains at (e.g.) $u \neq t$, would immediately be taken to show that it is possible *simpliciter* (i.e., without the possibility itself being linked to a time) that *S* obtains *simpliciter* (i.e., without the obtaining being linked to a time either distinct from or the same as *t*). However, if it is possible *simpliciter* that *S* obtains, then presumably it is possible *simpliciter* that *S* obtains at any given time; so we can coherently think that even granted that *S* does not obtain at *t*, it is possible that *S* obtains at *t*: just as possible as that *S* obtains at some other time concerning which it is not yet known whether *S* does obtain at it or not.

Thirdly, there is the idea of defining a possibility of being / not-being by reference to the temporal maximum to which the subject can be/not-be.[10] Possibilities—i.e., in fact, capacities—for different maxima are specifically different possibilities. Now, if we exercise one capacity, we do one thing; and if we exercise a different capacity, we do something else. For example, if we exercise our capacity to swim, we swim; and if we exercise our capacity to read, we read: these are different kinds of activities. Consequently, if there is a capacity or possibility which is the possibility of *being F to the temporal maximum M*, and there is also a possibility which is the possibility of *being F to the different maximum N*, then the exercises of these possibilities are different kinds of being *F*. And similarly for not-being *F*. In particular, since Aristotle is persuaded that "always" indicates a maximum, he is committed to the doctrine that the being/not-being of that which is/is-not *always* is different in kind from the being/not-being of that which is/is not for some *lesser* maximum. These two main ways of being/not-being *F* differ from each other not as more and less of one and the same condition or activity, but as distinct and in fact mutually exclusive kinds of being/not-being *F*.[11] Thus, we should not express the

[10] The generalised treatment of *can* φ in terms of maximal φ-ing seems to be unique to this *De caelo* argument.

[11] The second of the two main ways subdivides into various kinds. Where *F* is a substantial form, so that something's (the matter's) being *F* entails the existence of a member of the kind in question, there will be different temporal maxima for different living species. *De gen. et corr.* 2.10, 336b10–15, gives examples. However, there are also cases, e.g., when "*F*" = "sitting" or "standing," where it seems built into the notion of *F* that being *F*, where it applies, will cease at some point though not after any particular laid down interval.

difference by means of different quantification over times, i.e., as the difference between e.g.: "$(\forall t)(Fa \text{ at } t)$" and "$(\exists t)(Fb \text{ at } t)$ & $(\exists t)$ $(\sim Fb \text{ at } t)$."[12] It is the difference, rather, between "*a* is (*F* always)" and "*b* is (*F*, but not always)." From now on these different temporal "qualities" of being and not-being will be expressed by superscripts on the predicate.

(4) What feeds the idea of maximal being, and in particular the idea that "always'" indicates a maximum of time, is the notion of a natural life-span (αἰών). For etymological reasons Aristotle thinks this term particularly appropriate to the entire duration of the everlasting[13] heavens or the everlasting cosmos as a whole. The discussion occurs in *De caelo* 1.10. Aristotle has been considering the mode of being of the mysterious incorporeal, non-temporal, things "beyond the cosmos" (279a11–17). Of them, he says:

> ... they carry out their entire life span (διατελεῖ τὸν ἅπαντα αἰῶνα)[14] unalterable and impassible, being possessed of a vital activity (ζωή) that is the best and the most self-sufficient (*De caelo* 1.9, 279a20–22)

He then goes on:

> In fact, it was by divine inspiration that the ancients voiced this word [*sc.* αἰών]. For the completeness (τὸ τέλος) that comprehends the time of each thing's life so that that time is naturally all within it, is what for each thing is called its life span (αἰών). For the same reason, both the completeness of the whole heaven, and the completeness that comprehends all time and ⟨time's⟩ limitlessness (τὴν ἀπειρίαν), is ⟨called⟩ αἰών, immortal and divine: αἰών that gets its name from αἰεὶ εἶναι [= always being]. On this the being and living of other things depends, more strictly in some cases and obscurely in others.[15] (279a22–30; cf. 2.1, 283b26–29)

[12] Cf. Denyer (2000) sections 4–5.

[13] This translates ἀΐδιος, and means "everlasting in both temporal directions."

[14] I have translated as if this accusative is the object of διατελεῖ, but it may equally be the accusative of duration.

[15] Both translation and interpretation are debatable. The main queries concern the reference of ὅθεν at 29 and the meaning of the καὶ governing τὸ τὸν πάντα χρόνον ... περιέχον τέλος (26): is it conjunctive or epexegetic? If conjunctive, which seems more natural, then two completenesses are mentioned at 25–27, of which the second must be the αἰών of the entities beyond the heavens, which somehow embraces, without being identical with, τὸν πάντα χρόνον. Ὅθεν then refers to this, or perhaps to a generically conceived life-minus-temporal-finitude predicable both of the extra-mundane beings and of the everlasting heavens. If the controversial καὶ is epexegetic, only one non-temporally-finite completeness is at issue, that of the heaven, and ὅθεν refers to it. There is also the question whether the καὶ linking τὸν πάντα χρόνον and τὴν ἀπειρίαν (26)

This passage seems meant to defend his application of the term αἰών to the existence of the incorporeal and timeless[16] beings beyond the cosmos, by reasoning which I reconstruct as follows:

(a) Because their existence is a life of complete and all-inclusive perfection, it is correct to call it αἰών.
(b) One might, however, be surprised by this, since in ordinary use αἰών means "natural span," and for ordinary things, i.e., mortal things, the natural span is contained within the life of the cosmos, which extends beyond it in both temporal directions.
(c) But to understand the word αἰών we should not look to the *temporal finitude* of such mortal spans but to their *completeness* (on mortal terms).
(d) For in fact the term gets its meaning not primarily from these cases; its etymology ('always being') shows that semantic priority lies with a totality that comprehends all time.[17]
(e) Consequently, the ancient humans must have been inspired when they coined the word, given that the prior sort of case is so far beyond ordinary human experience.

Thus, the idea that "for all time" and "for unlimited time" indicate a temporal maximum or whole is one for which Aristotle has prepared the way quite carefully by the time he gets to the backbone argument of L_1, to the effect that what is/is-not^{always} cannot ever not-be/be. It is worth noting that his expressing this in terms of *unlimited* (or *infinite*; ἄπειρος) time (279a26; 281a34; 283a7–10; 283b29) is in striking contradiction with the *Physics* 3 doctrine of the unlimited (infinite): "it turns out that

---

is epexegetic or conjunctive. If the latter, the phrase may mean "every finite time and the limitlessness ⟨of time⟩." For discussion of various interpretations, with references, see Leggatt (1995) 205–206 *ad* 279a28.

[16] They are not merely not 'in time' in the sense established in *Physics* 4.12, 221b21–22, i.e. flanked by time before and after, hence of finite duration. Since they offer no purchase for change or motion they cannot even be everlasting (i.e., for Aristotle, having infinite duration in both directions), *pace* Leggatt (1995) 205 *ad* 279a22.

[17] The etymology shows *that* this kind of case is primary, but does not explain *why* it is. Presumably, the reason is that an αἰών that comprehends all time is an αἰὼν αἰώνων: a span that makes possible, by comprehending them, all finite temporal spans. If, instead (and contrary to the etymology), the latter determined the primary meaning, any move to present the timelessly complete life of incorporeal divine beings as an αἰών would be too farfetched or even nonsensical. (Such a move is endoxic, having Plato's authority; *Tim.* 37c6–d7.)

[the unlimited] is the opposite of what people say it is: it is not that of which no part is outside, but that of which some part is always outside" (206b33–207a2, tr. Hussey 1983). However, perhaps what the *Physics* discussion shows is not that Aristotle has there rejected the notion of an αἰών that comprehends all time,[18] but that he has clarified or re-shaped the notion of the *unlimited* so that, instead of suggesting an all-inclusive whole which is the paradigm of completeness, it now suggests something essentially always incomplete, since "that of which no part is outside is complete and whole" (207a8–9).[19]

(5) The idea that possibilities of being/non-being are defined by reference to temporal maxima has an interesting consequence, but one that seems to blast a hole in the argument as presented in section 2. Take someone whose possibility, capacity, or power for weight-lifting is defined by the limit of 50 kilos: he can certainly lift, e.g., 25 of the units, but if he does he is not then exercising a concurrently possessed distinct possibility. For different maxima in the same range are mutually exclusive. So if the person lifts any weight below his current maximum, he is still thereby exercising his possibility of lifting that maximum. It is also true, I think, that he exercises that same 50kg possibility on every gram within every amount he lifts. And, clearly, it can happen that he actually exercises his maximally defined possibility, yet never exercises it to the full. Analogously, then, one who is exercising the power of living the standard human lifespan, say three score years and ten, is engaged in the business of living-three-score-years-and-ten during every decade, year, and minute of the duration of her life. That is the kind of life it is throughout, just as, on a higher level of generality, it is—throughout—the kind of life lived by a *mortal* animal. Clearly, the specific temporal maximality of the life as lived each moment as long as it lasts cannot ensure that the maximum is actually reached. A life may be cut short, or one may lose ahead of time the *conatus* to go on. It therefore looks as if it could be true that

[18] The temporal version of this notion does not enter into the treatment of time in *Phys.* 4, but that may be because Aristotle there is interested in time as a dimension for ordering and measuring finite events, so that consideration of a duration which is the αἰών or life of the whole universe would be beside the point.

[19] At *Phys.* 3.6, 207a15–16, Aristotle criticises Melissus for speaking of the unlimited as a whole, by contrast with Parmenides who held that the whole of being is limited. Aristotle might have ranged his earlier self alongside Melissus in this respect; see Waterlow (1982) 76–77. Bogen and McGuire (1986–1987) sections xxv–xxxv, argue that there is no inconsistency between the *De caelo* and *Phys.* 3 on the unlimited.

something is exercising the grand possibility of being^{always} and yet could in the midst of time cease to be, leaving the maximum unaccomplished. And similarly for not-being^{always}.

Aristotle does not trouble about this gap in his argument. Presumably, he takes himself to be entitled to assume that the temporal maxima *will* be played out to the full. It may be that his method of demonstrating καθόλου rather than by appeal to specifically physical principles is designed to deal only with ideal cases, abstracting from possibilities of interference or failure. (One would not interrupt a geometrical demonstration of the path traced by a rolling cone by pointing out that something might get in the way.) We might wonder what can be learned about reality from a method designed to bring out the attributes of ideal objects. Aristotle would surely answer that in the case which most concerns him and those with whom he is disputing—namely, the being of the cosmos—the ideal is necessarily what is physically actual. The cosmos is so good and so physically all-embracing that nothing outside or inside it could cut it short or make it weary of being. Certainly Plato would have agreed.[20]

(6) In any case, let it be that Aristotle has established that what is^{always} cannot not-be, and that what is-not^{always} cannot be. (He has thereby also established that what can be and can not-be, both is $^{\text{not always}}$ and is-not $^{\text{not always}}$ at different times.) He now proceeds to draw some conclusions about beginning and perishing: what is^{always} is absolutely (ἁπλῶς) (a) imperishable and (b) ungenerable (281b20–282a4). Parallel conclusions are drawn at 282a21–22 for what is-not^{always}. We also get the obvious corollary that nothing generable or perishable is everlasting (ἀΐδιος), i.e., is^{always} (282a21–22). But we have not yet been given a valid reason for accepting that the attributes *phtharton* and *genêton* are *twins*, that if either belongs to a given subject, so necessarily does the other. Yet this was the main point to be proved, if we consider *De caelo* 1.12 from the perspective of 1.10. Aristotle reverts to this *agendum* at 282a25–26:[21]

> If something (a) is *agenêton* and (b) is—must it be everlasting [i.e., must it be^{always}]? And likewise if something (a) is *aphtharton* and (b) is—⟨must it be everlasting⟩?

[20] Cf. van Rijen (1989) 95–98. He, like Waterlow, makes sense of the argument and sees the gap as covered by the assumption that only things composed of perishable matter (*Meta.* 8.1, 1042b6) are subject to interruption or repression of exercise of possibilities, along with the assumption that nothing that can be^{always} F is thus composed.

[21] At this point the discussion I labelled "L_2" is under way; it probably starts at 282a14, but for a stretch it incorporates some L_1 argumentation: see 281a21–26.

But before examining how he addresses these two questions, we must face the fact that, now, at 282a27–30, he announces that he is using "*agenêton*" and "*aphtharton*" (and therefore by implication their contradictories) in their central senses (τὰ κυρίως λεγόμενα), and immediately assigns *different* central senses from the ones established back in chapter 11, 281a3–6. In that earlier place, the central senses were emphatically modal (cf. τὸ ἀδύνατον καὶ μὴ δυνάμενον, 5), whereas here they are as follows: "'*agenêton*' applies to that which is now and of which it was not previously true to say 'it is-not,' while '*aphtharton*' applies to that which is now and of which it will not later be true to say 'it is-not.'"

(7) This different stipulation of central senses for the same pair of terms surely tells us that the ensuing passage, and perhaps some of what precedes it, belong to a different composition from that of the textually earlier stipulation. Thus, we have evidence that the entire stretch of argument constituting chapters 11 and 12 of *De caelo* 1 incorporates a piece that was originally written separately. There are no obvious boundaries marking off this piece, and we should surely assume it was Aristotle himself who worked it into the resulting whole. That what we have is a synthesis of different compositions would help explain the inconsequential character of some of the argumentation.[22]

Should we worry that the discrepancy between the two sets of central senses of "*agenêton*" and "*aphtharton*" leaves the entire argument broken-backed? I think not desperately, for a combination of two reasons. First, I suggest, the apparently assertoric set proposed at the textually later point is actually no less modal than the overtly modal set proposed earlier, though modal in a different way. The suggestion is that when Aristotle says that the *aphtharton* is that which now is and of which it will not later be true to say that it is not, he means that whatever is correctly said to be *aphtharton* is *of a nature such that* it both is now and will falsify any future statement that it is not, and correspondingly for the *agenêton*. The upshot is that (granted the impossibility of interference) the *aphtharton* cannot perish and the *agenêton* cannot have begun. Thus, these terms

[22] On the inconsequentiality, see Williams (1966) 209–210, also Moraux (1965) lxxxii, n. 1 *ad fin.* (lxxxiv). Williams and Moraux (writing independently) both conjecture different compositions. See Moraux (1966) lxxxii, n. 1, for evidence drawn from the *Organon* and from Aristotle's use of letters as variables at 282a14–21. There may be further evidence in the discrepancy, noted by Williams (1966) 207, between 281b28 and 283b13 (on whether there is a δύναμις τοῦ γεγόναι).

are as strong in the second stipulation of senses as in the first. However, if we apply, as I think we should, the "*of a nature such that*" interpretation across the board, then "*phtharton*" will mean "such that it now is and later is not," and correspondingly for "*genêton*"; which is to imply that the *phtharton* must cease and the *genêton* must have begun. And at this point we might worry that the modality of this pair of terms is stronger now than when they were introduced earlier, since then they were contradictories of "*cannot* be, then not be" and "*cannot* not be, then be." This worry can be assuaged to some extent if not completely. As we are about to see more fully (this is the second reason mentioned above), the terms of Aristotle's earlier argument, given certain assumptions, enable him to move not merely from "is/is-not" to "can be/not-be," but also in the reverse direction. If this liberty could be extended to allow a parallel inference from "can begin/perish" to "begins/perishes" then the modality of "*genêton*" and "*phtharton*" in the first stipulation is no weaker than in the second.[23] Now admittedly there is no logically watertight justification, as far as I can see, for extending the liberty in this way. The best one can say, which no doubt is rather lame, is that the earlier argument created a logical atmosphere favorable to sliding from "can φ" to "φ."[24]

(8) We can now return to the point where Aristotle at last begins squarely to address the main question, i.e., whether *genêton* and *phtharton*, and likewise their contradictories, are mutually entailing. He leads up by asking: "If something (a) is *agenêton* and (b) is—must it be everlasting (i.e., must it be^{always})? And likewise if something (a) is *aphtharton* and (b) is—⟨must it be everlasting⟩?" (282a25–26). The relevance of this to the preceding context is as follows. He has been arguing that what is^{always} and what is-not^{always} (these both count as everlasting) are neither *genêton* nor *phtharton* (282a21–22; 281b25–30; 281b34–282a4). That reasoning will also have made it obvious that whatever is both *agenêton* and *aphtharton* is in the category of the everlasting. Thus, it is quite natural to ask next

[23] On this interpretation, one must take the contrast between ἐνεργείᾳ and δυνάμει at 282b19–20 as distinguishing the present being of something that is *genêton* from the merely future being of a *genêton* that has not been generated yet.

[24] The convergence in meaning of the two lots of central senses does not make it less anomalous that Aristotle separately prescribes them both: why do this if the textually second prescription neither verbally repeats the first, nor substantially improves on it? Thus, the inference to different compositions remains in force.

whether just one or other of the two attributes, being *agenêton* and being *aphtharton*, is sufficient for everlastingness. Aristotle states that the answer is "Yes," given and only given the assumption that "*agenêton*" and "*aphtharton*" entail one another.[25] And in the course of this he argues that on the same proviso "*phtharton*" and "*genêton*" entail one another too (282a30–282b7).[26]

However, Aristotle next states categorically that "*agenêton*" and "*aphtharton*" do entail one other, on the ground (now categorical) that "*phtharton*" and "*genêton*" entail one another (282b7–9). The latter, he says (282b9–10), is "obvious from what was said earlier"; even so, he now proceeds to spell the matter out more fully (282b10–22). Finally, having established (as he thinks) the inter-entailment of *genêton* and *phtharton*, he proves the inter-entailment of their contradictories (282b23–283a3).

(9) But how has Aristotle obtained the premise that *genêton* and *phtharton* entail each other? The backward reference at 282b9–10 is to 282a4–21. The argument there, like its explication at 282b10–22, relies on a set of relationships which can be represented as a sort of square of opposition. But before looking at what the relationships are supposed to prove, we must face what to some commentators has seemed a difficulty about the terms. The problem is that at 282a4–21 (cf. 282b10–13) Aristotle passes to and fro between two apparently quite different sets of terms without properly distinguishing them. One is overtly modal, the other is not:

I	A—can be^{always}	B—can not-be^{always}
	D—can not-be$^{\text{not-always}}$	C—can be$^{\text{not-always}}$
II	A*—is^{always}	B*—is-not^{always}
	D*—is-not $^{\text{not-always}}$	C*—is $^{\text{not-always}}$

[25] 282b2 should have a period after αὐτῶν, I believe. The point is that it is clear from the definition just laid down (282a27–30) that, given the assumption, *aphtharton* and *agenêton* severally entail "everlasting."

[26] This is because "*agenêton*" and "*genêton*," and "*aphtharton*" and "*phtharton*," are contradictory pairs: they "do not belong to the same thing, but one or other belongs to every given thing" (282b26–28). The first γάρ of 282b2 (καὶ γὰρ) is puzzling if, with the editors, we retain it (it is missing from E^1, and, as D.J. Allan is inclined to think, from Simplicius' text), since the gist ("*phtharton* entails *genêton*") does not support the point about what the definition makes clear (see previous note). Nor should it be supporting the truth of the assumption itself, since Aristotle is about to use the latter in showing that, it granted, *phtharton* entails *genêton* and *vice versa*. Might the γάρ be an element of καὶ γὰρ = "Yes, and ..." (Denniston 1996, 109–110) even though this is continuous prose?

Let us look first at square II.[27] It is assumed to offer an exhaustive division of everything in the relevant universe of discourse. A* and B* are contraries (cf. 282a 6, ἐναντίον), while C* and D* are called the "negations" (ἀποφάσεις; a10) of A* and B* respectively. From many other contexts in Aristotle's works, we might have expected a "negation" to be the contradictory of that of which it is negation, but this is not so here. True, A* and C*, and likewise B* and D*, are mutually exclusive, because even if "not-always" is in some sense weaker than "always", it is not weaker as "some" than "all" (in Aristotelian logic), or "proper part" than "whole," where the stronger includes the weaker. But the pairs A* and C*, B* and D*, are also not contradictories because neither pair is exhaustive of all options. One exhausts only the options for being, the other for not-being. In so far as C* and D* each entail "is some of the time and is-not some of the time," they entail each other. Aristotle sees C* and D* as "intermediate" between the contraries A* and B* (282a18–19).

Aristotle moves to and fro legitimately between squares I and II because the counterpart terms are mutually entailing (so the relationships within both squares are the same). The problematic entailments are, of course, those running from square I to corresponding points on square II, so let us focus on them. (a) We have already seen[28] that something to which A applies, cannot not-be for any time; hence, given that it must fall under some category or other in square II, it must fall under A*. Likewise with B and B*. Now (b) consider something, *x*, to which C applies; and suppose first that *x* is, then that it is-not. If *x* is, then it falls under either A* or C*; but it cannot fall under A*, given (as would normally be assumed) that A* entails A, for then by the earlier argument *x* would be something that cannot not-be for any time. If *x* is-not, then it falls under B* or D*, but it cannot fall under B*, given that B* entails B, for the parallel reason. Hence *x* falls under C* or D*, since it either is or is not. But either way, since C* and D* entail one another, *x* falls under C* and D*. By similar reasoning, whatever is D must fall under C* and D*.

(10) We may now return to "*genêton* ↔ *phtharton*," and the question of how this is meant to be established. At 282b8–12, Aristotle writes:

---

[27] The sets of terms are not set out in squares in our mss, but it is convenient to refer to them in this way.

[28] Section 2, above.

> ... *genêton* and *phtharton* are implied by one another. This too is clear from what was said earlier. For between what is^{always} and what is-not^{always} there is that which is accompanied by neither (ᾧ μηδέτερον ἀκολουθεῖ),[29] and this is what is *genêton* and *phtharton* ...

But what was said earlier showed only that C* and D* (= C and D) are between the extremes. No doubt Aristotle has been assuming all along that A* and B* (= A and B) and C* and D* are all instantiated in the universe, and that this is necessarily so. Hence, C* and D* are necessarily both instantiated. However, on a straightforward understanding, C* and D* each imply no more than variation either way between being and non-being. Thus, C* and D* can each be instantiated both by something that is and then is-not (perishes) and by something that is-not, then is (begins); but nothing guarantees that those two modes of instantiation necessarily coincide in one subject. To obtain that result, Aristotle needs a further assumption: that whatever falls under C* or D* is and then is-not, or *vice versa*, for a *determinate*, i.e., finite, time in each case (282b12–14). Given this, it is, of course, easy to show that anything under C* or D* must, when it is, have begun and be going to cease, and must, when it is-not, have ceased and be going to begin again (282b20–23).

But is it really, as I have just claimed, a further assumption that C* and D* import finite durations of being and not-being? One might think this constraint is already lurking in the principle whereby Aristotle obtained the result that what can be/not-be^{always} cannot not-be/be. That was the principle that being and not-being are exercises of capacities defined by temporal maxima. We might think, given the analogy of maximal powers for weight-lifting and so on, that what has a capacity for less than temporally total being/not-being must have a capacity specifiable by a finite number of units. But we are not forced to think this. The basic idea is that something goes on being/not-being for as long as it can. Now suppose that something begins to be. There is nothing contrary to logic about supposing that it has and exercises the capacity to go on and on without stopping. We can perfectly well think of that as the maximal being-capacity for this type of thing.[30]

---

[29] I.e., neither 'what is^{always}' nor 'what is-not^{always}'. Thus also J.L. Stocks in Barnes (1985), and Guthrie (1939). Leggatt (1995) takes μηδέτερον at b11 to refer to *to aphtharton* and *to agenêton*, which last appeared at b7.

[30] No doubt such a capacity is not for what could be called an αἰών of being, since a span, even of the whole of time, is understood as in some way fully bounded.

We have seen Aristotle allowing uncountable or immeasurable maxima in certain cases, since unlimited time in a sense implying 'all time' was treated as a maximum. (Not that this got any support from the analogy with weight-lifting powers; could anything lift all the heavy material there is, including its own?) So why should Aristotle not permit unlimited time in the sense of unlimited-in-just-one-direction to function as maximal in certain cases? But then the principle of temporal maxima would not allow him to draw the desired conclusion that *genêton* ↔ *phtharton*. It would legitimize beginners that never will cease and ceasers that never began. So just on the verge of drawing the desired conclusion, he shifts to emphasize *determinacy* of time, whereas at the beginning he had emphasized maximality. He now says:

> ... *genêton* and *phtharton* are implied by one another. This too is clear from what was said earlier. For between what is^{always} and what is-not^{always} there is that which is accompanied by neither, and this is what is *genêton* and *phtharton*. For it can both be and not-be, each for a determinate time (ὡρισμένον χρόνον); I mean that the being and the not-being are each for a given amount of time (ποσόν τινα χρόνον). (282b9–14)

No mention here of maximality. It is as if mere maximality has been quietly forgotten; or, alternatively, it is as if we are now to realize that at the beginning and all along maximality was understood as entailing determinacy.[31] We have seen (section 4, above) how Aristotle (starting from his doctrine that the heavenly bodies are animate [cf. *De caelo* 2.12, 292a18–21] and everlasting) configures the concept of αἰών (determinate span of life) so that all of time counts as an αἰών. On a more epistemic note, one might also point out that "always" provides as logical and informative an answer to "How long can it last?" as any phrase expressing a finite time, such as "one day" or "70 years." However, the same might be said of the answer "For ever—once it has started." One might try to exclude such an answer on the ground that it fails to pick out a single length of time: it can apply to an item that *x* started in the 14th Olympiad, and to another that started in the 30th; but even if we grant that one of those one-way-infinite durations is shorter than the other—the one that started later—why should *that* rule these cases out of consideration? Ordinary finite natural spans are different lengths for different kinds of beings.

[31] Cf. χρόνον τινὰ ὡρίσθαι at 281a29.

(11) For Aristotle, the question of one-way infinite being/not-being comes up rather as an afterthought, after he has concluded that *genêton* ↔ *phtharton* and, therefore, that *agenêton* ↔ *aphtharton*. He writes:

> But to say that there is no reason why something that comes to be should not be *aphtharton*, nor why something *agenêton* should not be and ⟨then⟩ be destroyed (with one-off [ἅπαξ] processes of coming to be in the one case and of perishing in the other),[32] is to withdraw one of the givens,[33] namely: anything can act or be acted on, be or not-be, either for unlimited time or for a determinate amount of time (ποσόν τινα ὡρισμένον χρόνον)—and the reason why something can for an *un*limited time is this: unlimited time is in a way determinate (ὥρισταί πως), as being the time than which none is greater. But[34] that which is unlimited ⟨merely⟩ *in some direction* (τὸ πῇ ἄπειρον) is neither unlimited nor determinate.[35]
>
> (283a4–10)

To this we can reply that whereas a stretch of being or not-being that is unlimited *in some direction* is not unlimited absolutely (ἁπλῶς, the opposite adverb to πῇ), still it does not absolutely lack unlimitedness, and arguably it does not absolutely lack determinacy, since, as we have seen, perhaps specifying such a duration does give information on how long something can and does last. To insist that any such duration is neither unlimited nor determinate in the way or ways relevant to the preceding argument is just to stipulate. Perhaps Aristotle senses the weakness of this, because he now launches a series of arguments for "*agenêton* ↔ *aphtharton*" (Moraux (1965) *ad loc.* calls them "supplementary") which seem to be designed to manage without the sheer *assumption* that what is/is-not$^{\text{not-always}}$ is/is-not for a *finite* span.

(12) But if we return to the point reached just before the postscript (i.e., to 283a3), it does look as if the whole argument up to now has depended on that assumption. The problem we have just touched on is that the assumption is far from obvious to anyone not already prepared to interpret all duration in terms of natural spans, and yet is necessary for establishing that *genêton* ↔ *phtharton*. However, a fact still more

---

[32] By contrast with the perpetual repetition implied by his preferred model.

[33] Cf. 281a28–b1.

[34] Keeping the δὲ of the mss, *pace* Moraux (1965).

[35] 283a8 has been correctly understood by Bogen and McGuire (1986–1987) section xxii, and by Verdenius (1969) 275, except that Verdenius distinguishes, as I would not, between ὡρισμένον (7) and ὥρισται (9), thinking the first means "determinate," the second "defined." Williams (1966) 205 also sees equivocation here.

exasperating for anyone hoping to find in this passage a decent argument over all is that the assumption by itself is sufficient basis for asserting that *genêton* $\leftrightarrow$ *phtharton*: and evidently so. Grant that what is/is-not$^{\text{not-always}}$ is/is-not for a *finite* span, and you have already granted that what begins must end and what ends must re-begin. So expecting the opponent to "give" that assumption (283a6) is asking him to surrender on the spot. This is one embarrassment. Another is that the whole L_2 apparatus of the squares of opposition marshalling contraries and negations now turns out to be unnecessary for the main task of *De caelo* 1.11–12. *If* the above assumption is granted, that task is thereby accomplished, and accomplished very visibly once we are no longer distracted by Aristotle's efforts with the squares.

(13) But before concluding that this material does not represent a perfectly coherent project (as we have already seen reason to suspect), we should examine another possibility of interpretation. In section 10, I said that C* and D*, straightforwardly understood as implying variation between being and not-being, are satisfiable by what is *genêton* and *aphtharton*, and by what is *phtharton* and *agenêton*. But a slightly more complex interpretation yields Aristotle's desired result *via* square of opposition II and without question-begging assumptions. Read C* and 'D* as, respectively, "is, then is-not" and "is-not, then is." Then invoke the principle, which has been at work all along, that being and not-being each divide exhaustively into the two modes of "always" and "not-always," and ask which mode is exemplified in the different limbs of C* and D* read as above. The answer for "is" in C* is already explicit: it is already presented as the "is" of "is, then is-not.". Bringing out the corresponding point for the "is-not" of C*, we write this as "is-not, then is," thus obtaining for the whole C* "is, then [is-not, then is]." This secures that the *phtharton* is *genêton*. Applying the procedure the other way round to D* secures that the *genêton* is *phtharton*.[36] On this reading, when Aristotle insists at 283a4–11 that the being and the not-being are either for ever or for a finite time, he is not clutching at an assumption extraneous to the second square of opposition, but is drawing on the meaning of its subcontraries.

[36] Since the procedure is indefinitely repeatable, the same subject begins and ceases to be (i.e., be *F*) infinitely many times. Thus, it seems C* and D* apply only where the subjects of the predications are everlasting, which would exclude such classic sublunary contingencies as accidents and human voluntary action (*De Interp.* 9, see esp. 19a12–21; *Eth. Eud.* 2.6.) See Sorabji (1980) 128–130, and Waterlow (1982) 61–62.

Unfortunately, there is really no evidence in the text for attributing to Aristotle the above analyses of C* and D*.[37] So it seems safest to adopt what I called the "straightforward" interpretation of C* and D*, a course which makes the inference to "*genêton* ↔ *phtharton*" a *non sequitur*.

(14) However, if this is our picture, we should not rest with viewing the failure of the whole complicated argument. We should also note something of abiding interest here. The argument reveals a fascinating moment in the development of modal concepts. The notion here forged of a "can be/not-be always" that excludes the possibility of not-being/being for any time at all, constitutes a powerful, albeit narrow, interpretation of impossibility and necessity. Particularly interesting is the fact that in this system contingency cannot be defined as a conjunction of possibilities, $\diamond p$ and $\diamond \sim p$, such that the conjuncts are severally entailed by, respectively, p and $\sim p$.[38] This is because in this system's square(s) of opposition the subcontraries entail each other, whereas in a regular modal square they are only mutually compatible.

Thus, here we have a *non-synthetic* concept of the neither-necessary-nor-impossible: i.e., a modality not composed out of two mutually independent elemental possibilities each consistent with, and following from, one of a pair of opposed necessities, the necessity of being and the necessity of not-being. Metaphysically, the possibility of being which belongs to something that has also the possibility of not-being is a different sort of possibility from (and alien to, exclusive of) that implied by necessary being. This is clear if we think of that which necessarily is as having its necessary being grounded in a possibility (or power) for being such that it is of the very nature of this *possibility* to exclude any possibility of not-being (and *vice versa* for what necessarily is-not). On these terms, that which contingently is cannot be defined or even coherently described as combining the possibility implied by necessary being (taken on its own) with that implied by necessary not-being (taken on its own).

[37] His understanding of the possibility of what is-not by reference to "another time" surely means that he interprets "can be / can not-be$^{\text{not-always}}$" as "can be, then not-be/can not-be, then be" (cf. πάλιν, 282b21). But nothing in the text encourages us to think that he took the further step of *analysing* what follows the "then" into another expression of the form "—, then—". In any case, such an analysis could not provide independent grounds for rejecting one-way infinities of being/not-being, since it allows instances of C*and D* that are/are-not, never having begun to be/not-be—and then are-not/are.

[38] Kneale and Kneale (1962) 85: "a statement of contingency is a disguised conjunctive statement."

When the two-way possibility that constitutes contingency is conceived of non-synthetically, the possibilities of being and of not-being involved are mutually entailing.

(15) It is logically possible, of course, that something that possesses some kind of non-synthetic possibility of being and of not-being, either is, or is-not, at all times. But our passage surely rests on the assumption that the only basis on which something could be/not-be at all times is its possession of what we may call a monolithic possibility of being, or one of not-being, i.e., one that excludes any possibility of the opposite. Or perhaps, rather, it is assumed that we could not rationally assert that something is/is-not at all times unless we presupposed for it the relevant monolithic possibility. (The monolithic possibility is itself explained in terms of temporally omni-comprehensive natural spans of being: we shall return to that notion at the end.) The main point now is this: in a scheme where omni-temporal being and not-being are grounded on the corresponding monolithic possibilities, that which contingently is, is at one time and not at another, and likewise *mutatis mutandis* that which contingently is-not. Hence, not only is it the case (a) that *the modality of contingency* with which we are concerned is not merely a conjunction, but a mutual entailment, of possibility of being and possibility of not-being. It is also the case (b) that, on the contingent level, *actual being* and *actual not-being* are mutually entailing (as holding at different times). As we have seen in section 10, each of these mutual entailers applies to what is *genêton* and to what is *phtharton*, but without its logically following that *genêton* ↔ *phtharton*.

At 282b14–22, Aristotle assembles the four terms "is^{always}," "is-not^{always}," "*genêton*," "*phtharton*" as if in a further square of opposition (the third in our account). Analogy may have misled him to the conclusion that *its* subcontraries ("intermediates", b15–17) entail each other—analogy with various other squares already invoked in his argument and in our commentary, where mutual entailment of subcontraries indeed obtains. We can certainly understand why he hoped to validate a category that necessarily unites the *genêton* and the *phtharton*. If each of these is a modular element that could sometimes occur without the other, one teaming up with *cannot-end*, the other with *cannot-have-begun*, then, at that rate, the everlasting (ἀΐδιον) could turn out to be so by an external, or brute, conjunction of the modular element *cannot-have-begun* with the modular element *cannot-end*.

A treacherous analogy, then, should perhaps carry blame for Aristotle's confidence that *genêton* ↔ *phtharton*, and that this is demonstrably so. Certainly, a sense that demonstration is *required* must be what explains the windings of his argument in *De caelo* 1.12. On the other hand, his sense there that demonstration has been *achieved* is due not only to the work he has sunk into elaborating the squares of opposition, but also to his assumption, based on considerations in chapter 11 (281a7–27), that being and not-being are for temporal maxima. This carries with it the claims (a) that one sort of temporal maximum is the complete plenitude of time; and (b) that the only alternative sort of temporal maximum is a finite length of time. The exhaustiveness of this division is supported by the notion that something's duration is its natural and fully bounded span. Given (a) and (b), little work is needed to derive the conclusion that what comes to be must perish, and what perishes must have begun. In fact, it follows trivially.

Bibliography

Allan, D.J. (1936) *Aristotelis De caelo* (Oxford).

Barnes, J. (1985) ed., *The Complete Works of Aristotle: The Revised Oxford Translation* (Princeton).

Bogen, J. and McGuire, J.E. (1986–1987) "Aristotle's Great Clock: Necessity, Possibility and the Motion of the Cosmos in *De caelo* 1.12," *Philosophy Research Archives* 12: 387–448.

Bolton, R. (2009) "Two Standards for Inquiry in Aristotle's *De Caelo*," in A.C. Bowen and C. Wildberg (edd.), *New Perspectives on Aristotle's* De caelo (Leiden) 51–82.

Denniston, J.D. (1996) *The Greek Particles*, 2nd edn revised by K. Dover (London/Indianapolis).

Denyer, N. (2000) "Never Will and Cannot," *Aristotelian Society Supplementary Vol.* 74: 163–178.

Guthrie, W.K.C. (1939) *Aristotle, On the Heavens* (Cambridge, MA).

Hintikka, J. (1957) "Necessity, Universality, and Time in Aristotle," *Ajatus* 20: 65–90.

——, (1973) *Time and Necessity, Studies in Aristotle's Theory of Modality* (Oxford).

Hussey, E. (1983) *Aristotle's* Physics *Books III and IV* (Oxford).

Judson, L. (1983) "Eternity and Necessity in *De caelo* 1.12" (a discussion of Waterlow 1982), *Oxford Studies in Ancient Philosophy* 1: 217–255.

Kneale, W. and Kneale, M. (1962) *The Development of Logic* (Oxford).

Leggatt, S. (1995) *Aristotle, On the Heavens I and II* (Warminster).

Moraux, P. (1965) *Aristote, Du ciel.* (Paris).

van Rijen, J. (1989) *Aspects of Aristotle's Logic of Modalities* (Dordrecht).
Sorabji, R.R.K. (1980) *Necessity, Cause and Blame* (London).
Verdenius, W.J. (1969) "Critical and Exegetical Notes on *De caelo*," in I. Düring (ed.) *Naturphilosophie bei Aristoteles und Theophrast* (Heidelberg) 268–284.
Waterlow, S. (1982) *Passage and Possibility: A Study of Aristotle's Modal Concepts* (Oxford).
Williams, C.J.F. (1965–1966) "Aristotle and Corruptibility," *Religious Studies* 1: 95–107 and 203–215.

TWO STANDARDS FOR INQUIRY IN ARISTOTLE'S *DE CAELO*

ROBERT BOLTON*

> One must not look for the same accuracy in λόγοι as in what comes through perception.
>
> (*Pol.* 7.7, 1328a19–21.)

1. *Aristotle's scientific method in the* De caelo

It has been clear since ancient times that the *De caelo* raises puzzling questions concerning Aristotle's scientific method. Already in Simplicius, for instance, we find clear attempts to identify peculiar features of Aristotle's method in the study of the heavens (Diels 1882, 292.3–20; Heiberg 1894, 510.19–31). Most, if not all, of the questions arise most prominently in connection with passages found in books 1–2, a fact that perhaps itself calls for an explanation. But for the investigation of these questions it is most useful to begin with what is arguably the most important remark on method in the *De caelo*, which comes in a passage in book 3, chapter 7. There Aristotle is discussing how the basic sublunar elements—earth, air, fire, and water—are generated from each other. He criticizes at some length on methodological grounds the view expressed in Plato's *Timaeus* that there is one exception to the general rule of the transformation of these elements into each other, namely, in the case of earth. Earth alone, Plato says, cannot be transformed into the other elements (54c–56d). This results from the need, in order to achieve the greatest *perfection* in the cosmos, as Plato puts it, to use two different types of basic triangles, or triangular surfaces, not constructible from each other, in order to build up the regular geometrical solids which, on Plato's account, compose the different elements (53c–55d).[1] In response, Aristotle says this:

* Rutgers University.

[1] Earth alone, on Plato's story, is built up out of isosceles right triangles while the

> It is not reasonable (εὔλογον) for it to turn out that one element alone [earth] has no part in the transformation [of the elements into each other]. Neither is it apparent on the basis of perception; rather [on this count] all [the elements] change equally into each other. As a result, [the Platonists] offer accounts which concern the φαινόμενα while their accounts are not in agreement with the φαινόμενα. The reason for this is that they have not acquired their first principles in the proper manner; rather, they were determined to bring everything into conformity with certain fixed ideas. For surely the first principles that concern sensible things must conform to the sensible, those that concern eternal things to the eternal, those that concern perishable things to the perishable; in general, principles must be conformable to their subjects. But because of their affection for their principles, they proceed like people who are defending their theses in reasoned arguments (ἐν τοῖς λόγοις). For holding their principles fixed as true they abide any consequence [of them], not seeing that it is necessary to judge some types of principles in the light of their consequences, in particular of what is ultimate. The ultimate thing [to judge a principle by] in the case of productive knowledge is the product; in the case of natural science, it is what is, in each case, authoritatively apparent from perception. (*De caelo* 3.7, 306a3 ff.)

Here Aristotle says quite plainly that the *final* test of the adequacy of any theory in *natural* science is whether the appropriate empirical implications or consequences of the theory are in fact properly apparent to perception.[2] Aristotle's complaint is that the Platonists discount the perceptually obvious transformation of earth into the other elements, for instance, no doubt, when earth or some earthy matter such as wood burns and is seen to change to fire and air. In *Tim.* 49b–c, Plato himself acknowledges that we do suppose that we *see* earth changing into other elements, but he then argues (53b–c) that this needs to be explained away in the

---

other elements are all constructed from scalene right triangles and thus, he holds, are transformable into each other (*Tim.* 54aff).

[2] There are disputes in the literature as to exactly what Aristotle means here by the proper or authoritative perceptual data in each case. Some have suggested that Aristotle has in mind only the types of perceptual data that in the *De anima* he describes as free from error or, more carefully, as least subject to error, namely, data that have to do with the so-called special sensibles (ἴδια αἰσθητά), such as color in the case of sight (2.6 and 3.3, 428b18). The proposal seems to be that Aristotle wants to specify an incorrigible, or a practically incorrigible, empirical base for the final validation of claims in natural science. See Irwin (1988) 33–34 with 314–315; Taylor (1990) 137–142. Cf. Owen (1986) 243. But when Aristotle refers to the perceptual data that are authoritative ἀεί (306a17), he does not have in mind one type of always inerrant data but rather the data that are "in each case" authoritative. See notes further below.

light of what is required for the most perfect cosmos.[3] It is very likely that it is just this move by Plato in the *Timaeus* that leads to Aristotle's complaint in *De caelo* 3.7 that Plato is too determined to hold on to certain "fixed ideas" (δόξας ὡρισμένας).[4]

This same approach is evident in many other passages in the *De caelo* where Aristotle is clearly relying on the same methodological rule stated in 3.7. In 2.13, 293a17 ff., for instance, Aristotle objects to the heliocentric (or pyrocentric) view of the universe adopted by some Pythagoreans, with its assumption of an imperceptible "counter Earth" which accompanies the Earth in its motion, because

> they do not look for accounts and causes that relate to the φαινόμενα but rather they drag and try to adjust the φαινόμενα in relation to certain of their own doctrines and opinions. (293a25 ff.)

Here Aristotle again objects to the sort of holistic approach which would permit one to adjust any φαινόμενα by reference to theory. In *Meta.* 1.5, Aristotle discusses further the Pythagorean view and says that they posited the invisible counter Earth, in particular, in order to bring the number of the moving heavenly bodies up to the perfect number ten (986a3 ff.). So here we have another case of the same type of reliance on dogmatic mathematical or numerological starting points for which Aristotle criticizes Plato in *De caelo* 3.7. As becomes clear in *De caelo* 2.14, 296a24–b6, where Aristotle responds in detail to the view that the Earth moves, either around a central fire or otherwise, the φαινόμενα that the Pythagorean proposals conflict with include observable data concerning the natural motion of bits of earth towards the center of the Earth and the movements of the fixed stars in relation to the Earth. These perceptual data, Aristotle argues, require a geocentric view.

A later passage in 2.14, 297b23 ff. provides us with yet another clear use of the rule of 3.7. Here Aristotle argues for a spherical shape for the Earth and also for its relatively small size, on the ground that this accounts for the appropriate perceptual phenomena (φαινόμενα κατὰ τὴν αἴσθησιν, 297b23 f.). These phenomena include the special changing patterns of light and shade on the Moon during a lunar eclipse,

---

3 In this type of case, the main visible perceptual data do not have to do with special sensibles, such as colors, but rather with what the *De anima* calls incidental sensibles (2.6, 418a20 ff.), such as earth and its transformation into other elements. So, Aristotle cannot mean here to restrict the ultimate authoritative perceptual data for judging a given theory in natural science to those that concern special sensibles alone.

4 See also *De caelo* 2.7, 289a19 ff. where changes of the type in question are mentioned.

which are explained, Aristotle says, by the spherical shape of the Earth (297b24–30); and the particular observable changes in the visibility of certain stars when we move about the Earth, which are explained according to Aristotle by the relatively small size of the Earth (297b30 ff.). In 2.12, 292a1 ff., Aristotle famously cites his own observation, as it seems, of the occultation of Mars by the Moon (on the evening of 4 May, 357 BC by some modern calculations) in support of the claim that Mars is further away from the Earth than the Moon. He refers there also to extensive Egyptian and Babylonian records of similar observations.[5]

One further feature of Aristotle's uses of perceptual data in the *De caelo* is worth special attention. As final evidence for the Earth's small size in 2.14, Aristotle cites the calculations of "the mathematicians" (298a15 ff.). This duplicates an earlier reference in 2.14, 297a2 ff. to the φαινόμενα reported in "what the mathematicians say concerning astronomy." Included here are reports of the relative movements and positions of the heavenly bodies (297a4 ff.). In 2.10 also, Aristotle cites various similar data from the astronomical works of "mathematicians" (291a29–32 with b9). These references to mathematicians as assemblers of relevant *observable* data for the confirmation of theories in astronomy are duplicated in other important texts outside the *De caelo* which offer general pronouncements on proper method in science. One is in *De part. an.* 1.1 where Aristotle contends that in general the natural scientist must proceed "just as the mathematicians do in their proofs in astronomy … first looking to the φαινόμενα … and then stating their causes and 'why'" (639b7 ff. with 640a13–15). An even more general rule of this same type is set down in very similar terms in *An. prior.* 1.30, where Aristotle says this:

[5] In none of these further cases, where Aristotle clearly relies on perceptual data to settle various questions definitively, are the perceptual φαινόμενα restricted to special sensibles. These passages, then, all cast further doubt on the contention that Aristotle is looking for some incorrigible empirical base for the validation of his views in astronomy, since the other types of perceptual data on which he standardly relies are not in the least, for him, incorrigible. So, if Aristotle does indeed suppose, as *De caelo* 3.7 indicates, that there is some *foundational* base of authoritative perceptual φαινόμενα such that theoretical proposals must ultimately be judged in any given case by reference to these types of φαινόμενα but not *vice versa*, these foundational φαινόμενα will not be epistemically incorrigible.

> Most of the principles in each science are unique to it. Therefore, it is the role of experience (ἐμπειρία) to yield the principles of each science. I mean, for instance, that experience in astronomy yields the principles of astronomical science, since it was only when the φαινόμενα were adequately grasped that the *demonstrations* [of the φαινόμενα] in astronomy [and thus the principles, on which genuine demonstrations must be based] were discovered. The same is true of any art or science whatever. (46a17–22: cf. *An. post.* 1.13, 78b34–79a6 and 2.19, 100a6–b5; *De an.* 1.1, 402b21 ff.)

From these closely connected passages, then, it is very clear that Aristotle regards all of the various types of empirical data concerning the movements of the heavenly bodies and related phenomena, assembled by the mathematicians (as he calls them) who have attempted to explain these data, as paradigms of the sorts of perceptual data that he takes to constitute the *necessary* and the *ultimate* authority for testing and judging the merits of any proposal in astronomy or any other science or discipline.[6]

However, though this strict empirical standard of confirmation is clearly and repeatedly articulated in *De caelo* 3.7 and elsewhere, and though it is followed with some obvious success, for instance in 2.13–14, much, indeed most, of what Aristotle claims and argues for in *De caelo* 1–2 is not defended on any such basis. In 1.1 for instance, Aristotle opens the work with a defense of the claim that the magnitudes with which natural science deals come in only one, two, or three dimensions so that the three dimensional magnitude is the complete or perfect magnitude. His argument for this goes as follows:

[6] In taking this stance, we should remember, Aristotle is directly controverting Plato's strong opposition in *Resp.* 7, 528eff. to the use of just such empirical data in proper theoretical astronomy. There Plato declares that there is no ἐπιστήμη of any perceptible things (αἰσθητά, 529b). So, if it is indeed correct, as has often been suggested, that Aristotle's views on scientific method, particularly in the *Analytics*, follow Plato's in that they are strongly influenced by the practices of mathematicians in his day, it is important also to see that the assembling of empirical data by astronomers of the sort we have just been describing counts for Aristotle himself as a necessary starting point for the proper practice of one branch of mathematics. As he says more explicitly in *An. post.* 1.13, the facts (τὸ ὅτι) or the φαινόμενα that one branch of mathematics, namely "mathematical astronomy," is required to explain are those provided by "perceivers" (αἰσθητικοί, 78b34–79a6). So, this one branch of mathematics has as a part of its very content theorems, or conclusions of demonstrations, supplied and suppliable only by "perceivers" (cf. *Meta.* 12.8, 1073b3–8, 32 ff.) In *An. prior.* 1.30, Aristotle says that this same pattern holds for "any art or science whatever" and, thus, by implication, for any branch of mathematics (46a21–22).

> Besides these three there is no other magnitude because "three are all" and "in three ways" is "in all ways." For just as the Pythagoreans also say, the whole world and all things are delimited by three, since end, middle, and beginning give the number of the whole and they give the number three. This is why, having taken it from nature as if it were one of her laws, we also use this number in the worship of the gods. And we also make attributions in this manner since we speak of two things or two people as *both* not *all*, and we use the latter appellation [*all*] first of three ... So, since all things and the whole and the complete [or the perfect (τέλειον)] do not differ from one another in form [i.e., in threeness] but, if at all, in their matter and in the things of which they are said, only a body will be a complete [or perfect] magnitude since it alone is determined by three and this is all.
>
> (268a8 ff.)

Needless to say, this argument, if one may call it that, makes no reference at all to any authoritative perceptual phenomena. Nevertheless, it is the only argument that Aristotle offers for the first main thesis that he tries to defend in the *De caelo*. Many other arguments, if not most, in the first two books follow a similar pattern. For instance, in 1.2 Aristotle supports the claim that there are just three types of *simple* underived locomotion in the universe—namely, away from the center, towards the center, and around the center—by an explicit appeal to his earlier contention, drawn from Pythagorean sources, that completeness or perfection is achieved "in three" (268b24 f.). Then, at 269a2 ff., on the basis of his conclusion that circular movement is simple, since the circle itself is simple (268b19), and of the further contention that every simple movement must belong by nature to some simple body, i.e., one with its own unique underived natural direction of motion (b26 f.), Aristotle then infers that there must *be* a simple body, different from the four traditional mundane elements, a body whose natural motion is circular (cf. 269a30 ff.). Here again, certain main premises of Aristotle's argument, such as the premise that the circle is simple and the premise that every simple movement must belong by nature to some actual simple body, are not in the least defended by appeal to any perceptual phenomena. Later, Aristotle himself refers to related premises only as a basis that one may posit (ὑπολαβεῖν) for reasonable (εὔλογος) argument (1.3, 270a12 ff.; cf. 269b18–23).

In *De caelo* 2.2, 284b6 ff., Aristotle argues further for various other remarkable cosmological conclusions, such as, for instance, that by certain calculations we, i.e., the Greeks, live in the lower left-hand part of the universe (285b24), largely on the basis of what he treats simply as a reasonable extension of the Pythagorean doctrine that there is a natural right and left in the universe.[7] He extends this to include a natural up and down and front and back, as in certain animals (284b18 ff.). In 2.4, 286b10 ff., Aristotle argues for the sphericity of the whole universe on the ground that this is "most appropriate" (οἰκειότατον) for it, and since the sphere is by nature the primary solid. This latter claim is in turn defended at 286b11 ff., on the ground that the circle is the primary plane figure (because its boundary line alone is one and complete and what is one and complete is prior to what is many and incomplete), and since "as the circle is among plane figures so is the sphere among solids" (286b25 f.). This is Aristotle's first main argument here for the sphericity of the universe. He then argues further (286b27 ff.) for the primacy of the sphere and, thus, for the sphericity of the universe on the ground that this is supported by the geometrical account of the generation of bodies from plane surfaces found in the *Timaeus*. His later argument continues in a similar vein (286b33–287a10).[8]

The most striking thing about these various arguments, and many others like them that one could cite, is not simply their failure to pay any heed at all to the strict empirical standard for final confirmation on which Aristotle later insists, though that is striking enough. It is rather that these arguments all rely on materials of a Pythagorean and Platonist character, based on numerology, on claims concerning the primacy, completeness, simplicity and perfection of certain mathematical entities and the like, materials of just the type that Aristotle explicitly rules out of court and dismisses as dogmatic "fixed ideas" in the passages where he lays down and employs his strict empirical standard. The rather stark question, then, which his procedures pose for us is just this: How can Aristotle use, in many passages in the *De caelo*, apparently without qualms, a mode of argument which in other passages he clearly denounces on grounds of methodological principle?

---

[7] On Aristotle's use of such directional principles in cosmology, see also the paper by J. Lennox in this volume below.

[8] On the problem of the sphericity of the universe see also P. Pellegrin's paper in this volume.

2. *Two standards of evaluation*

Now we might simply regard this problem as providing supporting evidence for the view of many modern commentators that in general the *De caelo* is an amalgam of incompatible doctrines, in methodology as well as in positive astronomy, perhaps because it contains strata from different periods of Aristotle's development.[9] But this type of view, I take it, should only be adopted as a last resort in case we are unable to find a basically consistent account of what the text presents to us. As a starting point, in attempting to reach such an account of the methodology of the *De caelo*, it will be useful to look again at our main passage in 3.7 where we have seen Aristotle set out his strict empirical standard for confirmation in astronomy (306a3 ff.). If we look carefully, we see that Aristotle does not in fact present there just this one empirical standard. Rather, he complains about the Platonists' doctrine concerning the transformation of the elements on two counts. He says first (306a3) that this view is not reasonable (εὔλογον), and only secondly does he say that it conflicts with the relevant perceptual data.[10] We can see that Aristotle does mean to introduce a dual standard here, since he goes on to give us an explicit example of what he has in mind in saying that the Platonists' doctrine is not εὔλογον.

> But, in fact, in the case of the bodies that *are* resolved [into each other on the Platonic theory], the stranding (παραιώρησις) of the triangles is not reasonable (εὔλογος). This [stranding] results from their transformation into each other because each [elemental] body is composed of a different number of triangles. (306a20–22)

As commentators have pointed out, on Plato's account of the transformation of water into air, for instance, certain triangular surfaces are left over which, at least temporarily, are not the surfaces of any body.[11] But it is just not εὔλογον, Aristotle complains, that there should be surfaces left dangling which are not the surfaces of anything. Here Aristotle's complaint is not that this goes counter to observation since, as the *Timaeus* indicates, the basic triangular surfaces on Plato's account are imperceptible (see 56b–c). Rather, Aristotle's complaint is that it just makes no sense to

[9] For views of this type, with further references, see Moraux (1965) cvi, ff.

[10] Cf. Simplicius (Heiberg 1894, 642.7–14), who says that Aristotle argues διχῶς against Plato, ἀπὸ τοῦ λόγου and ἀπὸ τῆς αἰσθήσεως. As we shall see in detail, this is exactly right.

[11] So already Simplicius (Heiberg 1894, 647.3 ff.).

commit oneself to surfaces that are not the *surfaces of* anything. In other passages also in the *De caelo*, Aristotle often draws a clear contrast between what is εὔλογον and what is directly evident from perception. In *De caelo* 2.11, for instance, Aristotle contrasts εὐλόγως argumentation for the conclusion that the shape of each heavenly body is spherical (291b11 ff.) with other arguments for this conclusion based on what is shown "through the things in the province of sight (διὰ τῶν περὶ τὴν ὄψιν)" (291b17 ff.). In 2.12, again, Aristotle contrasts one view that "would be held to be εὔλογον" on certain grounds (291b31 ff.) with another opposite view based on "what has come to us by sight" (292a3 ff.). How, then, are we to understand the nature of Aristotle's contrast between these two different sources of evidence—(1) the perceptually apparent and (2) the reasonable?

The appeal by Aristotle to what is or is not reasonable (εὔλογον) is very common indeed, especially in the *De caelo* but also in his other scientific works.[12] It is important to see, further, that Aristotle uses a family of related locutions to introduce this type of appeal. For instance, in his continuing discussion in *De caelo* 3.8 of the Platonic account of elemental transformation, Aristotle argues that this account is unreasonable (ἄλογον) since, given their shapes, Plato's basic elemental solids cannot all be combined together to produce larger visible bodies without leaving gaps of void or empty space between the basic elemental solids that make up the larger bodies (306b3–9). Aristotle does not explain just why he thinks this is unreasonable, but presumably he has it in mind that this means that the larger bodies would be constituted in part out of nothing, an idea which earlier in 3.2 he rejects out of hand (302a5). Also, as in 3.7, Aristotle goes on (306b9 ff.) to contrast this one type of difficulty for the Platonic theory—that it unreasonably constructs bodies in part out of void—with another type of difficulty, namely, that *observation* shows that air and water in particular fill up completely and without remainder the vessels which contain them, no matter what shape the vessels may have. He describes the former type of objection here as κατὰ λόγον (306b16) by contrast with the latter type of objection which is based on the appeal to perceptual data. So here, the expression κατὰ λόγον ("in accord with reason") is used as a substitute for εὔλογον ("reasonable").

[12] See Bonitz (1870) 297a42 ff. There is a still useful survey of Aristotle's uses of εὔλογος in the study by LeBlond (1938). LeBlond does not, however, sufficiently investigate Aristotle's parallel uses of other related locutions.

There are many other passages where the equivalence of these two modes of expression is even more directly expressed. This occurs for instance in *De caelo* 2.12, 292b25 ff., where Aristotle initially uses εὔλογος (b28) and then shifts to κατὰ λόγον (b31; cf. 2.10, 291a 32 and b3; 3.8, 306b9–16; *Phys* 3.7, 207a33 and b1).[13] In line with this, the *contrast* which we have seen in *De caelo* 3.7 and elsewhere between the appeal to what is εὔλογον and the appeal to what is supported by the perceptual φαινόμενα is matched, for instance in *De caelo* 2.14, by a parallel contrast between support which is κατὰ τὸν λόγον (297b17) and support by direct reference to what is perceptually apparent (297b23 ff.). This type of contrast, moreover, between what is κατὰ τὸν λόγον and what is κατὰ τὴν αἴσθησιν is itself also found in various forms in very many passages in the physical treatises and elsewhere (see, e.g., *Phys.* 8.8, 262a18–19; *De gen. an.* 3.10, 760b27–33) Bonitz (1870, 435a45 ff.) lists many of these in his *Index* under λόγος, but some that he misses deserve special attention here if we want to understand the historical background and the significance of Aristotle's contrast between what is εὔλογον or κατὰ λόγον and what is κατὰ τὴν αἴσθησιν.

We may begin with *Top.* 1.11, 105a3 ff. where Aristotle contrasts two types of questions, those which may be dealt with by λόγος or reasoned dialectical argument and others which can and should simply be settled by perception, such as whether snow is white or not. To see the background and significance of this contrast, which we have also found to be prominent in the *De caelo*, it is useful to consider several related passages in the *Metaphysics*. In 1.6, 987b27 ff., Aristotle claims that Plato was the first to posit numbers and forms as separate from perceptible things. This, Aristotle says, resulted from Plato's mode of inquiry ἐν τοῖς λόγοις. Here Aristotle uses the very same language which we earlier saw him use in *De caelo* 3.7, where he also describes Plato as defending his theses ἐν τοῖς λόγοις in disregard of the relevant perceptual data (306a3 ff.). Plato was the first to use this mode of inquiry, says Aristotle, since ear-

[13] In *De gen. et corr.* 2.3, after he has identified the four possible pairings of the elementary qualities—hot, cold, wet, and dry—Aristotle says this:

> It is κατὰ λόγον that these [four possible pairings] go with the [four] evidently simple bodies—fire, air, water, and earth. For fire is hot and dry, air is hot and wet (since air is like vapor), water is cold and wet, and earth is cold and dry. Thus, these [qualitative] differences are allotted εὐλόγως to the primary bodies and the number of these is κατὰ λόγον. (330b1–7)

See also Bonitz (18770) 297b7 ff.

lier thinkers were innocent of dialectic (987b29–33; cf. 9.8 1050b35). As Ross and other commentators point out, in his reference to Plato here Aristotle seems clearly to have in mind the well known passage in the *Phaedo* (99d–100a) where Socrates dissociates himself from the perceptual or empirical inquiries of his physicist predecessors and turns instead to inquiry in λόγοι (Ross 1924, 1.172–173). Ross himself translates the word λόγος in Aristotle's report in 1.6, 987b31 as "definition." But in *Phaedo* 99dff, Socrates does not turn from an empirical to a definitional inquiry. He turns rather to a particular mode of reasoned *argument* based on the use of hypothesis, which involves no special focus on definitions (100a, 101d–e). As we have just seen, Aristotle's discussion in 1.6 indicates that he thinks of the Platonic use of λόγος in the *Phaedo* as a use of dialectical argument, and λόγος in this passage should clearly be understood in this sense.[14] Earlier, when he wants to talk particularly about Socrates' interest in definitions, Aristotle uses the terms ὁρισμός or ὅρος, not λόγος (*Meta.* 987b3,6).

This, then, helps to set the historical background for Aristotle's own contrast in *Top.* 1.11 between λόγος (reasoned dialectical argument) and perception as alternate ways to try to settle things; and it shows us the influence of the *Phaedo* in fixing this contrast in these terms as an important one in subsequent discussions of methodology.[15] In later passages in the *Metaphysics*, as Ross also points out, Aristotle returns to the discussion of the special Platonic mode of inquiry in λόγοι, and he describes it generally in two further closely related ways:

(a) as inquiry using λόγοι that are καθόλου (13.8, 1084b23–25; cf. *An. post.* 1.24, 85a 37 ff.), and equivalently
(b) as inquiry that is λογικῶς (12.1 1069a26–28) or which uses argumentation that is λογική. (14.1 1087b18–21)

This last characterization of Platonist procedure is repeated in an important methodological passage in *De gen. and corr.* 1.2, 316a3 ff., where Aristotle contrasts the Platonists' λογικῶς mode of inquiry based on extended λόγοι or reasoned arguments (ἐκ τῶν πόλλων λόγων), with the

[14] For the use of hypothesis as employed in *Phaedo* 99dff, as a standard device in dialectical argument, see *Top.* 1.18, 108b12–19; 3.6, 119b35 ff.

[15] Plato himself is no doubt influenced by earlier uses of this type of contrast by, for instance, Parmenides who already contrasts reliance on perceptual experience with reliance on λόγος (Diels and Kranz 1956, 28B7). Another traditional version of this distinction is preserved in Aristotle's *Eth. Nic.* 10.1, 1172a34 ff.

proper φυσικῶς mode of inquiry for a natural scientist which, he says, is based fundamentally on extensive experience of the observable facts of nature (316a5–14). In these passages, then, we have two further closely related descriptions of that particular mode of support which, following the *Phaedo*, Aristotle himself wants to contrast with the empirical mode which he says in the *De caelo*, *An. prior.*, and *De gen. et corr.* is proper and ultimate for the natural scientist. In sum, this alternate mode of support makes use either of what is εὔλογον or of what is κατὰ λόγον or κατὰ τὸν λόγον καθόλου or of what is λογικῶς.

Aristotle's use of the expression λογικόν is of particular interest. In *Top.* 5.1, he introduces the term λογικόν to characterize that type of problem which is particularly suitable for treatment in λόγοι, that is by reasoned dialectical argument—which he earlier opposed to perception (1.11, 105a3 ff.). He says: "A problem is λογικόν with respect to which there arise many καλοὶ λόγοι" (129a30–31). So, a λογικῶς treatment of a problem, according to the *Topics*, is a treatment of it in "nice arguments" as opposed to a treatment of it by direct or ultimate reference to the data of perception. This, again, clearly reflects the Platonic precedent in the *Phaedo*.[16] Also, as is well known, Aristotle often contrasts in the physical treatises a mode of inquiry which is λογικῶς with one which is φυσικῶς, i.e., special to the natural scientist (see Bonitz 1870, 432b7; cf. *De caelo* 1.7, 275b12). In *Eth. Nic.* 7.3, φυσικῶς argumentation is that which "one must hear from the natural scientists" (1147a24, b8–9). In line with what we have already noted, in *De caelo* 1.12, 283b17 this contrast is alternately framed as a contrast between φυσικῶς inquiry and καθόλου inquiry. More generally, in *An. post.* 1.22 for instance, Aristotle contrasts λογικῶς procedure or argumentation with ἀναλυτικῶς (*scientific*) argumentation or procedure (84a7 ff.). He says in *An. post.* 2.8 that a λογικός proof cannot be a demonstration, i.e., a scientific proof (93a14–15). As we have seen in *De caelo* 3.7, a part of what is required for proper scientific φυσικῶς procedure is the final confirmation of theoretical proposals by reference to the appropriate perceptual φαινόμενα. This, as Aristotle says there, is what is ultimate and authoritative in φυσική (306a16).

[16] In *Top.* 1.14, Aristotle divides problems into ἠθικά, φυσικά, and λογικά. The later tradition came to treat these types of problems as belonging to three separate branches of theoretical philosophy. But as we can see from *Topics* 5.1, as quoted above, for Aristotle these types of problems are not mutually exclusive. A "physical" problem may also be λογικόν if it is especially suitable for treatment by λόγοι. Or it may not be, as in the case of the question whether snow is white or not. See, further, on this matter, below.

In *De gen. et corr.* 1.2, he again emphasizes, as we have noted, that it is φυσικῶς inquiry by contrast with λογικῶς inquiry that achieves success by finding principles that account for the authoritative observable phenomena (316a5–14). So, as we can now clearly see, Aristotle's appeal in *De caelo* 3.7 not only to what is proper and ultimate for φυσική but also to an alternate mode of procedure that validates results by appealing to what is εὔλογον, itself follows a *pattern* that is very common in his scientific works generally, one that we can detect not only by Aristotle's use of the word εὔλογον but also by his use of the other closely related expressions such as κατὰ λόγον, λογικῶς, and καθόλου. So, the dual track for inquiry and confirmation that we find in *De caelo* 3.7 is in all these various guises very common throughout Aristotle's scientific works.

The main importance of this result for our present purposes is this. In all of the texts from *De caelo* 1–2 that we have considered earlier, where Aristotle evidently ignores his strict empirical standard for confirmation and seems rather to make more use of characteristically Pythagorean and Platonic modes of argument, he himself describes his results and his argumentation either as εὔλογον or κατὰ λόγον or by use of one of the other related expressions that we have been investigating. In 1.1–2, for instance, where Aristotle defends certain theses by reference to the Pythagorean numerological doctrine of the completeness and perfection of the number three, he describes his results as κατὰ τὸν λόγον (268b5) or as κατὰ λόγον (268b25). In 1.2–3 Aristotle indicates that his arguments for the existence and character of the fifth body are based on what is reasonable (εὔλογον, 270a12). In 2.2 Aristotle's extension of the Pythagorean doctrine that there is a natural right and left in the universe to include a natural up and down and front and back, by analogy with certain animals, is recommended as εὔλογον (284b19, 23). The arguments in 2.4 mentioned above for the appropriateness of a spherical shape for the universe are based largely on καθόλου considerations (286b12), and on appeals to what is εὐλογώτατον (286b34).[17] The first two books of the *De caelo* are in fact noteworthy among Aristotle's physical writings for the frequency with which he makes use of this mode of argumentation.[18]

[17] For further discussion of the particular details of these arguments, see the contribution by P. Pellegrin to this volume.

[18] See the tables in LeBlond for data on the frequency of Aristotle's uses of εὔλογος, but not of the other expressions. For a use of λογικός in *De caelo*, see 1.7, 275b12.

So, rather than inexplicably ignoring or abandoning strict φυσικῶς procedure in these texts, Aristotle instead is *consciously* using the alternate mode of procedure and argument that he regards himself as also entitled to employ in his inquiries in natural science, a procedure which has a historical antecedent, at least, in Plato's turn in the *Phaedo* away from empirical data to the consideration of things in λόγοι or reasonable dialectical arguments. This makes it crucial for us, if we want to understand what Aristotle thinks he is doing in these texts where he employs εὐλόγως or λογικῶς argumentation, to try to say exactly what this procedure is for him, why he thinks it is necessary in astronomy, and what he thinks its merits and limitations are. This is of added importance in the study of the *De caelo*, since perhaps the two most problematic chapters in the work, 1.12 and 2.12, rely very heavily on this mode of argument. In 1.12 Aristotle argues against the one-way eternity of the universe—into the past or into the future but not both. At the end of the chapter he describes his main mode of argument as καθόλου as opposed to φυσικῶς (283b17–18; cf. 1.10, 280a32). In 2.12 Aristotle tries to account for the pattern of the differences in the number and character of the movements of the different heavenly bodies by viewing the actions of these bodies analogically as "of such a sort as those of animals and plants" (292b2; cf. 292a20). Here, as we have seen, he explicitly relies on what "would be held to be εὔλογον" (291b31) or what "would be held to be nothing παράλογον" (292a21) as opposed to what would be apparent from close observation (292a14ff., with 2.3, 286a3–7).[19]

[19] It is important to stress, as commentators often have, the highly analogical character of Aristotle's argument in *De caelo* 2.12 where he asks us to consider the heavenly bodies as of such a sort as living things (see recently, e.g., Leggatt 1995, 247–250). This itself makes it unlikely that Aristotle means here to be importing scientific principles from his biology as principles that have strict scientific standing here. We shall see further evidence for this below. Some of the main materials used in his biology on which Aristotle draws in developing his analogical comparison of the heavenly bodies with animals, in 2.12 and more especially earlier in 2.2, are fully discussed in J. Lennox contribution to this volume. See also the account of these and related matters in Morison (2002). There is not space here to discuss the proposal of some scholars that Aristotle uses ὡς at 292a20 to mean only 'as if,' so that he does not mean to actually commit himself in 2.12 to the view that the stars, and so forth are living things (see Leggatt 1995, 248–249 with references there.) As we shall see, the tentativeness of Aristotle's discussion here can in any case be established on other grounds.

3. *What is εὐλόγως procedure?*

What is it, then, to inquire or to argue by the standard of what is εὔλογον—or κατὰ τὸν λόγον (or καθόλου or λογικῶς) as opposed to what is κατὰ τὴν αἴσθησιν—or, more generally, what is φυσικῶς or ἀναλυτικῶς? The fact that there are these different standards for discussion or inquiry in Aristotle's scientific works generally, not simply in the *De caelo*, is one that has received increasing attention recently. For many decades, this fact was much obscured mainly due to the dominant influence of the view (pioneered in various versions and with various qualifications by Aubenque, Owen, Wieland, and others) that for Aristotle (as for Plato) the proper method of inquiry in science or philosophy is some form of dialectic. Resistance to this view began some years ago now with special attention to Aristotle's biological works, where it was argued that we can discern two modes of inquiry, one indeed dialectical and the other employing the alternate procedure we have seen described in *De caelo* 3.7, *An. prior.* 1.30, etc., which we can in general call ἀναλυτικός or scientific.[20] This approach by extension provides us with one possible way of trying to understand the two modes of inquiry in the *De caelo*. On this approach, the mode of inquiry that Aristotle labels as εὐλόγως, κατὰ τὸν λόγον, καθόλου, λογικῶς, etc. is just his dialectical mode of inquiry. The other is the more strictly scientific analytic procedure that Aristotle often contrasts with dialectic and describes as κατ' ἀλήθειαν as opposed to dialectic which, he says, proceeds κατὰ δόξαν (*Top.* 8.13, 162b30 f.). This scientific procedure, as we have seen, will involve final reliance on the empirical test discussed in *De caelo* 3.7 and *An. prior.* 1.30, and also reliance on the other requirements for reaching strict scientific knowledge laid out in the *Analytics*. Anything, including appropriate causal first principles, reached directly by these procedures or by appropriate use of what has itself been reached by these procedures will be reached by this strictly scientific mode of inquiry.

However, there is another way which commentators have often proposed for understanding the two modes of inquiry in the *De caelo* in particular. This alternate approach fits closely with some recent suggestions for understanding Aristotle's methodology in his *Metaphysics*, particularly in book 7. There we also find Aristotle employing a mode of inquiry which he labels λογικῶς by contrast with another which he does

[20] See Bolton (1987), for this approach, with discussion of the earlier literature.

not so explicitly label (see 7.4, 1029b13). On one recent proposal, λογικῶς inquiry in book 7 should be understood to be inquiry that restricts its argumentative base to what is found in Aristotle's so-called "logical" works, i.e., to the *Organon*. More importantly, this procedure also restricts itself there to what is for Aristotle *true* but *more general in scope* than what is specific to any given science or discipline. The other mode of inquiry, by contrast, draws rather on material specific to a given science or to a particular matter at hand. The purpose of this distinction, it is further suggested, is mainly pedagogical and reflects Aristotle's view, so it is claimed, that one should teach philosophy, including natural philosophy or science, by starting with the *Organon*, which presents truth that is more general in scope than what is special to any specific scientific inquiry, and only then move to the appropriate more special material.[21] A closely related approach to this has in fact often been proposed for understanding the *De caelo*. According to this approach Aristotle's procedures of argument, in certain passages in the *De caelo*, are meant to be elementary and exoteric and, thus, rhetorical in character, while in other passages his procedures are meant to be esoteric and more strictly scientific. This type of proposal for the *De caelo* goes back in one form or another to Diels and other scholars in the late 19th century, and it has been advocated in various versions by Jaeger, Wieland, and others more recently.[22]

Does either of these two types of proposals, we can now ask, fit or help us to understand the distinction of standards or modes of inquiry that we find in the *De caelo*? To begin, as it turns out, Aristotle himself in fact tells us the reason why he employs two modes of inquiry in the *De caelo*, and his reason is not drawn from pedagogy or, more generally, from the needs of elementary or exoteric presentation. In 2.3 286a3 ff., Aristotle

---

[21] This type of proposal is fully developed by Burnyeat (2001), esp. chapters 3 and 5, with further references there. The alternate suggestion offered here, that λογικῶς inquiry is standardly dialectical inquiry, is alluded to by Burnyeat (2001) 19, following Simplicius (Diels, 1895, 440a19 ff.), but basically ignored. See also F. Lewis *apud* Burnyeat (2001) 2 note 1.

[22] These scholars frequently do not agree on what material is to be assigned to what mode or on what the original source or sources of the material in the different modes may be. For references, see note 9 above. One should also contrast these two approaches both with that of Lloyd (1996), who proposes that in various passages in the *De caelo*, such as those where we have seen him avowedly proceed εὐλόγως, etc., Aristotle is engaged in "bluff" because he feels he is "seriously out of his depth" (161, 182), and with that of Falcon (2005, 85 ff.), who argues that in such passages Aristotle is signaling us that he is dealing with matters which he regards as in the end unintelligible to us.

points out that his investigation there into the question as to why different heavenly bodies have different patterns of movement is not based on the use of perceptual observations but rather on an alternate mode of investigation because the heavens are so far away from us, particularly in the sense that we can know very little of their character from perception (αἴσθησις) or close observation. In 2.12, 292a14 ff., where he returns to related questions, Aristotle repeats and expands on this point and offers it as the reason why he is required to rely in his inquiry on procedures that yield "what should be held to be nothing unreasonable (οὐδὲν ἄλογον)" (292a17–18). He says:

> On these topics, then, it is well to seek for still greater understanding, even though we have little to go on and are placed at such a great distance from the matters which concern the events themselves. Nevertheless, if we look at things on the basis of the following sorts of considerations, what is now puzzling should be held to be nothing unreasonable (οὐδὲν ἄλογον). We think of the stars as if we were thinking only of their bodies—of units with a certain arrangement, but wholly inanimate. But we ought to conceive of them as partaking in action and life, since in this way what results will be held to be not at all beyond reason (οὐθὲν παράλογον). (292a14–22)

So, the explanation of why Aristotle takes himself to be required at times to make use of, and to be content with, an appeal to what is εὔλογον or οὐδὲν ἄλογον does not primarily have to do with the needs of pedagogy or of elementary or exoteric presentation to others. As he indicates, Aristotle thinks he needs to use this mode or standard of argument in order *to discover and to confirm for himself in the best way that he can* the truth in astronomy, given the limits on the usability there of the alternate procedure for discovery and confirmation based on collecting the appropriate empirical data and finding their causes. In a closely related passage in *Meteor.* 1.7, Aristotle makes much the same point: "Concerning things not manifest to perception (αἴσθησις), we consider ourselves to have sufficiently proved our case in a reasonable way (κατὰ τὸν λόγον) if we have brought things back to what is possible" (344a5–7). Here Aristotle clearly does not mean that, on matters where the strict empirical scientific standard for confirmation is unavailable, all that is needed is a proposal that is in the modern sense *logically* possible, i.e., logically consistent and free of internal contradiction. He means rather that what is needed is a proposal that is coherent or epistemically possible, one that fits well enough with what is in general credible for us (see 344a7 ff.: cf. *De caelo* 2.1, 283b26 ff. and Heiberg 1882, 292.3–20).

In addition, it is also clear from the opening lines of *De caelo* 2.12, 291b24 ff. that Aristotle regards εὐλόγως procedure as distinctly inferior to the alternate more scientific level of procedure. He describes it as providing "little advance" (μικρᾶς εὐπορίας, 291b27). This is a point that Aristotle also makes in similar terms elsewhere, for instance in *De part. an.* 1.5, 644b22 ff., in his famous defense of the study of biology. There Aristotle again draws particular attention to the limits on our ability to advance our knowledge in astronomy due to the extreme paucity of appropriate perceptual data. He sharply contrasts this with our situation in biology where the relevant perceptual data are easily obtained; and thus, he says, we know *more* and we know *better* in biology than we do in astronomy (645a1 ff.). So, the fact that we must rely so much on εὐλόγως or λογικῶς argumentation in astronomy marks a deficiency in our current ability to fully master that subject in the best way, not any need to start from the *Organon* or from what is elementary or exoteric in character.

This explains then, we can now see, why Aristotle, in *De caelo* 3.7, can criticize the Platonists for proceeding ἐν τοῖς λόγοις (306a12–13) in using what is in fact comparable to some of his own εὐλόγως or λογικῶς argumentation, and for ignoring the more empirical scientific mode of reasoning that he there recommends *when the latter is available and goes counter to the former*. In such a case, exclusive reliance on εὔλογος argumentation is indeed reliance on dogmatic fixed ideas. On the other hand, in other places such as *De caelo* 2.12, Aristotle himself can rely exclusively on such argumentation when the superior mode is completely unavailable. Thus, at the outset of 2.5, 287b28 ff., Aristotle feels the need to preempt himself from criticism for relying exclusively there on the less exact and less conclusive mode of argument to defend the view that the outer heavens move *forward* rather than *backwards*. His argument (288a2 ff.) is that forward is more honorable (τιμιωτέρα) than backwards and thus better, and that nature always does what is better (288a2 ff.). This is why the outer heavens move forward.[23] Aristotle realizes, and quite rightly so as we might well agree, that this argument is more than a little weak. So he says, recognizing this, that the potential critic of this argument should take account of the mode of credibility of the type of argument used, "whether it is ἀνθρωπίνως or something stronger"

[23] See also *De caelo* 2.8, 290a29–b11 for Aristotle's invocation of such a teleological principle as something εὔλογον.

(287b33 f.). The merely "human" mode of argument, he indicates, may be the only one available "for now" while we wait for "more strict necessities" (ἀκριβεστέρας ἀνάγκας, b34). As earlier, the "necessities" Aristotle has in mind here are of course not logical but epistemic. What he awaits is a more strict and reliable guarantee of truth.

So, it seems very clear that Aristotle would not himself feel entitled to criticize so sharply the grounds for the Platonist λογικῶς account of elemental transformation as he does in *De caelo* 3.7–8, but for the fact that the more exact and conclusive procedure, which controverts the Platonist account, is readily available. On the other hand, however, in 1.3, 270b4 ff., Aristotle is clearly pleased to announce with regard to his view that the outer heaven is eternal and changeless that "λόγος bears witness to the φαινόμενα and the φαινόμενα bear witness to λόγος" (270b4–5). Similarly, in *Meteor.* 2.5 Aristotle rejects a certain view as "impossible both κατὰ τὰ φαινόμενα and κατὰ τὸν λόγον" (362b14; cf. *De part. an.* 3.4, 666a19). In these cases, since the results of the two modes of inquiry concur and are not in conflict, Aristotle is more than happy to avail himself of both.

In these various passages, then, to return to our main question, Aristotle seems not at all to have primarily in mind any doctrine which requires one, specifically for elementary or exoteric or pedagogical purposes, to use εὐλόγως or λογικῶς procedure first, and then to move to more genuinely scientific procedure for more advanced or esoteric purposes. Rather, his distinction of levels and his practice in the use of them has a more primary epistemological basis. He wants to take the fullest advantage of whatever viable procedures he can to discover and to confirm the truth for himself in astronomy, with proper caution concerning the relative soundness of the different procedures he uses for *this* purpose. His views on this point no doubt have pedagogical implications, but one of these is not, as we have seen, that one should always start with λογικῶς or ἀνθρωπίνως argumentation in order to achieve progress at the level of general truth by that means, and then advance to further more special truth by a more special scientific procedure. Sometimes, as in *De caelo* 3.7, the available εὐλόγως or λογικῶς argumentation is completely misguided and yields no truth and the scientific argumentation is all that we should or can profitably use. In other cases, εὐλόγως or λογικῶς argumentation is used, sometimes on very specific points as in *De caelo* 2.12, but only with the greatest caution, not for its superior value for elementary or exoteric purposes but because, unfortunately, nothing more accurate or secure is available for now. The extent of the use of such argumentation in *De caelo* 1–2 reflects the degree to which the topics discussed there fall into this category. But, in any case, it is important to see that the primary purpose of the distinction of levels in the *De caelo* is, broadly speaking, epistemological in the sense just described.

4. *Other discussions of the two standards*

We need now to consider two further passages in Aristotle's scientific works outside the *De caelo* which are especially important for a full understanding of his distinction of standards and its uses in science. The first is a familiar passage found in *Phys.* 1.2. After outlining in 1.1 the proper procedures for use in inquiry in natural science (περὶ φύσεως), Aristotle introduces in 1.2 another alternate procedure of investigation that is not within the proper scope of natural science (οὐ περὶ φύσεως), a procedure which can be successfully used even with, or by, those who lack or reject the basis for proper φυσικῶς or scientific inquiry (184b25 ff.). This alternate procedure, Aristotle indicates, is based on common (κοινή) knowledge, available even to those who lack all strict scientific knowledge. This procedure, he says further, is like the one used in discussing any thesis (θέσις) put forward on the basis of λόγος (λόγου ἕνεκα) such as for instance "the Heraclitean thesis" (185a5–7). Here Aristotle uses language, and the same example, which he also uses in a passage in *Top.* 1.11, where he explicitly identifies a θέσις as a paradoxical belief of a famous philosopher such as "the belief of Heraclitus that all things are in change," a belief which might be held, he says, "because it is supported by λόγος" (διὰ τὸ λόγον ἔχειν, 104b19–28). Such a θέσις, introduced on this basis, Aristotle indicates, is a proper subject of dialectical inquiry. He also says in *Phys.* 1.2 that the alternate procedure which he has in mind to φυσικῶς procedure is the one to be used to expose eristic arguments such as Antiphon's attempt to square the circle (185a5–20). Aristotle's terminology here is virtually identical to that which he employs in *Soph. elen.* 9 and 11 to distinguish proper scientific argumentation from argumentation based on those common things (κοινά) which, he says, are matters of common understanding and not properly known only by experts. Again, as in *Phys.* 1.2, he says there that the latter mode of argument is the one to be used to expose eristic arguments, including Antiphon's attempt to square the circle (170a34–b11, 171b4–7, b34–172b4). *In Soph. elen.* 11, Aristotle very clearly identifies this latter mode of argument as dialectic

(172a27–36, cf. Rhet. 1.1, 1354a1–3).[24] In *Top.* 8.14 he uses the terms common (κοινή) and general (καθόλου) interchangeably to describe the typical basis for this type of argumentation (163b32, 37; 164a3, 7–11). The latter term, καθόλου, as we have seen, is one of those that Aristotle uses standardly to signal the introduction of λογικῶς or εὐλόγως argumentation. These texts, then, strongly support the view that, for Aristotle, this mode of inquiry and argumentation alternate to the scientific one is what he calls dialectic. This fits well also with the parallel contrast which we find in *De an.* 1.1 between the procedures of a natural scientist (φυσικός) and those of a dialectician (διαλεκτικός, 403a29).

In any case, it seems clear that in *Phys.* 1.2 we find another appeal to εὐλόγως or λογικῶς procedure by contrast with φυσικῶς procedure. But the reason why Aristotle feels the need to use λογικῶς procedure there is *not*, as in the *De caelo*, because the relevant empirical evidence needed for using φυσικῶς procedure is unavailable. His dispute in that passage is with the Eleatics who completely deny the existence of change. But in Aristotle's view, as he explicitly indicates in *Phys.* 1.2, the empirical evidence in support of the existence of change, indeed of natural change, is easily available and overwhelming (185a12f.). The problem here, as Aristotle says, is that the posit of the existence of natural change is a *first principle* of physics. And, so Aristotle argues, one cannot use φυσικῶς procedure—including direct appeals to the relevant empirical evidence that confirms the principles of physics—to refute those who deny these very principles without begging the question. So, for this refutation λογικῶς procedure is required. Here we find another dimension to, or another use of, εὐλόγως or λογικῶς argumentation in addition to the one that is prominent in the *De caelo*. Here the need for this procedure is not due to the fact that the proper empirical evidence is unavailable *tout court*, but due to the fact that the status of the opponent's position renders this evidence unusable in debate. Nevertheless, as in the *De caelo*, the use of this procedure is again required not for pedagogical reasons or out of any need to use the *Organon*, but because the empirical scientific procedure is unusable.

In addition to this it is also clear, particularly from the related passages in the *Topics* and *Sophistical Refutations* that argumentation based on what is καθόλου or *common*, since this information is accredited by people in general, can be used in discussion with people in general on

[24] These passages are discussed more fully in Bolton (1991).

any pertinent subject at all (see, e.g., *Soph. elen.* 11, 172a27ff.; *Top.* 1.2, 101a25ff.). So, there are indeed exoteric, rhetorical or dialectical, uses of λογικῶς or εὐλόγως procedure on non-scientific subjects for instance, in addition to the more scientific uses. To claim that the use of this procedure in the *De caelo* or other scientific texts is not for such exoteric purposes is not to deny that, for Aristotle, the procedure has such uses. But these other uses are non-scientific unlike the use in the *De caelo* which has a genuinely scientific objective.

There is a second well-known passage outside the *De caelo* worth considering here, which illustrates particularly well the limitations of εὐλόγως or λογικῶς procedure. This is found in *De gen. an.* 2.8. There again Aristotle contrasts two modes of argument for a conclusion that he obviously wants to accept, namely, that mules are sterile. The one mode of argument he calls λογικός, the other he simply says is "based on the facts about the kind" in question (ἐκ τῶν ὑπαρχόντων τῷ γένει). The former is introduced as follows:

> Now perhaps a λογική proof [that mules are sterile] would be held (δόξειεν) to be more persuasive than those just given. I call it λογική for this reason, because, in so far as it is more [or very] καθόλου, it is further [or very far] from the proper starting points (οἰκεῖαι ἀρχαί) [by contrast with the other mode of proof.] The proof is of this sort ... It is impossible for any offspring to be generated by mules. For, an offspring of a *different* kind [than mules] cannot be generated [by mules] since what is generated from a male and female of the same species is the same in species as they are. And a mule [or offspring of the *same* kind] cannot be generated [by mules] because a mule is the offspring of a horse and an ass ... [So, no offspring can be generated by mules.]
>
> However, this argument is exceedingly καθόλου and worthless (κενός). For these [καθόλου] arguments that are not based on proper starting points are worthless. Rather, they are *held* (δόκουσιν) to be based on the material facts (πράγματα) though they are not ... What is worthless here *is held* (δοκεῖ) to be important (εἶναί τι), but it is worth nothing (οὐδέν). It is not correct (ἀληθές), since many [hybrid] animals that are not generated from [parents of] the same species do generate offspring [of their own hybrid kind], as was said earlier. This procedure (τρόπος) must not be used for inquiry either in natural science (περὶ τῶν φυσικῶν) or elsewhere.
>
> (747b27–748a14)

Perhaps the most interesting thing about this passage, for present purposes, is that Aristotle appears here to completely reject the use of λογικῶς or καθόλου argumentation concerning topics in natural science, though, as we have seen, he often uses it without question in the *De caelo* and in other scientific treatises. However, Aristotle's procedure here in

De gen. an. 2.8 is in fact parallel to what we find in *De caelo* 3.7. Just as in that discussion, Aristotle here completely rejects a certain use of λογικῶς argumentation because it relies on plausible and accredited contentions that observation shows to be false. The λογικῶς argument here supposes that if mules or animals of any other kind are produced by parents of different species then they cannot also be produced by mules or by parents of the same species. That is, the argument assumes that for any type of animal there can be only one way in which that type of animal is produced. But, however plausible, this, Aristotle says, is simply false. Sometimes hybrid animals that are produced by parents of different species are fertile and do then produce offspring after their own hybrid kind. The empirical evidence, then, shows that there are two ways in which these types of animals are produced not just one (cf. 747a29–34; see *De gen. an.* 2.7 for some of Aristotle's examples.)

So, if we look closely at this passage, we can see that the καθόλου mode of procedure (τρόπος, 748a14) that Aristotle rejects for scientific inquiry need not be λογικῶς argumentation as such but, as in the *De caelo*, only λογικῶς argumentation used when the results of a more reliable empirical mode of inquiry are easily available and undermine or go counter to it. Aristotle claims here that the assumption that every type of animal is produced in just one way is "exceedingly general" (καθόλου λίαν). That is, he must mean in context, this assumption overgeneralizes from certain common cases while neglecting other relevant cases. So, when Aristotle here criticizes a λογική proof as one that is very καθόλου and, thus, too far removed from some of the particular relevant starting points, he does not mean, as is sometimes suggested, that the proof is more *abstract* or that it draws on a generalization that is wider in *scope* than does some alternate form of proof. There is a true generalization of the same scope as the one Aristotle rejects that also concerns all types of animals, namely, that they are sometimes produced in two ways, not always only in one way. Aristotle himself uses this in his own argument, so he can hardly be objecting to the use of any generalization with this scope. He means rather that the generalization used in the λογική proof is secured at too great a distance from all the relevant particular data and thus in ignorance of them, and that it is thereby more liable to error. So, this passage is important because like passages in the *De caelo* it emphasizes the risks of εὐλόγως or λογικῶς argumentation on empirical grounds. The "proper starting points," it should be noted, that the λογική proof here ignores (747b28–30), are not proper causal first principles but proper empirical data from which the attempt to find an appropriate explanation should start. (748a14–16) Aristotle's language here in describing the λογικῶς argument as "very far (πορρωτέρω) from the proper starting points" (747b29) should remind us of his similar remarks in *De caelo* 2.3 and 2.12 where he emphasizes that he relies on what is εὔλογον in astronomy precisely because we are far removed (πόρρωθεν) from the most relevant particular data (286a4–5, 292a14–17). So, λογικῶς procedure is described here as καθόλου, at least in part, because it operates at the level of over-generalization, in ignorance of, or in disregard of, at least some of the relevant observable particulars. In *De gen. et corr.* 1.2, Aristotle also emphasizes this characteristic of λογικῶς procedure, that it may overgeneralize in ignorance of relevant observable particulars (316a5–14). Thus it is that this procedure is always more risky and always more liable to error than proper φυσικῶς or scientific procedure.

Also, it seems clear, further, that Aristotle does regard the λογική proof that he produces here in *De gen. an.* 2.8 as dialectical. He describes it and also arguments of this type not simply as exceedingly general (καθόλου λίαν) because they do not take account of all the relevant particular facts proper to the subjects in question, but also as arguments that "are held (δοκοῦσιν) to be based on the material facts" (748a9,) and as "held (δοκεῖ) to be important" (748a11). So, for Aristotle here, it is not just any possible generality or generalization of a certain scope or level of abstraction that forms the basis for a λογική proof, but only one that has a certain level of general acceptability or plausibility. This explains why Aristotle can say at the beginning of our passage here (747b27) that "a λογική proof would be held (δόξειεν) to be more persuasive" than an alternate mode of proof. This would not follow simply from the fact that a λογική proof is wider in scope than some others. Rather, as we shall see below in more detail, Aristotle's point here seems to be that people in general typically operate in their thinking on many questions at the level of generalities or overgeneralizations—without attention to or knowledge of all the relevant particulars—so that if one can argue λογικῶς, from accredited generalities, one is more likely to get a persuasive argument, even though the risks of error in such arguments are much greater than in others (cf. *Top.* 1.14, 105b10–12 and the remark quoted at the head of this discussion.) So, this passage provides further strong support for the view that λογικῶς argumentation for Aristotle is just dialectical argumentation, which operates, typically, from what is generally accredited.

5. *Dialectic and λογικῶς procedure*

There is additional evidence to support the view that Aristotle's two standards or modes of inquiry in the *De caelo* are just the two that he often elsewhere marks out as *dialectical* and *scientific* (as in *Soph. elen.* 9, 170a34ff.). To begin with, Aristotle frequently defends the use of dialectic, or of reasoning from common or accredited opinions (ἔνδοξα), in dialectic or rhetoric, on epistemological grounds because, as he says for instance in *Rhet.* 1.1, "people have an adequate natural inclination toward the truth and so they turn out to attain truth for the most part" (1355a14ff.). Thus, as Aristotle says more specifically, for instance in *Eth. Nic.* 1.8 concerning the ἔνδοξα, or the views of the many and the wise about happiness: "It is reasonable that neither group is entirely mistaken, but right [in their views] on at least one point if not most" (1098b27–29). In *Eth. Eud.* 1.6, Aristotle uses this point to argue that "we must give some proof of what we say (ἀναγκαῖον δεικύναι πως)" by reference to what is as widely acceptable as possible (1216b26ff.). Also, however, in these and other passages, Aristotle draws attention, on epistemic grounds, to the limitations of reliance on common or noted opinions. In *Eth. Eud.* 1.6, he says that what is customarily held may well need correction in the light of what is "better known" (τὰ γνωριμώτερα, 1216b34). In *Meta.* 2.1, Aristotle says that while we do collectively attain the truth to a significant degree, our intellect (νοῦς) may still be blind to what is "most evident by nature" (993a30–b11). So, if we look in Aristotle for a mode of inquiry and of reasoning which he identifies and recommends for reaching credible results but which is still ultimately less reliable in science than strict scientific analytic procedure, the only thing that we find meeting this description is dialectic.

It would, of course, be a lengthy, and very tedious, exercise to try to show that every example of reasoning which, in the *De caelo* or elsewhere, Aristotle describes as εὐλόγως (or κατὰ τὸν λόγον, or λογικῶς, or καθόλου) fits his standards for dialectical argument. In fact, given the variety in the usage of such terms as these, both in ordinary Greek and in philosophical Greek, there is bound to be variation in Aristotle's own usage. The most that such a study could be expected to reveal is that in one very typical and common use, such as the one in focus here, all of these terms are used to designate dialectical argumentation. In further support of this thesis a few general observations may be worthwhile.

As we have seen, by contrast with those passages where Aristotle relies on empirical scientific φυσικῶς procedure, there are others in the *De caelo* where he argues εὐλόγως, etc. by use of the following devices:

(a) an appeal to longstanding tradition in, say, religion or mythology—to support, for instance, his views on the eternity, divinity and perfection of the heavenly bodies;
(b) an appeal to the testimony of recognized sages such as Pythagoras or Plato;
(c) an appeal to standard linguistic usage or to etymology;
(d) a very frequent use of analogy, involving, for instance, a projection from what we know from familiar close-to-hand terrestrial cases to what holds for unfamiliar far-away celestial cases; and
(e) an appeal to what holds καθόλου.

It is very easy to show from the *Topics* that each of these modes of argument forms an essential part of the standard equipment of dialectic. This is, of course, well recognized in the case of the appeal to tradition, or to recognized sages, or to facts about common linguistic usage and etymology. It is not so clearly recognized for the case of argument by analogy or for argument from what holds καθόλου. But, as for the case of analogy, in the *Topics* Aristotle emphasizes very strongly the importance in dialectic of argument based on what he calls "similarities" (τὰ ὅμοια). He says, for instance, in *Top.* 1.18 that "it is an accredited opinion (ἔνδοξον) that whatever holds of one of a group of similars, holds similarly of the rest" (108b13–14). Thus, he says, one can readily argue on this basis in dialectic so that the collection of lists of similarities and differences is most useful as preparation for dialectical argument (108a38 ff.; cf. 1.17, 108a7 ff.; 1.10, 104a12 ff., 8.1, 156b10 ff.).

Equally, and perhaps more importantly, Aristotle emphasizes in *Top.* 1.14 that dialectical premises should be "as general as possible" (μάλιστα καθόλου, 105b32). His main reason for this can be constructed from several passages. As he says right away in the opening lines of the *Topics* (1.1 100a18 ff.), dialectic is an art for reasoning about *any question at all*, by contrast with a special science or other discipline that has a restricted subject matter. Thus, as he says most explicitly in *Top.* 8.14, it is crucial, for the mastery of the art of dialectic, to develop techniques or lines of argument that have as general (καθόλου) or common (κοινόν) a range of application as possible, given the obvious difficulty of having relevant special points to hand for arguing on any question whatever (163b32–164a11, 164b16–19). This is particularly the case, he indicates, on out of the way or technical subjects where the special material is not generally accessible (164b16 ff.). This explains, then, why the material that comes to have the necessary widespread accreditation that makes it suitable for use in dialectical argument, especially on difficult scientific questions, will very often have a general or common character. It is not, as we have noted, that *anything*, or anything *true*, with wide range of application will have this kind of general credibility or that only such things will have this credibility. Rather, it is that if Aristotle wants to use a non-scientific mode of argument suitable for dealing with any subject at all, which he expects to have some legitimate pull on us or credibility for us as a group, that mode of argument must frequently, and especially in technical areas such as science, draw on what is καθόλου or κοινόν and can be used on a variety of subjects. This is required given our only feasible way of developing a facility for, and of using, a non-specialized technique of argument with universal coverage. Thus, in *Rhet.* 1.2, Aristotle specifically identifies dialectical and rhetorical lines of argument as "those that are common (κοινοί) in application to matters of right conduct, in natural science, in politics and to many things that differ in kind" (1358a10–14). As he goes on to make very clear (1358a17 ff.), these are not the only types of arguments usable in dialectic or rhetoric, but they do have a special prominence there in virtue of which dialectical argument can simply be referred to as καθόλου argument or as argument from κοινά.

This is exactly the way in which Aristotle introduces dialectic and its counterpart rhetoric in the opening lines of the *Rhetoric*, namely, as the technique that draws on "the things which it is common (κοινά) in a certain manner to everyone to understand and which are not drawn from special scientific knowledge" (1354a1–3). This is also the way Aristotle refers to dialectic in *Soph. elen.* 9, 170a34 ff. There he identifies the modes of argument that properly belong to dialectic in the following way:

> It is clear then that we [dialecticians] need not grasp the procedures (τόποι) for all types of refutations, but only for those that belong to dialectic. For these procedures are common (κοινοί) by comparison with [what belongs uniquely to] each [special] art and capacity. And [thus] it is the business of the scientist (ἐπιστμήμονος) to consider the type of refutation that accords [uniquely] with a particular science—whether it is apparent and not real, and if it is real, why. But it is the business

> of dialecticians to proceed from the common things that fall under no one art. For if we [dialecticians] grasp the accredited proofs (ἔνδοξοι συλλογισμοί) on each subject, we will have the basis for the refutations, since a refutation is a proof of a contradictory. (170a34–b2)

Here Aristotle identifies dialectic as that capacity that we possess for reaching reasoned results which draws on what we *commonly* understand and find credible independent of any mastery we may have of any *special* discipline or evidence. The manual for the art of dialectic which Aristotle constructs in the *Topics* and *Sophistical Refutations* is an attempt to codify the procedures that are involved in the skilled exercise of this natural capacity (see again *Rhet.* 1.1, 1354a6–11). These procedures are first described in the passage quoted above as employing the κοινοὶ τόποι and then, equivalently, as employing the ἔνδοξοι συλλογισμοί. This equivalence is appropriate, even though not all ἔνδοξα pertain to all or many subjects, because the technique of reasoning on this basis draws on our common mode of understanding, i.e., on the general capacity we have for reaching reasoned results that is applicable to anything at all by contrast with any science or other special mode of understanding. This, doubtless, is why Aristotle says at the beginning of the *Rhetoric* that dialectic is based on what it is common to everyone to understand. He also makes clear there his view of the importance of dialectic, so understood, as the basis for our normal rational everyday social interaction (a3–6, cf. *Top.* 1.2, 101a30–34).

There are, then, two connected ways in which dialectic simply is reasoning that concerns what is common or general. First, and most basically, it is the exercise of our non-specialized intellectual capacity for arguing reasonably about any subject at all. Secondly, given the only practical way this capacity can be developed into a skill, it characteristically, though not exclusively, involves using lines of argument (τόποι) which employ commonly intelligible premises that are applicable to many different kinds of things. It is important to see, moreover, that though dialectic does standardly reason from ἔνδοξα, i.e., from the recognized convictions or testimony of "the many and the wise," it is not essential to dialectical reasoning, as Aristotle describes it, that premises be *introduced as* the testimony of the many or the wise. He says only that this is "useful to add" when introducing such dialectical premises, not that this is standard or necessary (*Top.* 8.1, 156b20–23). He also does not fix on these types of premises for dialectic because by tradition these are the stipulated bases for dialectic. No one before Aristotle, including Plato, describes dialectic in this way, as employing as its premises the convictions of the many and the wise. Aristotle expects, rather, that opinions of these types will on their own typically have a common credibility, and he fixes on them in his own original attempt, as he sees it, to describe and codify the modes of argument that have this commonly felt reasonableness (*Soph. elen.* 34, 184a8–b8; cf. *Rhet.* 1.1, 1354a6–12). Further, he also says importantly that whatever is appropriately *similar* to, or *akin* to, the ἔνδοξα counts also as a dialectical premise since it will likely enjoy the same level of general credibility as do the ἔνδοξα (1.10, 104a12 ff.; 1.14, 105b3 ff.).[25]

Aristotle supposes, then, that the types of convictions that he calls ἔνδοξα will, typically, be ones that fall under our common mode of understanding so that we can avail ourselves of beliefs of these types and similar ones, as ready sources for information that has this standing, information to which we can expect people to respond on a common basis. The mode of discussion that operates on this common basis is what Aristotle calls dialectic.[26] We can see further from the way Aristotle ranks the types of belief that he calls ἔνδοξα at least a part of what his interest is in making use of them in the mode of reasoning that proceeds on the basis of what is commonly intelligible or reasonable. Since there are bound to be substantial disagreements in the use of such a mode of argument concerning what is or is not generally *reasonable* (εὔλογον), if we expect to make progress in discussion on this basis there must be some means of resolving these disagreements. To this end where, for instance, the views of "the wise" conflict with those of "the many" the former are disqualified by Aristotle from direct use as dialectical premises (*Top.* 1.10, 104a8–12). This, and other rules for dealing with disagreements in the use of the type of discussion which is based on our common understanding or on what we commonly find reasonable, would, of course, need to be backed up by Aristotle on appropriate epistemological grounds.[27]

[25] In addition to those ἔνδοξα which are the recognized views of the wise, Aristotle also includes among dialectical premises "the opinions which accord with the arts," since people would concede these as so described (*Top.* 1.10, 104a14–15, 33–37). Here, clearly, Aristotle does envisage the introduction of even unfamiliar views in a two-stage way, as the consensus of the experts. So, he does assume, reasonably enough, that our common understanding does underwrite such a procedure. But, still, he takes these premises to be different ones than the celebrated opinions (ἔνδοξα) of the wise (104a8–15).

[26] This, arguably, is one main unifying element in the conceptions of dialectic of Plato and Aristotle. Interest in arguments that proceed on a common basis is prominent already in the Socratic dialogues of Plato: see Bolton (1994). For further discussion of this topic in Aristotle, see now Smith (1997) 155 ff.

[27] These matters are further discussed in Bolton (1999).

In sum, then, all of the different types of arguments which Aristotle uses when he is employing the εὔλογος standard in the *De caelo*, do have, for him, an important place in dialectic. This gives a *unity* to εὐλόγως argumentation, in all its forms, when understood in this way, as dialectical argumentation. That is, all these modes of argument make appeal to what has some pull on us and has credibility and reasonableness for us generally, and legitimately so for Aristotle, independent of any connection with strict scientific results or canons of evidence.[28]

6. *Some final implications*

For our purposes here, the main interest of the proposal that Aristotle's non-scientific εὐλόγως or λογικῶς level or mode of argument in the *De caelo* is dialectic derives from the epistemic limitations that Aristotle standardly places on the value of dialectic in an area such as natural science where a superior mode of argument is available (see, e.g., *An. post.* 1.19, 81b18 ff.). If, as we have seen in the *De caelo*, dialectical argument on scientific questions may offer us only "little advance" or the best we can do "for now," and if its use often requires some apology for reliance on a very fallible "human" mode of credibility, then that must affect our view of Aristotle's commitment to and confidence in the arguments which he uses in places such as *De caelo* 1.12 and 2.12, where his mode of argument is largely of this type. In addition, if the fact that this mode of argument does still have "human" credibility makes it useful for Aristotle in leading us through the initial stages of scientific inquiry where, as he often emphasizes, we must start with "what is best known to us," we must also be prepared to see him, as he informs us he will, revising results reached on this basis in his later discussion (see *Phys.* 1.1, 184a16–24). A proper treatment of his use of εὐλόγως or λογικῶς procedure for any initial or expository purposes must take this into account. In any event, however, given the importance for current research on Aristotle generally of understanding the different standards by which he operates in his procedures of inquiry, the *De caelo*, with its special and nearly unique information concerning the need for these different standards and concerning their relative merits, is clearly worth special attention.

---

[28] It seems clear from *Top.* 8.5, 159b25 ff. that dialectic may also proceed for Aristotle in a restricted form, not, as standardly, from what is credible for us generally but rather from what is credible for some restricted group or individual.

It may be helpful to conclude on a historical note. In emphasizing the limitations of εὐλόγως procedure in *De caelo* 2.5, as we saw earlier, Aristotle describes this procedure as merely ἀνθρωπίνως (287b33 f.). This characterization involves, one may well suspect, a reference back to Socrates' description in *Apol.* 20d of the results that he reaches and sustains by his elenctic method as mere "human wisdom" (ἀνθρωπίνη σοφία) by contrast with more strict σοφία which he lacks and cannot reach by this method but which would definitely be preferable. Even more importantly, Aristotle's language here echoes Plato's own preemptive warning at *Tim.* 29c–d that his cosmology and astronomy there has only the status of a defective εἰκὼς μῦθος, of a "likely story," because those who have devised it have only an ἀνθρωπίνη φύσις, a merely "human nature."[29] In *De caelo* 2.5, then, Aristotle in effect puts his εὐλόγως argumentation in the *De caelo* on the same level as what Plato offers us in his cosmology in the *Timaeus*. The main difference, of course, is that unlike Plato, Aristotle thinks that it is in principle also possible to do better in cosmology. This may help us to understand, historically, why Aristotle felt the need to use and emphasize in the *De caelo* the distinction, which is entirely absent in Plato, between εὐλόγως and φυσικῶς procedure in the study of the natural world.

Bibliography

Bolton, R. (1987) "Definition and Scientific Method in Aristotle's *Posterior Analytics* and *Generation of Animals*," in A. Gotthelf and J. Lennox (edd.), *Philosophical Issues in Aristotle's Biology* (Cambridge), 120–166.

Bolton, R. (1991) "Aristotle's Method in Natural Science", in L. Judson (ed.), *Aristotle's Physics* (Oxford), 1–29.

——, (1994) "Aristotle's Account of the Socratic Elenchus," *Oxford Studies in Ancient Philosophy* 11: 121–152.

——, (1999) "The Epistemological Basis of Aristotelian Dialectic," (revised version) in M. Sim (ed.), *From Puzzles to Principles: Essays on Aristotle's Dialectic* (Lanham, MD), 58–105.

Bonitz, H. (1870) *Index Aristotelicus* (Berlin).

Burnyeat, M. (2001) *A Map of Metaphysics Zeta* (Pittsburgh).

Diels, H. (1882) *Simplicii in Aristotelis physicorum libros quattuor priores commentaria*, Commentaria in Aristotelem Graeca 9 (Berlin).

[29] In the *Meteorology* though not in the *De caelo* Aristotle does sometimes use εἰκώς in this manner, as equivalent to εὔλογος, e.g. at 346a30 and 372b22. But there he may feel less need to distance himself from the *Timaeus* than in the *De caelo*.

——, (1895) *Simplicii in Aristotelis physicorum libros quattuor posteriores commentaria*, Commentaria in Aristotelem Graeca 10 (Berlin).
Diels, H. and Kranz, W. (1956) *Die Fragmente der Vorsokratiker*, 3 vols (Zürich).
Falcon, A. (2005) *Aristotle and the Science of Nature* (Cambridge).
Heiberg, J.L. (1894) *Simplicii in Aristotelis de caelo commentaria*, Commentaria in Aristotelem Graeca 7 (Berlin).
Irwin, T. (1988) *Aristotle's First Principles* (Oxford).
LeBlond, J.M. (1938) *Εὔλογος et l'argument de convenence chez Aristote* (Paris).
Leggatt, S. (1995) *Aristotle: On the Heavens I and II* (Warminster).
Lennox, J.G. (2008) "*De caelo* 2.2 and Its Debt to *De incessu animalium*," in A.C. Bowen and C. Wildberg (edd.), *New Perspectives on Aristotle's* De caelo (Leiden), 187–214.
Lloyd, G.E.R. (1996) *Aristotelian Explorations* (Cambridge).
Moraux, P, (1965) *Aristote, Du ciel* (Paris).
Morison, B. (2002) On Location: Aristotle's Concept of Place (Oxford).
Owen, G.E.L. (1986) "*Tithenai ta phainomena*," in M. Nussbaum (ed), *Logic, Science, and Dialectic: Collected Papers in Greek Philosphy of G.E.L. Owen* (Ithaca, NY).
Pellegrin, P. (2008) "The Argument for the Sphericity of the Universe in Aristotle's *De caelo*: Astronomy and Physics," in A.C. Bowen and C. Wildberg (edd.) *New Perspectives on Aristotle's* De caelo (Leiden), 163–185.
Ross, W.D. (1924) *Aristotle's Metaphysics* (Oxford).
Smith, R. (1997) *Aristotle: Topics, Books I and VIII* (Oxford, 1997).
Taylor, C. (1990) "Aristotle's Epistemology," in S. Everson (ed.), *Epistemology* (Cambridge), 116–142.

NATURAL, UNNATURAL, AND PRETERNATURAL MOTIONS: CONTRARIETY AND THE ARGUMENT FOR THE ELEMENTS IN *DE CAELO* 1.2–4

R.J. Hankinson*

In chapters 2–4 of book 1 of his *De caelo*, Aristotle sets out to establish the existence of a fifth element, the aether, which is distinct from the standard four of Greek physics—earth, water, air, and fire—and which uniquely serves to provide the material for the heavenly bodies. His reasons for doing so are various. The heavens are by common consent divine and what is divine ought to be made of something better than the materials of the lower world (*De caelo* 1.3, 270b1–12); the terrestrial four elements inter-transmute and, hence, anything made of them is subject to destruction; but the heavens are eternal and unchanging, as long observation confirms (270b12–16); and the etymology of the word αἰθήρ from ἀεὶ θεῖν ("always running") confirms that the ancients too supposed it to be quite different in type from the terrestrial elements (270b16–25).[1]

But Aristotle also deploys a set of arguments of an altogether different nature in order to commend the aether's existence. Bodies which exist by nature, the elements, all have distinct, intrinsic, natural tendencies to move (*De caelo* 1.2, 268b14–16); indeed nature is by definition a source of motion and rest (*Phys.* 2.1, 192b13–14). But if a body has a natural tendency to move, a disposition, that is, which is uniformly

---

* University of Texas at Austin.
This paper has developed out of a long engagement with the arguments of *De caelo*, and a collaboration with Mohan Matthen which has lasted nearly as long. While I can now isolate no specific points in the interpretation presented for which he was originally responsible, some, perhaps many, of them are no doubt owed to him. The best, and also the least, I can do is express my general gratitude to Mohan for his companionship and support, both intellectual and otherwise, over the years. I would also like to thank Tim O'Keefe for reading over the typescript, making several very helpful suggestions, and saving me from a couple of logical errors.

[1] The etymology, of a typically fanciful and unconvincing Greek type, is owed to Plato's *Crat.* 410b. Aristotle (1.3, 270b24–25; cf. 3.3, 302b4–5) rejects the altogether more plausible Anaxagorean derivation of the word from αἴθειν (to gleam), since he wants to resist its assimilation to fire.

exemplified under the appropriate circumstances, the motions described by the bodies must be simple in the sense that there should be no jerkiness or deviation about them. But the only possible types of such uniform motions are those in a straight line and those along circular paths (*De caelo* 1.2, 268b16–20).[2]

Aristotle is not, as is sometimes claimed, committed to supposing that there must be an individual element associated with each of the possible types of simple motion, however they are to be individuated.[3] As Alexander puts it (*apud Simplicium*: Heiberg 1894, 13.20–14.3),[4] the geometrically-determined types of motion function for him as material causes of possible actual motions. In other words, if a motion is natural (in the sense of being the expression of an essential tendency of an elemental body),[5] then it must conform to one of the simple trajectories; but there is no converse necessity that every determinable simple trajectory must have some simple body whose nature it is to move along it. Thus, first impressions and those of many commentators notwithstanding, Aristotle does not suppose that there is a simple geometrical route to the conclusion that there must be some simple body whose nature it is to move in a circle (*De caelo* 1.2, 269a5–7).

Still, Aristotle does contend that

> (T1) all change in respect of place (which we call movement) is either straight, or circular, or a mixture of the two, since these are the only two simple motions. The reason for this is that these, namely, the straight and the circular, are the only simple magnitudes.[6] (*De caelo* 1.2, 268b16–20)

---

[2] Xenarchus, a Peripatetic of the first century BC, objected that the regular cylindrical helix was also simple: so why should an element not move spirally (Heiberg 1894, 13.22–14.21)? I discuss this and other Xenarchan objections to the Aristotelian system in Hankinson (2003).

[3] On this issue, see Hankinson (2009).

[4] Alexander's commentary on *De caelo* is lost, but it is extensively quoted and discussed in Simplicius' own commentary.

[5] This parenthesis is required, since Aristotle is happy to allow that the capacities for movement of animals are in a perfectly good sense natural (*De an.* 2.1; *Phys.* 2.1; but see Heiberg 1894, 242.3–11, where Simplicius glosses an Aristotelian argument by distinguishing between "things which move by nature [*sc.* in the elemental sense]" and "self-movers"). Of course, even animals can have tendencies to move *qua* composites of elements: it is in this sense, as Simplicius says, that a man falling off a roof falls with the motion appropriate to his (predominantly heavy) elemental composition (Heiberg 1894, 40.18–20; cf. 17.19–20), since compound bodies "move in respect of whatever [*sc.* elemental body] predominates": *De caelo* 1.2, 269a1–2, 5–6 (T5 below), 28–30. Cf. *De caelo* 1.4, 271a29–30; *De motu an.* 10, 703a25–28; Heiberg (1894) 24.14–20.

[6] The translations of the *De caelo* are taken from Hankinson and Matthen (2009) with minor alterations.

That might seem to suggest that every conceivable motion can be represented as some combination of straight and circular components, although it is doubtful whether Aristotle could actually have thought that.[7] Less improbable is the supposition that he thought that all actual motions in the world were so composed, including the voluntary movements of animals (cf. Heiberg 1894, 15.9–17 and T3 below). But for our purposes, what matters is the claim that is being made about the analysis of motions that have to do with elemental tendencies. If we observe some movement that appears to occur naturally (i.e., it seems to be intrinsic to the thing itself to move in that way; it is not simply being carried, pushed or pulled by something else), then that motion must reduce to some combination of simple, elemental motions—for if it does not, then the basic assumption that animates Aristotle's entire discussion, namely, that constant motile tendencies must be ascribed to intrinsic features of the elements which compose bodies, will have to be abandoned.

Now that last argument may appear to be circular—and so, in a sense it is. But Aristotle's cosmological method is to isolate that set of basic explanatory assumptions which best fits the facts overall; and to this end no part of the system as a whole can properly be considered independently of any other part of it. Ultimately, the account as a whole will be justified as a form of inference to the best explanation.[8]

Thus, it is a fundamental feature of Aristotle's account of the mechanisms of the world that all of their dynamics must, ultimately, be explained in terms of the continuous motions of stuffs whose nature it just is to move in that way. But there are (or so he thinks) only two basic types of motion, revolutionary and rectilinear (*De caelo* 1.2, 268b16–20); and of these, only the former is such that it can continue uninterruptedly for ever (*Phys.* 8.9) because it is the nature of this trajectory itself to have no beginning or end.

[7] The claim is repeated at *Phys.* 8.8, 261b28–29, and 8.9, 265a13–15. The geometry of conic sections was in its infancy when Aristotle composed *De caelo*; but it is hard to see that he could have supposed that, e.g., a parabola or an ellipse could be produced by superimposing straight and circular motions (although, as Galileo was to show, a parabolic arc can be produced as the vector sum of one trajectory of uniform velocity and another uniformly accelerating at an angle to it). At first sight, it looks as though *De caelo* 1.2, 268b16–20 suggests that all composites must be composites involving both straight and circular motion—but that certainly cannot be Aristotle's intention. See Hankinson and Matthen (2009) *ad loc.*

[8] I deal with this aspect of Aristotle's methodology in Hankinson (2009).

This last is true only if there are no straight trajectories of infinite length (if there were, then motions could continue for ever along them). But Aristotle argues at length in *Phys.* 3.4–8, and *De caelo* 1.5–6 that there can be no such things, both for geometrical and dynamical reasons. A particular line has to be a certain determinate length; but "the infinite ... is not that of which no part is outside, but that of which some part is always outside" (*Phys.* 3.6, 206b33–207a2): in other words, it is not of determinate length.

Dynamically speaking, nothing can traverse an infinite distance in a finite time, unless it travels infinitely fast—but the concept of infinite velocity is (Aristotle thinks) incoherent, since velocities are measured by ratios that hold between finite distances covered and finite times elapsed (*De caelo* 1.5, 272a7–273a6; *Phys.* 6.2, 238a20–30). Equally, nothing moving at a finite velocity along an infinite trajectory can ever complete its journey: there will always be more—indeed infinitely more—left to traverse. But natural motions are defined in terms of their endpoints; and for something to be of nature to move in a certain direction is for it to be able to arrive at its goal (cf. *De caelo* 1.6).

For similar reasons there cannot be a body of infinite magnitude, since if there were it would have to be either infinitely heavy or infinitely light; but the speed of a moving element is proportional to its weight, and there is (again) no ratio between the infinite and any finite quantity (*De caelo* 1.6, 273a21–274a13).

I do not intend to discuss directly the plausibility of any of these claims.[9] Rather I want to examine various features of the notions of contrariety, and of the natural and the unnatural, that Aristotle makes use of in his argument for the existence of the fifth celestial element. At first sight, his deployment of these notions seems to be confused and inconsistent, and his argument vitiated by fallacies of equivocation. Yet Simplicius was to argue (or was to supply some of the material for such an argument) that these impressions are unfounded, and that Aristotle does in fact have a consistent and consistently employed conception of both of these related notions. In what follows, I shall sketch Aristotle's arguments, present the case for the prosecution, and then attempt to rebut it using material derived from Simplicius and elsewhere; then finally I will try to assess the merits of the case as a whole.

---

[9] See Hankinson and Matthen (2009) *ad loc.*

1. *The six hypotheses and the argument for the aether*

> (T2) Circular motion, then, is that around the center, while straight is that upwards and downwards. By "upwards," I mean away from the center; by "downwards," towards the center. Consequently all simple movement must either be away from the center, towards the center, or around the center.
>
> (*De caelo* 1.2, 268b20–24)

Aristotle, as Alexander was to point out, distinguishes, from all the possible straight and circular motions, three which have orientations which are privileged in terms of the overall structure of space: this "was what he indicated with the article, when he said around *the* center" (*apud Simplicium*: Heiberg, 1894, 14.31–15.1). As Aristotle himself said at the outset of the chapter, its purpose was to investigate "the parts which the universe possesses in virtue of its form" (*De caelo* 1.2, 268b13);[10] and that form is, Aristotle thinks, spherical.[11] This is what determines the privileged motions. Consequently, says Alexander, not every motion which is geometrically circular is circular in this genuine sense:

> (T3) The motion of wheels in general, since it does not occur around the center of the totality, is not simple circular motion. For it has in a way an up and a down in each part, ⟨since any given part is⟩ at one time going up, at another down .... Moreover, ⟨motion⟩ rightwards and leftwards, forwards and backwards, whenever they are simple, occur either upwards or towards the center. For the movements of animals are no longer simple, since they

[10] For this translation, see Matthen and Hankinson (1993). The usual rendering "formally distinct parts" (Guthrie 1936, 11; cf. Moraux 1965, 3) does not capture the proper (and required) sense of the Greek.

[11] At this stage of the argument, this is simply assumed, although no doubt Aristotle thinks that it is a hypothesis of considerable empirical plausibility (cf. *De caelo* 1.5, 272a5–6: "we see the heaven revolving"). Later on, however, he will seek to establish it on the basis of further arguments, saying at 2.4, 287a11–12 that "it is apparent (φαίνεται), and has been established (ὑπόκειται), that the totality revolves in a circle." These arguments depend to some extent on the premise that there is a naturally-moving circular body, the *probandum* of *De caelo* 1.2–4; and so once again there is a whiff of circularity about the procedure. But Aristotle does invoke considerations independent of that assumption: the sphere is the most perfect form and, hence, most appropriate to the most divine body (*De caelo* 2.3, 286b10–287a11); it revolves in a circle (as perception shows: below, T4; n. 19), but there can be no void beyond it (established in *De caelo* 1.9; cf. *Phys.* 4.6–9); hence, it must be spherical in shape: *De caelo* 2.4, 287a11–22); and in any case, the system as a whole will stand or fall in terms of its overall coherence and explanatory plausibility.

> occur by way of flexing and extension of the limbs.[12] Consequently even lateral movements are rectilinear upward and downward ones: so says Alexander.[13] (*apud Simplicium*: Heiberg 1894, 15.2–13)

Any motion of a body can perhaps be viewed as some combination of the simple motions (although it need not actually be so). Consider any possible motion of any point within the celestial sphere: either it maintains a constant radial distance from the center, in which case its motion is (genuinely) circular; or it does not, in which case the radial distance either increases (and it moves upward), or decreases (and it moves downward).

But what about the motion of a wheel around an axis perpendicular to the Earth's surface (or, more properly, a radius of the cosmic sphere)? All of the points on its surface maintain a constant radial distance from the center; and so it seems to follow that, in this genuine sense, they are not moving at all, since they are moving neither towards, away from, nor around the center of the cosmos.[14] This initially paradoxical result will become crucial in what follows.

According to Simplicius, in order to establish the eternity, and elemental distinctness, of the heavens

> (T4) on the basis of motions, he [*sc.* Aristotle] adopts these six hypotheses: [H1] that there are two simple motions (circular and rectilinear);[15] [H2] that simple motion is of a simple body; [H3] that the motion of a simple body is simple;[16] [H4] that there is ⟨only⟩ one natural motion for each

[12] Although these flexings and extensions are themselves products of lateral motions, and the extremities describe arcs of circles and rotations (*De motu an.* 1, 698a18–b1).

[13] Translations of Simplicius, *In de caelo* are taken from Hankinson (2002) with minor alterations.

[14] Matters are slightly complicated by the fact that every point on the circumference of the spinning horizontal wheel describes a trajectory on the surface of a cosmic sphere (a cosmic sphere is a geometrical object, a sphere of arbitrary radius centered on the center of the universe). Of course, not every point describes a great circle on that sphere or even the smallest segment of one. So, not every point has a circular motion as this is defined either. But this need not bother us any further.

[15] See *De caelo* 1.2, 268b16–20.

[16] These (asserted at 269a3–4: T12 below) express distinct theses: [H2] states that simple motions belong only to simple bodies, while [H3] holds that simple bodies move (naturally) only with simple motions. The conjunction of [H2] and [H3] thus expresses the claim that all and only the (natural) motions of simple bodies are simple. But they do not entail that simple motions need be paired one-one with simple bodies (i.e., they are compatible with more than one body moving with the same simple motion).

> body;[17] [H5] that for one thing there is one contrary;[18] and [H6] that the heavens move in a circle, as perception confirms.[19]
>
> (Heiberg 1894, 12.6–11)

I discuss the nature of these "hypotheses" elsewhere;[20] the crucial two for our purposes are [H4] and [H5].

It is helpful to lay out Aristotle's arguments in a formal manner. Here is the one which involves [H4]:[21]

[A]

1	(1) if *x* is a simple body, then *x* has only one natural motion (Ass.; [H4])
1	(2) if *x* moves upwards naturally, it can have no other natural motion (1)
1	(3) if *x* moves upwards naturally, it cannot move in a circle naturally (2)
4	(4) fire and air move upwards naturally (Ass.)

hence,

1, 4	(5) fire and air cannot move in a circle naturally (2, 3)
1	(6) if *x* moves downwards naturally, it can have no other natural motion (1)
1	(7) if *x* moves downwards naturally, it cannot move in a circle naturally (6)
8	(8) earth and water move downwards naturally (Ass.)

hence,

1, 8	(9) earth and water cannot move in a circle naturally (7, 8)

hence

1, 4, 8	(10) if *x* moves in a circle naturally, *x* must be something other than fire, air, earth and water (4,9)

---

[17] *De caelo* 1.2, 269a8–9; 1.3, 269b33–34; cf. 1.8, 276a29.

[18] First expressed at *De caelo* 269a10, 14; see also *Meta*. 10.5, 1055b30 (cf. 10.4, 1055a19–20). For a slight refinement, see (E1) and note 44 below. The principle is anticipated in Plato, *Prot*. 332d.

[19] Cf. *De caelo* 1.5, 272a5–6; 2.4, 287a11–12 (note 11, above).

[20] Hankinson (2009).

[21] The logically pedantic (such as myself) might object (rightly) that the argument as presented is not formally valid; but that validity is easily, and I think uncontroversially, secured. The numbers in the left-hand column (e.g., T1) refer to the assumptions upon which each line of the proof depends; those on the right refer to the lines of proof from which the line in question immediately follows. The system is basically that of Lemmon (1978), although I have not usually thought it worthwhile to abbreviate the rule of inference used at each stage on the right hand side as well. Note: "Ass." abbreviates "assumption".

What is at once noticeable about that argument when expressed formally is how economical it is. It makes use only of [H4], along with two empirically-based assumptions ((4) and (8)) about the motions of the terrestrial elements, which commended themselves widely to reflective Greeks:[22] it just looks as though heavy things fall and light things rise; and it is at least defensible (and explanatorily economical) to suppose that they do so because they are, at bottom, composed of some small stock of basic elements whose nature it is to do so.[23]

Moreover, [H4] itself has much to be said for it as a conceptual truth, if it is interpreted as a conditional as it is in (A1) (i.e., if it is not assumed that there are such things as simple bodies so defined): for what it is to be simple in the appropriate sense is just to be endowed with one and only one defining internal tendency. This is a stipulation, to be sure—but there is nothing methodologically unsavoury about that. The task of physics is precisely to see whether it makes empirical sense to suppose that anything in fact satisfies this stipulative definition.[24]

Equally, (A10) is conditional in form. It remains to be established whether there is a body with such a nature; but if there is, it cannot be any of the terrestrial four.

The same goes, in parallel, for the argument linking simplicity of motion with a certain geometrical conception of simplicity of magnitude enunciated in T1, which may expressed as follows:

[B]

1 (1) if *x* is a natural motion [i.e. the motion of a natural body], then *x* is simple (Ass.)
2 (2) if *x* is a simple motion, then *x* follows a simple trajectory (Ass.)
3 (3) the only three simple trajectories are upwards, downwards, and in a circle (Ass.)

hence,

2, 3 (4) the only three simple motions are upwards, downwards, and in a circle (2, 3)

---

[22] Not to all of them, of course: atomists explicitly reject both the type of element-theory involved and the idea that there are privileged natural directions of motion, as does Aristotle's successor Strato (see note 39 below).

[23] All of these assumptions are not only questionable, they are false. Moreover, there are further questions as to why, and how, Aristotle arrives at the need for four, rather than two, sublunary elements (see Hankinson 2009, §VIII). But we may ignore these concerns here.

[24] See Hankinson (2009) esp. §§II–III.

hence,

1, 2, 3 (5) if *x* is a natural motion, then *x* is either upwards, downwards, or in a circle (1, 4)

Once again, it is important that (B5) does not assert categorically that there *are* natural motions; it only says what they must be like if there are to be any. The connection between natural motion and simplicity enunciated in (B1) and (B2) again is conceptual; but it relies on an intuitive principle of causality, to the effect that a single causal influence can manifest itself only in a single, particular, determinate way.[25] One might take issue with (B3), as Xenarchus did (note 2, above), but it too is defensible.[26] And even if the defence fails, it may be replaced by a weaker principle which will still do the work required of it, namely, of defining the parameters within which a motion may be considered to be both natural and simple: for again, (B5) does not assert that, if there are natural motions, they must follow all of the possible available simple trajectories.[27]

The conclusion is drawn (and the conditional nature of the reasoning emphasized) in the following lines:

> (T5) (a) If, then, there is simple motion, and circular motion is simple, and if it is both the case that the motion of a simple body is simple and that simple motion is of a simple body (since if it were of a compound, it would be in respect of what predominates), there must be some simple body which naturally moves with a circular motion in accordance with its own nature. (b) For while it might move with the motion of another, distinct body by force, it could not do so naturally, given that there is only one natural motion for each of the simple bodies.
>
> (*De caelo* 1.2, 269a2–9)

In (a) Aristotle expresses two of Simplicius' six hypotheses, [H2] and [H3] (although as [H3] does no work in the argument, it is perhaps better to think of [H2] and [H3] as forming for Aristotle a single compound assumption) and part of a third ([H1]. (b) also invokes [H4] (see further below, 100). This apparently yields the following simple argument:

---

[25] Plato makes use of such principles in the *Phaedo* 100b–102b, namely, that the same effect cannot be produced by opposite causes and that the same cause cannot produce opposite effects; see Hankinson (1998) ch. 3.1b.

[26] Heiberg (1894) 13.22–14.21; see Hankinson (2003).

[27] As Alexander put it (*apud Simplicium*: Heiberg 1894, 13.12–20, 29–14.3), the simple trajectories are material causes of the natural (simple) motions (see Hankinson 2003, §II and 2009, §IV).

[C]

1 (1) if *x* is a simple motion, then *x* belongs to a simple body ([H2])
2 (2) circular motion is simple (part of [H1])

hence,

1, 2 (3) circular motion belongs to a simple body (1, 2)[28]

But that seems clearly inadequate, since as (b) implicitly allows, something might move circularly by force; and if it does so, its motion is by definition not natural for it; and if its motion is not natural for it, then it is not, again definitionally, a simple body which is so inclined to move. For while it has been part of the arguments' structures so far that naturalness of motion implies simplicity, the converse has never been asserted.

It seems as though there is an ambiguity lurking in the terminology here. If we treat simplicity of motion purely as a matter of what geometrically-determined trajectory is followed, namely, either towards, away from, or around the center (cf. T2: call that "geometrical simplicity" or "simplicity$_G$"), then not all simple$_G$ motions will necessarily be the motions of simple bodies. For the compounds move "in respect of what predominates" (cf. note 5), and hence (C1) will be false. Conversely, (C1) is plausible only for a stronger sense of "simple," where simplicity of motion is treated as being a defining characteristic of simple bodies (call that "physical simplicity" or "simplicity$_P$"), in which case it becomes a stipulation. Hence, not all simple$_G$ motions will be motions which are also simple$_P$.

But (C2) seemed to be a purely geometrical determination; and this is true even if "circular motion" properly so called is restricted to motion about the center of the cosmos; for nothing as yet has determined that such a motion is natural for anything (recall that simplicity of trajectory does not entail the existence of an associated simple motion). In effect, (C1) needs to interpret simplicity of motion strongly, as simplicity$_P$. But (C2) is only (thus far at any rate) plausible in the weaker sense of simplicity$_G$.

[28] This formulation is a trifle equivocal as it stands. (C2) should really be interpreted as a universally-quantified conditional ("take anything you like, if it is a circular motion then it is simple"), from which the most that follows (without a further categorical premise to the effect that "there is circular motion") is itself conditional: "if there is circular motion, then it is simple." This will be of importance later on—but I prefer to leave the ambiguity standing at the moment.

Hence, it seems that if we are to conclude that some specific circular motion belongs to a simple body in line with (C3), then we need to have further reasons to suppose that this is the case. We cannot simply infer it from the fact that its motion is circular. Rather, we need to know that it is the *right* sort of circular motion; and it seems we cannot do that, in this argument at any rate, without begging the question at issue.

But there is a better way of reconstructing that argument, one which does not trade on the ambiguity between simplicity$_P$ and simplicity$_G$. This version of the argument also employs [H6]:

[D]

1 (1) the heavens move in a circle (Ass., [H6])
2 (2) natural circular motion is simple$_P$ (Ass.)

hence,

1, 2 (3) if the heavens move with a natural motion, that motion is simple$_P$ (1, 2)
4 (4) if x is a simple$_P$ motion, x belongs to a simple body (Ass., [H2])

hence,

4 (5) if the heavenly motion is simple$_P$, heavenly motion belongs to a simple body (4)

hence,

1, 2, 4 (6) if the heavens move with a natural motion, heavenly motion belongs to a simple body (3, 5)

That argument does not equivocate on the notion of simplicity in the manner of [C]; rather it consistently construes simplicity as simplicity$_P$. Of course, the conclusion is only conditional; it requires further considerations to establish the truth of its antecedent and, hence, to detach the consequent. But Aristotle does in fact supply such considerations. A page or so later, he argues that

> **(T6)** if movement in a circle is natural for anything, it is clear that it would be one of the simple and primary bodies which moves naturally in a circle, just as fire does upwards and earth downwards. But if the things that move in a circle by circular movement do so unnaturally, it would be bizarre and downright absurd for this to be the only continuous and eternal motion, seeing that it is unnatural: for it is evident in other contexts that the unnatural things are the most swiftly destroyed.
>
> (*De caelo* 1.2, 269b3–10)

That the heavenly bodies do revolve eternally is, Aristotle thinks, a reasonable inference from observation: "in all of past time according to the records handed down from one ⟨generation⟩ to another, neither the whole of the outermost heaven nor any proper part of it has ever apparently changed" (*De caelo* 1.3, 270b13–16).[29] Furthermore, if the heavens really do move as they do unnaturally, then they must be forced to do so by something else; and that something else can only be, for Aristotle, something which itself has such a movement by nature.[30] Thus, in one way or another, the antecedent of (D6) will be satisfied: either directly (the heavens really move themselves), or indirectly (there is something which moves naturally in a circle, and which moves the heavens). Hence, either way (D6) will imply that there is indeed "some simple body which naturally moves in a circular motion in accordance with its own nature" (above T5(a)).

2. *The hypothesis of contrariety*

So far, then, so methodologically salubrious. Let us now turn to the arguments involving the remaining hypothesis, [H5], the contrariety principle; for it is here, according to the prosecution, where the serious trouble begins.

---

[29] According to Simplicius (Heiberg 1894, 117.24–30), this is true for Babylonian records going back 1,440,000 years (by contrast the Egyptians can manage a mere 630,000).

[30] At this point it will be objected that, for Aristotle, the ultimate cause of all motion, including circular motion, is an *unmoved* mover. But, *pace* Guthrie (1936, 79 n. c), there is no real hint of the doctrine of the unmoved mover in *De caelo*. Thus, while 2.6, 288a27–b4 does distinguish between the mover and the thing moved in the case of the heavens, it is at least possible that the "mover" here is the internal nature or essence of the moving body in virtue of which it moves; equally, there is no need to suppose that other passages (1.7, 275b25–29; 1.8, 277b9–12; 2.12) imply the doctrine of *Meta.* 12.8 (esp. 1074a31–38) and *Phys.* 8.4–6 (on this issue, see Elders 1965, 27–33, although his is an excessively developmental and stratigraphic approach). Moreover, an unmoved mover is, properly speaking, a mover *for a particular body*; and for that body, the motion induced by its mover is natural. One might also take issue with the contention that only something which has a natural tendency to move in a circle can induce an unnatural circular motion in something else: "Why could an unnatural circular motion not be induced by a confluence of various natural non-circular motions?" (this question was posed by Tim O'Keefe). Part of the answer is supplied in note 7 above. But even so, one might point out that you can generate circular motion out of oscillating linear motions plus suitable constraints, as in any piston-engine; and although there is no reason to suppose

In chapter 5 of his "philosophical lexicon" (*Meta.* 5), Aristotle discusses the senses of the term "opposite" (ἀντικείμενον), one species of which is "contrary" (ἐναντίον):

> (T7) Said to be contraries are (a) those items, of things different in type,[31] which cannot apply at the same time to the same thing; (b) those items which differ the most of things of the same type; (c) items which differ the most in the same subject; (d) items which differ the most in regard to the same capacity (δύναμις); or (e) are such as to exhibit the greatest difference, either *simpliciter*, or generically, or specifically.
> (*Meta.* 5.10, 1018a26–31)

That is not entirely lucid—it is not clear what if anything (e) adds to what has gone before; nor how (b) differs from (c)[32]—but it establishes that for Aristotle there is no fixed or even preferred meaning of the term ἐναντίον. This in turn suggests that the same thing may have more than one contrary, depending on the context. At *Top.* 2.7, 113a14–19, Aristotle seems to say as much.[33]

On the other hand, common to (b)–(e) is the idea that contraries so called are in some sense polar opposites, at opposite ends of a continuous scale; which tends to imply [H5]. This is supported by Aristotle's remark at *Meta.* 10.4, 1055a3–5:

> (T8) Since things which differ from one another can do so either more or less, there is also such a thing as the greatest difference, and this I call "contrariety" (ἐναντίωσις). (*Meta.* 10.4, 1055a3–5)

---

that Aristotle knew that, he was surely familiar with the centrifugal circular action of a sling-shot. Ultimately, here as elsewhere, Aristotle's assumptions will stand and fall not on *a priori* considerations of conceptual consequence, but on how well they can account for the phenomena of the world as it apparently is.

[31] The word translated "type" here is γένος. But the sense cannot be that of things generically different in any strong sense. Rather, Aristotle simply means any two properties *F* and *G* which are such that, if *F* holds of something *G* cannot (and trivially of course *vice versa*), properties such as red and green for example (which clearly do fall within the same genus, namely, that of color). Indeed, if they are not in some way generically related, it is hard to see how *F* and *G* could be incompatible (other than in the weak sense of being predicates which only apply to types which never intersect, e.g., "happy" and "prime"). However, it is clear that γενός in the description of type (b) does have something like its technical sense; and it is hard not to think that Aristotle has been at least irritatingly (although not uncharacteristically) careless about terminology here.

[32] See Kirwan (1971) 152, Ross (1923) *ad loc. Meta.* 10.3, 1054b31–1055a1 distinguishes between contraries "which are different in genus, and those which are in the same category of predication and, hence, in the same genus and the same in genus."

[33] His subject is that of actions and dispositions in which there is more than one variable. His point is simply that the contrary of, e.g., doing harm to one's enemies may be either doing harm to one's friends or doing good to one's enemies (cf. *De int.* 14).

There then follows a rather obscure argument designed to support this claim, namely, that contrariety is "complete difference" (*Meta.* 10.4 1055a5–19; see T10 below), the upshot of which is that

> (T9) it is clear that it is not possible for there to be more than one contrary to one thing, since there is nothing more extreme than the extreme, and nor are there more than two extremes of a single interval.
> (*Meta.* 10.4, 1055a19–22)

So [H5] is explicitly asserted here. Yet T7(a) seems clearly compatible with there being many contraries in that sense—the same thing cannot be both red and green; but then it cannot be both red and blue either; so in sense (a), both blue and green are contraries of red (as well as of each other).

Yet for all that, it seems that for Aristotle (a) is a degenerate, or peripheral, sense of contrariety. And even though he does say elsewhere that "privation (στέρησις) may be allowed to count as a contrary" (*Phys.* 5.1, 225b3–4), the point is linguistic rather than substantial: "not-light," the privation of light, may *refer* to "dark," which is properly speaking its contrary. The point is reinforced in *Meta.* 10.4:

> (T10) The primary contrariety is between a state (ἕξις) and its privation; but not every kind of privation (for "privation" is said in many ways), but whatever form of it which is complete. (*Meta.* 10.4, 1055a23–26)

Moreover, even if something can count as the contrary of *F* just in case it is not (fully) *F*, that need not compromise [H5], if the contrary of the positive condition is considered to be the complete range of privative possibilities, in other words, if the contrary refers to the whole set of not-*F* states collectively and not distributively. Thus, being blue and being green are not separate contraries of being red, but rather different ways in which something may satisfy the privative contrary condition to redness, namely, that of non-redness: the contrary condition as such is not differentiated.

But even so, it is clear from T10 and elsewhere that properly speaking contraries are co-ordinate. If A is the contrary of B, then B is the contrary of A. While that works logically in the case of simple privatives, it is clear that "redness" and "non-redness" are not conceptually speaking on all fours: non-redness is an umbrella-concept in the way that redness is not. Equally substances cannot have contraries except accidentally (*Cat.* 5, 3b24–27; 4a29–32), nor can definite quantities (*Cat.* 3b27–32).

Furthermore,

> (T11) every contrariety must be a sort of privation, while perhaps not

> every privation is a contrary. ... Every contrariety has a privation as one of the contraries, but not all of them in the same way. ... Hence, for some of them there is an intermediate, for instance a man who is neither good nor bad, while for other there is not: it must be either odd or even. ... Consequently it is evident that one of the contraries is always predicated by way of privation; but it is enough if the primary and generic contraries ⟨exhibit this feature⟩, such as one and many;[34] for the others can be reduced to these. (*Meta.* 10.4, 1055b14–29)

And a little earlier, Aristotle explicitly distinguishes between contradiction and contrariety: "contradiction is not the same as contrariety; and privation is a form of contradiction" (*Meta.* 10.4, 1055b3–4; cf. *An. prior.* 2.8, 59b8–60a14; 2.11, 62a11–19). Again the upshot is that not all expressions of privation (and hence not all contradictories) will properly express contrariety. However, as T11 also maintains, it is not an entailment of two things being genuine contraries that there be some intermediate state between them, as the case of oddness and evenness shows: there is no continuum of states between two extremes, but rather everything in the genus is parcelled out bivalently between the two species. On the other hand, strictly speaking, intermediates themselves cannot be contraries—"there is no evident case of an intermediate which is contrary, nor could there be by definition" (*Meta.* 10.5, 12–15)—something which will become important later on.[35]

Finally, consider the following from the *Topics*:

> (T12) Further, you must see whether there is a contrary in the same species. This investigation is multifarious: first you should see whether the contrary is also in the same genus, the genus itself having no contrary (for contraries must be in the same genus if the genus has no contrary). But if the genus does have a contrary, you must see whether the contrary is in the contrary genus: for the contrary must be in the contrary genus, if the genus has a contrary. (*Top.* 4.3, 123b1–8)

Whatever the obscurities of that[36] this too suggests that of each contrary there is only one (genuine) contrary; and that, pretty clearly, has emerged as the dominant Aristotelian idea, whatever the indeterminacies of T7. With all of this in view, let us now return to Aristotle's arguments in the *De caelo*.

---

[34] That is, "many" $=_{df}$ not one (for the domain of positive integers).

[35] Aristotle discusses the nature of intermediacy between contraries at some length in *Meta.* 10.7.

[36] Some of them are occasioned by the fact that the *Topics* is a manual of practical argument, a primer of strategies to be used in probing the weak spots of an opponent's position.

3. *Fallacies of equivocation?*

Consider the following argument, which Aristotle does not express directly, but which nevertheless relies exclusively on assumptions to be found in T13 below, as indicated:

[E]

1	(1) for each thing there is (at most)[37] one contrary (Ass.: 269a10, 14 [H5])
2	(2) natural motion is the contrary of unnatural motion (Ass.: 269a9–10)[38]
3	(3) upwards and downwards are contraries of one another (Ass.: 269a14–15)

hence,

3	(4) motion upwards is the contrary of motion downwards (4)
1, 3	(5) the only contrary of motion upwards is motion downwards (1, 4)
1, 3	(6) if *x* moves upwards naturally, its only contrary motion will be motion downwards (5)
1, 2,3	(7) if *x* moves upwards naturally, its only unnatural motion will be downwards (2, 6)
1, 3	(8) if *x* moves downwards naturally, its only contrary motion will be motion upwards (4)
1, 2, 3	(9) if *x* moves downwards naturally, its only unnatural motion will be upwards (2, 8)

Note the wording of (E6): it does not merely contend that the only way for a naturally upwardly-mobile body to move downwards is unnaturally, but makes instead the altogether stronger claim, in line with the dominant account of contrariety discerned in section 2, that the only sort of unnatural motion for it will be that downwards. The same goes *mutatis mutandis* for (E8). It follows from this that, e.g., sideways motion will not (as such) be unnatural for it. Even so, it may contain either a natural or an unnatural component, according to whether it tends to bring the body closer to, or further away from, its natural place (cf. T3 above). (The notion of natural place is part of Aristotle's general conceptual repertoire in physics and cosmology; suffice it to say here that the natural places of

[37] See note 44, below.

[38] (2) might of course be derived from a prior, more general, principle, namely,
(2*) the natural is the contrary of the unnatural;
but I present the argument here using only premises which Aristotle explicitly assumes.

the terrestrial elements are those regions in which they are naturally at rest, and towards which they strive if they are removed from them).[39]

Now consider the following text which supplied the material for argument [E]:

> (T13) If unnatural ⟨motion⟩ is contrary to natural ⟨motion⟩, and for one thing there is one contrary, then since circular motion is simple, if it is not the natural motion of the moving body then it must be unnatural for it. So if fire, or something else of this kind, moves in a circle [*sc.* unnaturally], then its natural motion will be contrary to the circular. But for one thing there is one contrary; and upwards and downwards are contraries of one another. But if it is some other body which moves in a circle [sc. unnaturally], there will be some other motion natural to it. But this is impossible, since if it is upwards it would be fire or air, and if downwards water or earth. (*De caelo* 1.2, 269a9–18)

Natural motion is the expression of the body's essential motile tendency; unnatural motion, properly so called, is movement that actively *resists* that tendency. Thus, not all motions which are not natural are unnatural in this central sense.

T13 yields the following argument, which makes use of some of the earlier premises:

[F]

1 (1) if *x* moves in a circle, then *x* moves either naturally or unnaturally (Ass.: cf. C2, E1, E2)
2 (2) if *x* moves in a circle unnaturally, there must be some other motion which is natural for it (Ass.: cf. E2)
3 (3) the natural motion of fire [and air] is upwards (Ass.)

hence,

[39] Cf. *Phys.* 3.5, 205a8–b1; 4.1, 208b8–12; 4.4, 211a4–6, 212a21–30; 4.5, 212b29–213a11; *De caelo* 1.3, 269b18–29; 4.3–5. Strictly speaking, for Aristotle an object's place is the innermost surface of the surrounding envelope of material (*Phys.* 4.4, 212a2–14), and so, place, conceived in this way, is a two-dimensional surface. But such a place clearly circumscribes a region, and so this way of talking is in this sense harmless. Furthermore, precision demands that we speak of Aristotelian elements moving towards the extremities of their places (a piece of earth will always have the tendency to fall, even if it is supported by more earth in *its* natural place: see e.g., *De caelo* 2.14, 298a8–b17); but again, this is not important for our purposes here (although it does introduce further difficulties of its own). The Peripatetic Strato (note 22, above) rejected the notion, at least in Aristotelian form, that everything has weight (i.e., a natural tendency downwards) and that there is no corresponding upward tendency (Wehrli 1950, Frr. 50–52). Xenarchus certainly modified this notion too (see Hankinson 2003).

3 (E6) (4) the only unnatural motion of fire [and air] is downwards (3, E6)
5 (5) the natural motion of earth [and water] is downwards (Ass.)

hence,

3 (E8) (6) the only unnatural motion of earth [and water] is upwards (3, E8)

hence,

2, 3 (E6, E8) (7) if x moves in a circle unnaturally, x is neither fire nor earth [nor air nor water] (4, 6)

but

2 (B1, B4) (8) if x moves in a circle unnaturally, it must move naturally either upward or downward (2, B5)
2, 3 (B1, B4) (9) if x moves upward naturally, it must be fire [or air] (3, 8)
2, 5 (B1, B4) (10) if x moves downward naturally, it must be earth [or water] (5, 8)

hence,

2, 3, 5 (B1, B4) (11) if x moves in a circle unnaturally, it must be either earth or fire [or air ore water] (8, 9, 10)

hence,

2, 3, 5 (E6, E8, B1, B4) (12) x cannot move in a circle unnaturally (7, 11: *reductio*)

hence,

1, 2, 3, 5 (E6, E8, B1, B4) (13) if x moves in a circle, then x does so naturally (1, 12)

But, given that there *is* something which moves in a circle, and given [H4] (cf. T5(b)), then there will be an element, the aether, whose (sole) natural tendency will be for circular motion.[40] Thus (F13), along with [H6], allows us to satisfy the antecedent of the conditional in (D6); the arguments are no longer simply hypothetical.

But, according to the prosecution, this is precisely where the problems begin to surface. T5(b) appears to allow that something *could* move in a circle forcibly; but now (F13) seems unequivocally to have ruled that out.[41] What is more, this conclusion seems much too strong for

40 Of course, [H4] could be controverted, as it was by Xenarchus (see Hankinson 2003).

41 This apparent self-contradiction is not *in itself* an objection. It is of course perfectly dialectically respectable to assert that p, only if q, and then to deny q: the antecedent of the conditional in *modus tollens* is not first asserted and then denied.

any intuitive construal of the premises which are supposed to entail it. The difficulty again seems to involve an equivocation. (F1) commends itself as a logical truth; but it will be so only if "unnatural" is read simply as the contradictory of "natural," i.e., as "non-natural" (henceforth "unnatural$_N$"). For only in that case will its consequent exhaust all the available logical possibilities and, hence, be necessarily true, which in turn entails (by a standard theorem of propositional logic) the truth of the conditional itself.

But then it will not necessarily be the case that if a motion is unnatural (in this sense), then it must be the natural motion of something else, since that only follows if "natural" and "unnatural" are used in their strong, contrary senses, in which "unnatural" means "contrary to the nature of" or "counter-natural" (henceforth "unnatural$_C$"). But read in this way, (F1) is no longer a logical truth. Again, the crucial moves in the argument, the alleged entailments of (F4) and (F7) by (F3) and (F6), work only if "unnatural" is read as unnatural$_C$.

Things get worse when we consider that Aristotle himself, at any rate when he wrote the *Meteorologica*, did not endorse the conclusion of this argument either. In that text, he holds that the body of fire, and the upper air as well, are carried around by contact with innermost heavenly sphere (*Meteor.* 1.3, 340b32–41a12).[42] And this seems in flat contradiction with (F12), not to mention (F7).

At this point one might invoke developmental hypotheses. *De caelo*, it may be said, is an early, immature work (after all, it seems largely if not wholly innocent of the doctrine of the unmoved mover),[43] and there is nothing to prevent Aristotle from having at one time supposed that fire could never revolve under any circumstances, on the basis of the flawed argument [F], but later recanting and allowing the sphere of fire to be moved by the aether.

That may be right (although I am inclined to doubt it). But developmental hypotheses are generally evidentially underdetermined, and usually philosophically boring. And if Aristotle had changed his mind about so fundamental a feature of his system, one might have expected him to

---

[42] Actually, it is difficult to determine whether in the physics of *Meteor.* the sphere of fire, now called "the inflammable" (τὸ ὑπέκκαυμμα) and explicitly said *not* to consist of fire as such, really is distinct from the upper air (*Meteor.* 1.3, 340b21–30). But this complication does not affect our argument. On the issue of the movement of the firesphere, see Leggatt (1995) 178–179.

[43] See note 30, above.

signal the fact more explicitly in the *Meteor.* (although it must be said that Aristotle is not one for public recantation).

But in any case all is not yet lost. One premise of T13 has not been considered in the previous analysis, namely, that asserting the *simplicity* of circular motion (269a10–11; cf. 269a3: T5(a), (C2), (D2)). As Simplicius puts it:

> (T14) Ingeniously he says that the reason why, if circular motion is not natural then it is unnatural, is that it is simple. For if it were not simple, it would be possible for it to be neither contrary nor unnatural, but simply non-natural (οὐ κατὰ φύσιν). (Heiberg 1894, 19.20–24)

Simplicius introduces the expression "οὐ κατὰ φύσιν" (as opposed to "παρὰ φύσιν") precisely in order to signal the distinction between unnatural$_C$ (to which the latter expression should properly refer) and unnatural$_N$ (which the new coinage is designed to pick out). Effectively, he argues that Aristotle is entitled to take the disjunction of (F1) as involving the strong notion of unnaturalness$_C$ and yet still as being exhaustive because of his prior acceptance of the claim that circular motion is simple.

Here we need to recall the ambiguity diagnosed earlier between senses of "simple" (above, [C] and [D]), and the establishment in (D6) and T6 above of the conclusion that there is something which has a natural revolving tendency, and which is ultimately responsible for the phenomena associated with the heavenly bodies. The conclusion that fire (for example) does not move in a circle unnaturally (F7), then, is only meant to establish that, on this view, fire does not move in a circle unnaturally$_C$, against its will as it were. For all that it may still move in a circle, but not naturally.

And this sits perfectly well with the considerations brought up by Alexander (T3 above), and those deriving from argument [E]. Every motion in a straight line, no matter what its orientation, will involve at least a component of motion either upwards or downwards, away from or towards the center. The only motion not to do so is that around the center—ignoring for the moment the rotation of horizontal wheels (pages 87–88, above)—but this is precisely the motion of the firesphere or ὑπέκκαυμα (see note 42, above). Genuine circular motion does not take the elements either any closer to or any further away from their natural positions, since they maintain the same radial distance from the center; hence, this revolving motion is neither natural nor unnatural$_C$ for them. To put it another way, *qua* element they are not moving at all. If "unnatural" is defined in this way for motion, and if the simple motions are defined by simple lines, it does indeed follow that there can be (at most)[44] one motion contrary to each of them.

Yet the prosecution will urge that this construal does not sit terribly well with the text as we have it. Indeed, it looks as though T13 fulfils some genuine argumentative function for Aristotle only if it establishes that fire cannot move in a circle *at all* (for why should it matter that it only cannot do so *against* its nature?); and yet that is precisely what the analysis just urged seeks to deny. Moreover a few lines later, he argues that

> (T15) From these things it is clear that there is some bodily substance other than the ones which exist around here[45] which is more divine and prior to all of them; and ⟨it would be so⟩ even if one were to assume further (a) that all motion is either natural or unnatural, and (b) that what is unnatural for one is natural for another, as is the case for those upwards and downwards. For the latter is unnatural for fire; the former, for earth; and *vice versa* in the case of the natural. Consequently, (c) it is necessary that the circular motion too, since it is unnatural for these, is natural for some other body.
> (*De caelo* 1.2, 269a30–269b2)

Here (a) seems to demand reading "unnatural" as "unnatural$_{\mathrm{N}}$," if it is to have any intuitive plausibility, since to assert that all motions were either natural or unnatural$_{\mathrm{C}}$ would be to assert that there were in fact no motions which were neutral for any particular object. Moreover, the conclusion (c) also requires interpretation in accordance with unnaturalness$_{\mathrm{N}}$, since evidently circular motion is *not* unnatural$_{\mathrm{C}}$ for fire and earth. On the other hand, (b) seems plausible as a general claim only if it is taken in the sense of unnatural$_{\mathrm{C}}$, since there are all sorts of motions (oblique ones, for example) which are unnatural$_{\mathrm{N}}$ for every element, but natural for none. Once again, things look like a mess.

The argument may be formalized as follows:

[G]

1 (1) all motion is either natural or unnatural (but not both)[46](Ass.)

2 (2) for any *y*, if *m* is unnatural for *y*, then there is an *x* such that *m* is natural for *x* (Ass.)

[44] This qualification to [H5] is only implicit in Aristotle's text. But it is required both generally in that some things are not contraries at all and, hence, can have no contrary (page 96, above), and in particular by *De caelo* 1.4, which argues that there is no contrary to circular motion (see page 113, below).

[45] That is, the four terrestrial elements.

[46] The disjuncts not only need to be (a) jointly exhaustive, they must be (b) mutually exclusive as well. "Natural" and "unnatural," in whatever interpretation, satisfy (b).

hence,

2	(3) if circular motion is unnatural for earth, water, air and fire, there is an *x* such that circular motion is natural for *x* (2)
4	(4) the natural motions of earth and water are downward, of fire and air upward (Ass.)
5	(5) every body has at most one natural motion (Ass.: [H4], (A1))

hence,

4, 5	(6) circular motion is not natural for earth, water, air and fire (4, 5)

hence,

1, 4, 5	(7) circular motion is unnatural for earth, water, air and fire (1, 6)

hence,

1, 2, 4, 5	(8) there is an *x*, distinct from earth, water, air and fire, such that circular motion is natural for *x* (3, 7)

And *x* of course is the aether. So on the face of it Aristotle has his desired conclusion.

Yet equivocation threatens here too. (G1) only has any initial plausibility if "unnatural" is read as "unnatural$_N$"—compare the case of (F1) above—but the pivotal premise (G2) seems to commend itself only if "unnatural" is construed as "unnatural$_C$." Oblique motion is not natural for anything, although an object, e.g., a ball descending a Galilean inclined plane, may move along an oblique or curved, trajectory as a result of its natural motion. Here the motion is best seen for Aristotle (and perhaps for Galileo too) as a form of constrained natural motion. Further, it is not true that every unnatural$_N$ motion has at least some component that is natural or unnatural in the stronger sense—consider again the motion of horizontal spinning tops (page 88, above). Again, for Aristotle, (G7) can only possibly true in the sense of unnatural$_N$. Here then, if anywhere, it seems as if Aristotle is guilty of equivocation as well as of adopting implausible principles.

Whether they also satisfy (a) is precisely what is at issue in the distinction between unnatural$_C$ and unnatural$_N$. Simplicius discusses the interpretation of "unnatural" at Heiberg (1894) 21.1–25, 51.28–52.18 (see also page 106ff., below). Simplicius (Heiberg 1894, 231.1–19) also discusses one of Aristotle's arguments (*De caelo* 1.7, 274b29–32) against the possibility of there being an element infinite in size, in terms which show that he is construing "forcible" (as applied to motion) to mean unnatural$_C$ (cf. *De caelo* 1.7, 276a6–16 and Heiberg 1894, 243.24–244.33).

Against this, one might note that [G] is a supplementary argument, one not necessarily uttered *in propria persona*. The concessive "even if" of T15 suggests that it takes off from considerations not endorsed by Aristotle, which in turn implies that in his view not all motion is necessarily either natural or unnatural. In other words, the proper sense of "unnatural" is unnatural$_C$. Equally, if we allow for the sake of argument that all motion *is* either natural or unnatural$_C$ (i.e., that *any* motion that is not natural for something actively counteracts its natural tendencies) then [G] involves no equivocation, since there is *ex hypothesi* only one sense of "unnatural". That is, every motion which is unnatural$_N$ will *ipso facto* be unnatural$_C$.

Even so, it is hard to see why, just because some motion is strongly contrary to the nature of one thing, it must be natural for something else. This would entail among other things that Strato's view (note 39, above) that every physical object has weight and, hence, tends to fall but yet may still be forced upwards by extrusion, must be conceptually incoherent. But Strato's position surely cannot be ruled out on logical grounds alone.

Might Aristotle falsely have supposed that it could? The evidence is against this. The extrusion theory predates Strato (it is owed to Democritus) and Aristotle was well aware of it. At *De caelo* 4.2, 310a7–13, he argues against those "who make all matter the same: nothing is unqualifiedly light or upwardly mobile, but is either left behind or squeezed out," on the grounds that large quantities of air and fire would be heavier than small ones and, hence, would have less of a tendency to rise (relative to the other bodies). He thinks that this can be shown empirically to be false (cf. *De caelo* 1.8, 277a27–b9) but he does not hold it to be logically incoherent.[47]

The closest he comes elsewhere to explicitly endorsing the principle expressed by (G2) is at *De caelo* 1.9, 278b29–34 (cf. Heiberg 1894, 283.1–9). In the course of arguing that no body could be outside the sphere of the fixed stars, Aristotle says that, if it were, it would be in an unnatural place; but this unnatural place would have to be natural for something else. But, of course, this is not the same principle and neither is the claim that whatever can move naturally can also be moved unnaturally (*De caelo* 1.7, 275b23–25); and although Aristotle may have thought that they derived from a more general axiom relating the natural and the unnatural, he nowhere either says, implies, or even suggests this.

[47] See also Heiberg (1894) 267.10–269. 28, 569.5–8, 712.27–30.

I think that, to the extent that Aristotle would have endorsed (G2) (and given the concessive nature of the argument there is no strong reason to suppose that he did), he would have done so on physical, dynamical grounds. That is to say, if something *is* moving unnaturally in whatever sense of "unnatural," there must be some reason why it is doing so; and that reason can only be because it is being moved by something which *can* move naturally in that way. At *De caelo* 1.7, 275b25–29, Aristotle considers the case of a (putative) infinite body being moved unnaturally,[48] and concludes that, if it were to do so, it could do so only under its own steam, in which case it would either be an animal ("unnaturally" then means "unnaturally *qua* physical body")—which is impossible because there cannot be an infinite animal—or under the influence of something else for which that movement is natural. Thus, if we restrict ourselves to consideration of motions that are not produced by animals, if there is an unnatural motion—if some body moves with a motion which is not natural to it[49]—then it must be caused to do so by something which naturally moves that way.[50] And, hence, G2 is vindicated but by physical, rather than purely conceptual, premises.

4. *Preternatural motions*

So, the defence will claim, there is a consistently clear and simple doctrine lurking behind the admittedly confusing and baroque facade of Aristotle's texts. Properly so called, unnatural (unnatural$_C$) motion is motion which fights against the natural tendency of the body in question. Given that definition, there will be some (not many) motions which are neither natural$_C$ nor unnatural$_C$, such as that of the firesphere (see note 42, above). This is the preferred interpretation of Simplicius; and he coins the term "preternatural" (ὑπὲρ φύσιν)[51] to describe, e.g., the circular motion of the firesphere:

[48] The precise interpretation of the text here is disputed; see Guthrie (1936) 64–65 note *c*, Hankinson and Matthen (2009) *ad loc*. But these disputes do not greatly affect the issue here.

[49] Again, we need not distinguish between unnatural$_C$ and unnatural$_N$.

[50] But see note 30, above.

[51] I am not certain that it is Simplicius' coinage, but I know of no earlier example of it. Note that it is not simply equivalent to "non-natural" (οὐ κατὰ φύσιν). For a thing's movement to be preternatural (or supernatural as "ὑπὲρ φύσιν" is sometimes rendered, though that has too many connotations of non-physical explanation), it must be both non-natural and the product of the natural motion of something superior to it (see T16).

(T16) This type of motion is neither natural nor unnatural but rather preternatural, or unnatural in that it is in accordance with the nature of something else more powerful that provides living motion ⟨for it⟩ in accordance with the highest degrees of life. And something unnatural in such a way is not a contrary; for it obtains neither in respect of contrary qualities (as the upward and downward do), nor do ⟨such motions⟩ conflict with one another, since the natural is rather preserved by the preternatural. (Heiberg 1894, 51.23–28; cf. 21.25–27)

Earlier, Simplicius has characterized this sense of non-natural motion as "something other than natural, since it is ⟨the motion⟩ of something greater which controls it" (21.22–23), i.e., the revolving aether. Indeed, it might be said that, properly speaking (and consistent with the suggestions made earlier), the firesphere was not really moving at all (at least *qua* fire); and Simplicius indeed suggests that when revolving at the periphery fire is really at rest (21.26–32).[52]

Here Simplicius is in stark disagreement with Xenarchus who holds (followed by Philoponus) that fire moves upwards only when it is in an incomplete state (i.e., when it is not in its natural place). Aristotle himself characterizes the motion towards its natural place for a body as motion "towards its form" (*De caelo* 4.3, 310a33–34), but when it is fully actualized (i.e., when it is at the periphery), it moves naturally in a circle (Heiberg 1894, 20.10–17, 21.32–22.17). But we need not concern ourselves with those arguments now.[53] Suffice it to say that it appears that a consistent set of considerations can be elaborated on Aristotle's behalf in favour of the postulation of the fifth element. Whether he is ultimately justified in so doing is another matter, and one that depends in large part on the plausibility of [H4].[54]

Still, even if one is convinced that there are mutually coherent interpretations of all of these arguments, and that Aristotle need not equivocate fatally on the senses of "natural" and "unnatural," one might still legitimately wonder whether Aristotle himself really had the entire business straight. It certainly would have helped if he had adopted something like

[52] This provides an implicit answer to an objection of Galileo:

On the other hand they [*sc.* the Peripatetics] describe as preternatural to them that circular motion with which they [*sc.* fire and the upper air] are incessantly moved, forgetting somehow what Aristotle has said many times, that nothing forcible can last a long time. (Besomi and Helbing 1998, 50; cf. Drake, 1967, 47)

On this interpretation, these movements are *not* forcible, since they are not contrary to the elements' natural tendencies (see note 55, below).

[53] They are discussed in Hankinson (2003).

[54] On this, see Hankinson (2009).

Simplicius' distinctions of terminology (even if he had simply used the phrase οὐ κατὰ φύσιν for unnatural$_N$). But terminological inexactitudes of this sort are surprisingly common for a philosopher of Aristotle's (often justified) pretensions to rigor; and a partial reason (if not an excuse) for it in this case may derive from the broad range of meaning of the Greek preposition "παρά," which may as readily mean "alongside" as "against" (a distinction which neatly parallels the ambiguity in question).

Consider the case of *De caelo* 1.2, 269b3–10 (T6 above), where Aristotle holds that it would be "bizarre and downright absurd" if the only continuous and potentially (indeed for him actually) eternal motion was unnatural for the thing moving thus, "for in other cases the unnatural motions are evidently the ones which are the most quickly destroyed." Here the requisite sense of "unnatural" seems clearly to be unnatural$_C$: it is only when something is actively fighting against the movement that it is undergoing, i.e., when it is being constrained actively against its nature, that such a state of affairs should have any tendency to rapid dissipation.[55]

On the other hand, what follows is apparently grist to the prosecution's mill:

> (T17) Consequently, if it were fire that moved thus, as some say, this motion would be no less unnatural for it than that downwards, since we can see that the motion of fire is away from the center in a straight line.
> (*De caelo* 1.2, 269b10–13)

The contention that circular motion would be just as unnatural for fire as downward motion is can only be true for the unnatural$_N$ sense of "unnatural": it cannot be a natural motion for it. But if all that Aristotle is seeking to prove here is that fire cannot move *naturally* in a circle (relying on [H4]), this would not matter (except from the point of view of terminological infelicity): if fire does move in a circle, it does so not according to its nature; hence, there must be something else, whose nature it is, which is moving it (page 106, above).[56]

---

[55] Thus, if this is correct, Galileo's point (note 52, above) against the Aristotelians is misplaced.

[56] But see *De caelo* 3.2, 300a20–27: each body has many unnatural motions. This is an unsatisfactory argument designed to show that every body must have one natural motion; since things evidently move, and they do so either naturally or unnaturally; but if unnaturally, forcibly, and if forcibly, *against* their nature. It is harder to see how this can be acquitted of trading on the equivocations mentioned. This passage is mentioned by Leggatt (1995) 179, to show that Aristotle did on occasion incline to the sense of unnatural$_N$.

Much the same seems to go for the following argument from chapter 3:

> (T18) (a) It is impossible that the body which moves in a circle should have either weight or lightness, since it can move towards the center or away from the center neither naturally nor unnaturally. (b) For natural movement in a straight line is not possible for it, since there is one ⟨natural movement⟩ for each of the bodies, and so it would itself be one of those which move in one of these ways. (c) But if it moved unnaturally, then if the downward movement were unnatural, upward movement would be natural; and if the upward were unnatural, the downward would be natural. For we laid it down that, in the case of contraries where one of them is unnatural, the other is natural. (*De caelo* 1.3, 269b29–270a3)

It seems as though T18 seeks to show that the heavens cannot move either upwards or downwards at all (Aristotle certainly takes that to be the case). But that should only follow if the disjunction in (a) is indeed exhaustive and, hence, that "unnaturally" is taken in the broad, inclusive sense as "unnaturally$_{N}$." But the contentions of (c) only work if "unnatural" is construed as "unnatural$_{C}$"—for it is only in this sense that the fact that the body has a particular motion unnaturally (whether upwards or downwards) entitles us to conclude that its converse would have to be natural for it (by the contrariety principle) and, hence, that it would have to be *per impossibile* one of the sublunary bodies.

But once again, this is too quick. Aristotle is not trying to establish here the general conclusion that the aether cannot move up or down at all (although he certainly accepts it), but only that it has neither weight nor lightness. For if aether did, it would have to have either motion upwards or motion downwards as one of its natural motions; but then it would be one of the sublunary elements—"heavy" and "light" just mean having these intrinsic motile tendencies (*De caelo* 1.3, 269b20–26; cf. *De caelo* 4.1). Thus "unnatural" in (c) may perfectly consistently be interpreted as having narrow scope.

But what about the conclusion a few lines later that "it cannot be moved in respect of place by being violently impelled either up or down, since it can move neither naturally nor unnaturally with any motion foreign to it" (270a8–11)? Again, it looks as though this is supposed to be a blanket prohibition on its moving up or down at all—and if that is supposed to be arrived at by argument, then the argument does indeed trade on an equivocation. But if all that is being said is that, if it does move up or down, then it does not do so either by force or naturally, then the argument can once again be read consistently (albeit at the expense of making its conclusion weaker).[57] It is not clear, here, whether defence or prosecution has the better case.

5. *Xenarchus' objections, Simplicius' replies*

Xenarchus thought that the argument of T18 and the others canvassed were designed to show on purely logical grounds that none of the sublunary elements could be such as to move naturally in a circle, since, if they did so, one thing would have to have more than one contrary—i.e., one unnatural motion (downwards, say, if it is fire) to two distinct natural ones (upwards and around). I reject that reading of the argument. But even though on my interpretation T18 is not designed to supply an argument for [H4], it none the less depends on it. However,

> (T19) Xenarchus objects ... to the dictum that for one thing there is one contrary. For he says, (a) it is easy for people to move fire forcibly along a line of any form whatever, be it simple or complex. "And we say," he says, "(b) in the ethical treatises too that there are two opposites for each of the virtues, e.g., wickedness and ingenuousness to prudence, over-confidence and cowardice to courage, and similarly with the others."
>
> (*apud Simplicium*: Heiberg 1894, 55.25–31)

(a) apparently rebuts Aristotle's arguments construed as claiming that there is no way for fire to move in a circle at all (i.e., in the manner of my prosecution); while (b) takes us back to the principle of contrariety. Simplicius is not impressed:

> (T20) It should be said against (a) that it is necessary that the unnatural motions too be proper to each; for these too exist by nature and not as a result of contrivance ... . Moreover, what is unnatural must be such as to be natural for something else. In regard to (b) ⟨it should be said⟩ that since each of the virtues is a balance, the two ⟨opposites⟩ to each of them are opposed as a single imbalance to a balance. And in fact while one is excess, the other deficiency, common to both is imbalance.
>
> (Heiberg 1894, 55.31–56.7)

As regards (a), Simplicius simply asserts that by "unnatural" Aristotle means the determinately counter-natural motions (i.e., those which are unnatural$_C$); they are "by nature" since the nature of the body in question

---

[57] Tim O'Keefe ingeniously suggests that Aristotle may simply be assuming that it cannot be moved unnaturally because there is nothing in contact with it to force it to move unnaturally. But that does not square well with the argumentative content of T18.

actively resists their imposition; and he evidently supposes that Aristotle endorses the proposition (see 269b33–34, (G2); section 3, above).

The reply to (b) is more interesting, however, and takes us back to the distinction between types of contrariety discussed earlier (T7 above). Simplicius allows that Xenarchus is right to hold that Aristotle is committed to there being a sense in which the extremes are opposed to the mean. Indeed, Aristotle explicitly says as much at *Phys.* 5.1, 224b30–35.[58] Moreover, and more immediately germanely to our subject, at *Cat.* 14, 15b1, he remarks that rest is the contrary *simpliciter* of motion; but of course there are contrarieties within the domain of motion itself. This suggests that there may be a sense of generic contrariety such that there can be a contrary to a genus which is itself composed of two contrary species, arranged as it were orthogonally to it. This is compatible with, although distinct from, Simplicius' example, where rashness and cowardice are contraries to one another on the continuum of confidence in the face of danger, while both together, *qua* departures from the mean, are the contrary to courage as imbalance is to balance (Heiberg 1894, 56.9–11).

Simplicius' reply takes the form of insisting that, properly understood, these generic contrarieties do not compromise the contrariety principle [H5]. Courage as such is not contrarily opposed to cowardice; so it is not the case that there are two distinct contraries, courage and cowardice, for rashness.

Even so, as Xenarchus points out,

> (T21) If these things are true, it is not necessary for the heaven to be of a fifth body on the grounds that two things cannot be opposed to one, the circular ⟨motion⟩ of the fire and that downwards or upwards.[59] For the upward is opposed to the downward as excess to deficiency, while that which is common to both, namely, the rectilinear, ⟨is opposed to⟩ circular ⟨motion⟩ as imbalance to balance.
>
> (*apud Simplicium*: Heiberg 1894, 56.12–17)

---

[58] Cf. *Top.* 4.3, 123b27–30: "there is an objection, namely, that deficiency and excess are in the same genus (for both are in the genus of evil) while the moderate, being intermediate between them, is in the genus not of evil but of good." The objection is to the general claim just enunciated that "the intermediates are in the same genus as the extremes" (123b25–27), e.g., that the various shades of grey are intermediate between black and white.

[59] Heiberg's text reads, "the circular ⟨motion⟩ of the fire and that downwards to that upwards," which cannot be right. An alternative solution would be to read "the circular ⟨motion⟩ of the fire and that upwards to that downwards." But Simplicius' response (T22, below) favors the version I give (see Hankinson 2002, *ad loc.* for further details).

So Xenarchus here adopts the sort of analysis urged by Simplicius in T20. Simplicius, however, rejects this too:

> (T22) While what he says here is elegant, I do not think that it relates to the contrariety as stated by Aristotle. For he does not oppose ⟨motion⟩ upwards and downwards to that in a circle, but rather the circular and the downward to the upward. But in regard to the supposition which says that it is fire which is carried in a circle, given that he says that ⟨the motion⟩ of the fire here has the structure of something which naturally moves upwards, it is not absurd for Aristotle to say that circular motion is no less unnatural for it than downward. And this is how he previously opposed them. So, since two things are reasonably taken to be unnatural for each natural, he concludes that there are two opposites for a single thing,
> (Heiberg 1894, 56.17–25)

—which contradicts [H5]. But Xenarchus' contention calls for a little more analysis.

Simplicius argues that, if there are two natural motions for fire (circular and upward) but only one unnatural motion (downward), there will be two distinct contrary states opposed to one, but not in the manner exemplified by the oppositions of the ethics. That is, it will not be an example of a genus, itself containing two contrarily-opposed polar opposites, being itself opposed *in toto* to some contrary. Rather, Simplicius urges, we will have to take circular and upward motion, *ex hypothesi* the natural motions for fire, and oppose them together as a an *ensemble* to downward motion. This does indeed look awkward, since circular motion does not seem to be opposed to downward motion in the same straightforwardly topographical manner as upward motion is.

But Xenarchus (as Simplicius objects) explicitly rejects that form of opposition in T21. Rather, we are to oppose upward to downward motion in the standard Aristotelian fashion as contrary opposites determined by what is natural and unnatural$_C$ for fire *in its incomplete state*. Then, that set of motions which are appropriate only to fire in its incomplete state are in turn to be opposed, as generically unnatural if you will, to the natural circular motion of fire in its complete, actual condition. If this is right, rising and revolving are natural for fire in different senses (or rather for different senses of "fire")—but that is a strength (from an Aristotelian perspective) and not a weakness of Xenarchus' treatment.[60]

[60] Xenarchus' position is also compatible with Aristotle's strictures in T12 above. When looking for the contrary of something, see if the genus of which it is a species has itself a contrary. If it does, the contrary species should be in the contrary genus. If it does not, then it must be within the original genus. The sort of case that Aristotle has

We may conclude that, no matter how precisely the crucial and controversial components of these arguments are to be interpreted, Aristotle does not have a knock-down argument derived from his own conceptions of contrariety against fire's having (in certain circumstances) a natural tendency to revolve. But on the other hand, it seems that Xenarchus' argument does indeed allow that in a sense the revolving body is different from any of the sublunar four: for it is fire in a state of complete actuality. Aristotle's principles, then, will not (in concert with empirical evidence) yield the conclusion that the revolving body is utterly distinct in type from any of the sublunar four;[61] but they do show (if of course they are acceptable) that it must be different in form from them in a nontrivial sense. In a way, then, Aristotle's conclusion (G8) may be allowed to stand.

6. *No contrary to circular motion?*

With this in mind, let us finally—briefly—turn to the arguments mounted by Aristotle in *De caelo* 1.4 in favour of the view that there is no contrary to circular motion. For here Aristotle himself makes use of the opposition between straight and circular:

> (T23) That there is no other movement contrary to circular movement one may confirm in a variety of ways. First of all, we state that the rectilinear is most opposed to the circular, since concave and convex seem to be opposite not only to one another but also to the straight, when they are taken together and as a unity. Consequently, if anything were contrary to it, rectilinear motion necessarily would be particularly contrary to circular

in mind is that of justice (*Top.* 4.3, 123b8–38): it is a species of virtue and its contrary, injustice, is a species of vice. But vice and virtue, the genera, are themselves contraries. But the case of motions is not on all fours with this: even if the straight and the circular are opposed, upward and downward are not themselves opposed in such a way that one is a restriction of the straight and the other of the circular.

[61] As noted at the outset, he has other reasons for holding this as well. The heavens are apparently eternal and unchanging and, hence, should be composed of stuff whose nature it is not to change. But fire, even completely actualized fire, will still retain its capacity for inter-transmutation into the other elements (*De caelo* 1.3, 270b1–26). See also *De caelo* 1.10–12, Aristotle's extended argument for the eternity of the world. One of the major contentions of this argument is that Plato's *Timaeus* view of a world which was created but which will remain undestroyed by divine *fiat* (although in principle destructible) is incoherent, since it entails supposing that something can possess a genuine capacity which is unactualized for an infinite time (see Hankinson and Matthen 2009, *ad loc.*; Waterlow 1982).

> motion. But the rectilinear motions are opposed to one other in regard to place, since up and down is both a difference and a contrariety of place.
> (*De caelo* 1.4, 270b32–271a5)

The argument is a type of *modus tollens*. "The best possible candidate for contrary to circular motion would be rectilinear motion. But rectilinear motion cannot be contrary to circular, since it already contains its own contrarieties. Hence, circular motion has no contrary." It is easy enough to see how Xenarchus would handle that argument consistently with the views expressed in T19 and T21; but we may allow ourselves a refinement of it. If circular movement (in any direction) is generically the contrary of rectilinear movement (in any direction), then motion clockwise and anti-clockwise should be the specific contraries within the one genus, answering to upward and downward in the other.[62]

Aristotle is quite well aware that the natural candidate for the contrary to circular motion in one direction is circular motion in the other; and he is no less aware that his preferred model of the universe, the nested spheres of Eudoxus and Callipus, require there to be counter-rotating spheres. Indeed his own physical realization of the system, sketched in *Meta.* 12.8, needs even more of them, and precisely in order to counteract the motions of the original spheres. None the less, he feels constrained to argue that, strictly speaking, anti-clockwise motion is not contrary to clockwise motion (*De caelo* 1.4, 271a5–22), for contrariety of motions is determined by their having opposed starting and end-points;[63] but motion in a circle is always both from and to all of the same places (271a19–28). But while it is of course true that (complete) circular motions in either sense touch all of the same points, they do not do so in the same order; and it is easy enough to devise an ordering measure which will distinguish between them.[64]

[62] Cf. T12 and note 60, above.

[63] He adduces other considerations too, some of which are not easy to disentangle (see Hankinson and Matthen 2009, *ad loc.*). But their obscurities need not concern us here. Note that at *Phys.* 8.8, 262a8–12 he appears to say that motions in opposite directions along a circular trajectory *are* contraries. However, at 264b10–17 he qualifies this by saying that these motions are opposite (ἀντικείμενα) and not contrary (ἐναντία), presumably since contraries in this sense are determined by their having opposing termini at a maximum distance from one another (so Ross 1936, 716–717).

[64] It can be done with materials that Aristotle himself provides. At *De caelo* 1.4, 271a19–22, Aristotle writes, "And not even the circular movement from A to B is contrary to that from A to C; for the motion is from the same ⟨point⟩ to the same ⟨point⟩, while contrary motion is that from one contrary to the other." But the first motion takes the points mentioned in the order A B C A, while the second takes the in the order A C B

This raises the obvious question: Why should Aristotle be so concerned to abstract circular motion from the realms of contrariety? The answer to that is to be found not in logic or mathematics, but in physics. The final considerations of *De caelo* 1.4 are physical—and metaphysical—in nature:

> (T24) If circular movement is contrary to circular movement, one or other of them will be to no purpose ... for if they were of equal strength, there would be no motion for them, while if one motion prevailed, the other would not exist. Consequently, if one of them exists, the other body would be pointless and would not move with the motion proper to it. For we would say that a shoe for which there was no wearing was to no purpose; but god and nature do nothing to no purpose.[65]
>
> (*De caelo* 1.4, 271a22–33)

The idea behind this is that for motions to be genuinely contrary, they must not just tend in opposite directions: they must have a tendency to impede one another in the manner in which an admixture of heavy elements impedes the tendency of light ones to rise (cf. T5(a) above). But if that applied to the celestial bodies, it would involve a waste of effort. Aristotle is clearly here examining the supposition that one and the same heavenly body (celestial sphere) could be composed of distinct, distinctly-tending elements, fighting it out, as it were, within it, which surely would be a messy and uneconomical way of building a universe. It is a further question why one should suppose that the universe will be economically constructed.[66] Aristotle, or so I believe, takes the "nothing to no purpose" principle as a highly-confirmed empirical hypothesis regarding the workings of the world. But discussion of this lies beyond the scope of this paper.

What emerges from this is that Aristotle does not suppose that the counter-moving spheres of a Eudoxan-style system are opposed, precisely because none of them has any tendency to impede the motion of any other. Of course, the resultant motions of the planets will be complex vector-sums of the various compounded motions; but each of the

---

A. Thus, for the first, A is invariably the immediate predecessor of B, B of C, and C of A; while for the second, B is the immediate predecessor of A, C of B, and A of C. Of course, this can be generalized for any number of points.

[65] That "nature does nothing in vain" is a constant refrain of Aristotle's; but this is the only occasion upon which god gets equal billing with nature. See Hankinson and Matthen (2009) *ad loc.*

[66] Congruent issues surface in connection with the supposed desiderata of elegance and economy in theory-construction. Why *should* Ockham's Razor be a prerequisite for any well-groomed theory?

motions in and of itself continues perfectly smoothly, and without friction. And in this sense they are not contrarily opposed.

At this point, one might wonder why Aristotle could not simply have distinguished between different sorts of motive contrariety, between a general contrariety of direction (in which the clockwise spheres can be said to move contrariwise to the anti-clockwise ones) and contrariety of opposition, which is foreign to the revolving body (or bodies). That is all he needs: and it is perfectly compatible with his physical considerations. But here, finally, it is worth once more considering Aristotle's overall aim in all of these arguments. He is out to show that the heavens are totally different in essence from the elements of the terrestrial world: the latter are in a constant turbulent state of inter-transmutation, growth, diminution, change, and decay. The heavens, by contrast, simply continue for ever on their serene path. The important thing is to remove the matter of the heavens from any possibility of change and, hence, generation and destruction; and that is to be done by insisting that they are not composed of parts with conflicting, mutually destructive tendencies. It is in this sense—the most important sense—that they cannot exhibit contrariety.

7. *Conclusion*

I hope that it has emerged from the foregoing that the concepts Aristotle deploys in his argument for the fifth element need not be supposed to be radically ambiguous and equivocally deployed. In particular, Aristotle need not be found guilty of confounding two quite distinct senses of παρὰ φύσιν in order to take advantage, whether advertently or not, of the power of one and the plausibility of the other.

And yet the suspicion remains that, even if it can be shown that Aristotle was not seriously confused, none the less he may still have had less than a secure grip on the concepts in question. And even if the arguments of the defence in favor of a generous understanding of T15 find favour, it is hard to resist the conclusion that the similar manoeuvres undertaken to buttress T13 succeed only at the considerable expense of rendering the argument in its context irrelevant. It is not that no good Aristotelian argument (in the sense of one which consistently applies consistent Aristotelian theses) can be found here: it can.[67] But it depends

[67] It is supplied above at page 102 ff.

on a consideration of the uniqueness of fire's natural upward movement (and so if it moves in a circle, something else is moving it), which makes the detour through the consideration of the impossibility of fire's unnatural circular motion redundant. The best the defence can do here is to point to the principle invoked by Galileo (note 52) that for Aristotle unnatural processes cannot be extended without end; and to the notion that if fire really were being moved contrary to its nature, it would be exerting a constant retarding effect on the circular motion, a retarding effect which would not have any point to it.

But that, crucially, is not what Aristotle says in the context of T13;[68] and one surely would expect him to have done so if the argument was indeed intended to be taken in the manner the defence suggests. In summation, then, a good Aristotelian case can be made (and to some extent was made by later Aristotelians) for there being a consistent and philosophically interesting construal of these arguments and of their key concepts. Whether, or perhaps rather to what extent, it is reasonable to think that Aristotle himself had it all straight is a good deal less clear—and those are questions I leave for you, the jury.

Bibliography

Barnes, J. (1992) *Aristotle: Posterior Analytics*, 2nd edn (Oxford).
——, (1995) ed., *The Cambridge Companion to Aristotle* (Cambridge).
Besomi, O. and Helbing, M. (1998) edd. *Galileo. Dialogo sopra le due massimi sistemi del mondo Tolemaico e Copernicano: I. Testo* (Padua).
Drake, S. (1957) *The Discoveries and Opinions of Galileo* (New York).
——, (1967) *Dialogues Concerning the Two Principal World Systems*, 2nd edn (New York).
Elders, L. (1965) *Aristotle's Cosmology* (Leiden).
Guthrie, W.K.C. (1936) *Aristotle: On the Heavens* (London/Cambridge).
Hankinson, R.J. (1998) *Cause and Explanation in Ancient Greek Thought* (Oxford).
——, (2002) *Simplicius: On Aristotle's De caelo 1.1–4* (London/Ithaca, NY).
——, (2003) "Xenarchus, Alexander and Simplicius on Natural Motions, Bodies and Magnitudes," *Bulletin of the Institute for Classical Studies* 46: 19–42.
——, (2009) 'Mathematics and Physics in Aristotle's Theory of the Ether', in K. Algra and F.J. de Haas (edd.), *Mathematics and Physics in Antiquity* (Leiden).
Hankinson, R.J., and Matthen, M. (2009) *Aristotle: On the Heavens 1* (Oxford).

[68] *De caelo* 3.2, 300a20–27 is even more resistant to charity in this regard (cf. note 56).

Heiberg, J.L. (1894) *Simplicii in Aristotelis de caelo commentaria*, Commentaria in Aristotelem Graeca 7 (Berlin).
Kirwan, C. (1971) *Aristotle's Metaphysics: Books Γ, Δ, E* (Oxford).
Leggatt, S. (1995) *Aristotle: On the Heavens I and II* (Warminster).
Lemmon, E.J. (1978) *Beginning Logic* (Indianapolis).
Matthen, M. and Hankinson, R.J. (1993) "Aristotle's Universe: Its Form and Matter," *Synthese* 96: 417–435.
Moraux, P. (1965) *Aristote: Du ciel* (Paris).
Ross, W.D. (1923) *Aristotle's Metaphysics* (Oxford).
——, (1936) *Aristotle's Physics* (Oxford).
Sorabji, R.R.K. (1988) *Matter, Space and Motion: Theories in Antiquity and Their Sequel* (London).
——, (1988) ed., *Philoponus and the Rejection of Aristotelian Science* (Ithaca, NY).
Stocks, J.L. (1922) *The Works of Aristotle Translated into English: Vol. III. De Caelo, De Generatione et Corruptione* (Oxford).
Waterlow, S. (1982) *Passage and Possibility* (Oxford).
Wehrli, F. (1950) *Die Schule des Aristoteles: V. Straton von Lampsakos* (Basel).
Wildberg, C. (1987) *Philoponus: Against Aristotle on the Eternity of the World* (London).
——, (1988) *John Philoponus' Criticism of Aristotle's Theory of the Ether* (Berlin).
Wolff, M. (1988) "Philoponus and the Rise of Pre-Classical Dynamics," in Sorabji (1988), 84–120.

WHY DOES EARTH MOVE TO THE CENTER? AN EXAMINATION OF SOME EXPLANATORY STRATEGIES IN ARISTOTLE'S COSMOLOGY

MOHAN MATTHEN*

1. *Introduction*

Why, and how, does earth move to the center of Aristotle's universe—the Totality, as he calls it? In a famous passage in *Physics* 4.1, Aristotle says:

> The movements of the simple natural bodies, like fire and earth and their like, not only show that place is something, but also that it has some power. For, if it is not impeded, each body moves to its own place, one above and the other below. (208b8–12)

The thesis is that something about the movement of earth and fire to their proper places shows that place has power, and exists.

In an important article, Peter Machamer argued that Aristotle's statement had almost uniformly been misconstrued as implying that "there is some sort of attractive power of the place on the moving element" (1978, 377). This cannot be correct, Machamer points out, for a little more than half a page on, in the same chapter, Aristotle says:

> Of what could one make place a cause? None of the four causes is present in it: it is not a material cause of existent things, for nothing is made of it; nor is it a form and definition of things; it is not an end; it does not change existent things. (*Phys.* 4.1, 209a19–22)

If a place affected a moving body *either* in the way that the military leader exerts an attraction on his followers, i.e., by conducting himself in a way that they wish to emulate (cf. *Meta.* 12.10, 1075a11–24), *or* in the way that a gravitating mass does in Newton's theory, then clearly it would be a final or efficient cause. Aristotle's statement above implies that place is not like this.

How could it be? For Aristotle, a thing's place is "the innermost motionless boundary of what contains it" (*Phys* 4.5, 212b20–21). Places are

* University of Toronto.

defined relative to things in his system. This leads to two problems of interpretation so far as our present passage is concerned. Firstly, what is the *thing* that contains the proper place of earth and of fire? Secondly, how does the boundary of this thing act on earth or fire? To our way of thinking today, a two-dimensional surface cannot exert an influence all by itself—it has no mass and no other power by which to exert an influence. Only the thing to which this surface belongs can act on something else. Does Aristotle have a different view?[1] And if so, is it coherent? Machamer says: "What is needed is an account of the power of natural place which is somehow not causal" (1978, 378). This may be right in some sense, but it leaves us with the problem of what Aristotle means when he says that place has some power. One might add that we need also to validate Aristotle's inference that place exists.

These questions are important for a proper understanding of the theory of elements laid out in the *De caelo*. For in this work, one of Aristotle's main concerns is the investigation of the elements and their natural motions. My purpose here is to investigate the structure of Aristotle's account, with particular reference to the problem stated in the title: what is the proper account of how earth moves by nature to the center? I shall try to show how Aristotle's notion of natural place can best be understood in light of some of his ideas about how the Totality is to be defined.

2. *E-change*

To a mechanist, movement is a fundamental and unified kind of quantifiable change. In the differential calculus, the foundational mathematical treatment of change in classical mechanics, movement is defined as the first derivative of position with respect to time. On this way of thinking, circular motion belongs to the same broad category as linear change. To an extent, this is true for Aristotle as well (cf. *De caelo* 1.2, 269a2–4). But for him this category cross-cuts with another distinction that is vital for our purposes. From the point of view of this other distinction, circular and linear motion are fundamentally different from one another, and each is grouped with other kinds of change.

The heavenly body rotates; fire and earth move in straight lines upwards and downwards. In our mechanistic understanding, both move,

[1] Seemingly he agrees: see *De caelo* 3.7, 306a20–22, and Bolton (2009), section 2.

though on different kinds of paths. For Aristotle, however, there is a major difference between these movements. Circular motion of an extended body in place is not a change of place (*De caelo* 1.9, 278b30), for the "innermost motionless boundary of what contains" a rotating thing—its place—does not change; moreover, when that thing is spherical, it is always contiguous with that boundary. An important consequence of this is that circular motion is not motion *to* a place. As we shall see in a moment, something that moves to a place stops moving when it gets there. The heavenly body never stops; its activity is temporally homogeneous—at any given moment, it is in the same state as at any other moment.[2] In this respect, circular motion is not a *process* by which the potentiality to be somewhere is actualized. Rather, it is an *actuality* (ἐνέϱγεια). It imitates God's actuality, as does also the contemplative or theorizing activity of the human intellect, which, for us, is the highest good. Circular motion belongs, therefore, to a class of homogeneous activities that encompasses other things than motion.

The linear motion of earth and fire are, by contrast with actualities, end-oriented. In Aristotle's scheme, end-oriented change (or *e-change*) has a terminus.[3] For Aristotle, e-change is a passage from the *lack* of some quality *F* to the *possession* of that quality *F*. Consider natural change. A thing *x* that is naturally *F*, but not actually *F* (because it is impeded from becoming *F* or because it is immature and still progressing toward *F*), will possess a potentiality to achieve its natural state, *F*. (The terms "potentiality," "power," and "tendency" will be used interchangeably here.)

Let *Po*(*F*) be *x*'s potentiality to achieve the *static* condition, *F*. In virtue of *Po*(*F*), *x* will *also* possess a further potentiality *Po*(towards *F*) to undergo a certain process: for in order to achieve the static *F*-state it must first leave its non-*F* state and traverse the points that are qualitatively or quantitatively intermediate between this non-*F* state and the *F*-state that is natural to it. *Po*(towards *F*) is the *dynamic* potentiality for *change* or *process*. When *x* gets to, or becomes, *F*, it ceases to be in process; its impetus to change is exhausted. Something that is *F* no longer has the potentiality to undergo the process of achieving *F*, except in so far as it has a potentiality to be displaced from the *F*-condition.

---

[2] This is, of course, false on the definition of motion just sketched from the differential calculus, since each particle of a rotating body changes direction at all times, from which it follows that its state of motion is in constant change.

[3] The expression "e-change" coincides roughly with a technical use of ϰίνησις in Aristotle, *Phys.* 3.1, which contrasts with the use of ἐνέϱγεια in *Meta.* 9.6. But Aristotle does not always use ϰίνησις in this way; sometimes, he uses it in the way we use "movement."

It is the distinction between *Po*(*F*) and *Po*(towards *F*) that Aristotle has in mind when he says that "alteration is the actuality of the alterable *qua alterable*" (*Phys.* 3.3, 202b25–26). Since the actuality of *x qua Po*(*F*) is the static condition *F*, *change* (or alteration) must be defined as the actuality of *x qua* tendencies like *Po*(towards *F*). The latter is actualized when impediments to change are removed, or when *x* comes into contact with something that imparts to it an impetus toward *F*. *Po*(*F*) is actualized when something gets to *F* and is no longer changing. The potentiality for process toward *F* is causally subordinate to, and properly understood in terms of, the first potentiality, viz., the tendency to be in the *F*-state, where *F* is the end-point of change.

3. *E-change and the movement of earth*

Aristotle's framework for understanding the natural motion of earth is best understood within this schema for e-change. Consider first the terminus. He identifies earth (the element) as

> (A) that which is at rest at the center of the Totality.[4]

This is the *being* or *form* of earth—it is what makes earth what it is. When they get to their natural places, the sublunary elements actualize their form. "To be in a certain place, i.e., up, is the actuality (ἐνέϱγεια) of the light" (*Phys.* 8.4, 255b11); earth for its part actualizes its form by being at the center. To be in a certain place is, or at least is some part of, what it is to be earth. (It is also part of the form of earth that it is cold and dry—*De gen. et corr.* 2.3, 330b5.)

And now the associated e-change:

> (B) Earth moves toward the center if it is not impeded.

The natural motion of earth *follows from* (A), i.e., it is a *consequence* of the form of earth, but not part of this essence.[5] The form of earth is static; its natural motion is a realization of this static essence.

[4] Cf. *De caelo* 2.3, 286a20–21.

[5] In contrast, Gill (2009, p. 144 n. 10) takes the view that the actuality is "dynamic." "[A] thing is most actual when it is *acting* according to its nature. What needs to be explained is why things *stop* moving, not why they *start*," she says. It should be noted that the dynamic interpretation cannot be extended to the natures of living things. A human being is most actual when mature, not when it is developing. There is no need to explain why a child stops growing once it has attained maturity. It stops because it has attained its form.

Earth is heavy, Aristotle keeps saying. What does this mean? Aristotle consistently defines the heavy in terms of natural motion to the center (*De caelo* 1.3, 269b23; 4.1, 307b32; cf. Gill 2009). This means that earth is heavy in virtue of (B), not (A), and that heaviness is not its form, but rather the innate source of movement that follows from its essence by the *Po*(*F*)/*Po*(towards *F*) inference explained in section 2. Sheldon Cohen (1994) captures the thought well: "The change that restores it to its natural condition of stasis is natural in following from an intrinsic principle of rest" (156–157).

A clod of earth that is displaced from the center possesses a natural potentiality *Po*(at the center), which is actualized when it is stationary at the center. This corresponds to (A) above, the form of earth. But this stationary condition can be actualized only if it participates in a certain process; it must leave its non-central position and traverse the points that lie between this initial position and the center. Thus, it will possess a natural tendency, *Po*(towards the center), to actualize its form by moving to the center. This corresponds to (B). It is particularly important to note that this power of motion, *Po*(towards the center), is activated only when an element is not in its natural place. When it occupies that place, its nature is to stay there: for as Aristotle emphasizes in *Phys.* 2.1, nature is a cause both of movement and of rest. Heaviness is part of earth's *nature*—and a thing's nature is its source of motion. It is important here not to conflate nature and form or being.[6]

This is the point made in *De caelo* 4.3: "The account of a thing being borne to its proper place is similar to that of other comings-to-be and changes" (310a20–22). "Each thing borne to its own place is borne to its own form" (310a34–b1). And later:

> When air comes to be from water, that is, light from heavy, it moves toward the upper place. The moment it is light, it is no longer becoming light but *is* in that place. *It is evident, then, that it moves while it is potential and progressing toward its actuality there, and toward the quantity and quality proper to its actual state.* It is by the same cause that what already is and exists as earth and fire moves towards its own place, unless something prevents it. (*De caelo* 4.3, 311a1–7)

The idea expressed in the third sentence of this passage (italicized above) is complicated by the fact that Aristotle is thinking of evaporation—water actually changing into air as it moves upward. Thus, part of his point is

[6] Cohen (1994) thinks that in order to accommodate Aristotle's thought that the elements are not self-movers, one must identify nature and essence. I am sceptical.

that water is potentially air, and potentially possesses air's natural capacity to rest in an upper place. This would be to attribute an iterated potentiality to water—a potentiality to acquire a potentiality. However that might be, he is clearly saying that air's potentiality is actualized when it achieves its proper place, whether that state is being in a particular place (as in the case of an element displaced from its natural place), or having a certain quantity or quality (as in the case of water that is in the process of being transformed into air).

4. *Doubts about the dynamic interpretation*

Suppose for a moment that the static interpretation proposed in the previous section—the idea that the form of earth is its static condition at the center—is mistaken. Suppose that, as many commentators have contended (recently Benjamin Morison (2002) 51–53, and now Gill in this volume), earth is defined by the dynamic principle of movement toward the center, not by the static principle of rest there. How then would we explain the immobility of earth at the center? Gill asserts (2009, notes 7 and 10) that earth is held in place at the center by the many bits of earth vying for that place. It is at rest "not because it is programmed to stop, but because its downward progress is impeded by clods of earth approaching from other directions or by the large ball of earth already there." The problem with this, as I see things, is that it is not clear why a clod of earth held in place would be in *natural* rather than *imposed* rest. Like a car on the Don Valley Parkway—apologies for a Toronto joke—earth is immobile because other things crowd in, not on account of an innate principle.

The case of fire is slightly different. Gill (1991, 261) is right to say, "Fire is not programmed to stop at the periphery—it would proceed upward indefinitely *if it were not confined by the sphere of the moon*" (my emphasis). Here, the periphery she is talking about is that of the sublunary sphere. Fire is supposed to go to the periphery *of the universe*, but since the heavenly element occupies that spot, fire is prevented from going all the way out. Still, this does not argue in favour of a dynamic interpretation of the elementary essences. In her contribution to this volume, Gill acknowledges (note 10) that fire would *naturally rest* at the outer extremity, if only it could get there. This surely requires a static interpretation.

There is an interesting complication here because Aristotle says, "the heavy and the light are said both simply and relative to something else"

(*De caelo* 4.1, 308a7–8). I take his theory of *relative* heaviness and lightness in the following way: while each of the upwardly moving elements (fire and air) rests at the periphery, and each of downwardly moving elements (water and earth) rests at the center, fire and earth have a stronger impetus toward these places than their pair-mates.[7] Thus, if earth were to be removed, water would go to the center, but otherwise its path to the center is cut off, and it forms a shell around earth. Thus, I do not see a need to distinguish between the natural resting place of water and that of earth.

This pairing of sublunary elements makes sense of the treatment of book 1, where the basis for distinguishing water from earth and fire from air is skated over—the claim in 1.2 is that there are just three kinds of simple movement for the elements: away from the center, towards the center, and around the center (268b23–24). In my view, Aristotle's main motivation for positing four sublunary elements instead of two comes not from considerations of natural place and natural movement, but from the qualitative considerations—four elements are needed to accommodate all compatible pairings of hot-cold and dry-wet (*De gen. et corr.* 2.2). The relative heaviness/lightness view is needed in order to bring the center/periphery view of book 1 into line with the tangible qualities view just mentioned. And it is a bonus that this version of the four-element theory explains the ocean's being on top of earth. (On the other hand, fire is not observed in the upper region, but seems rather to be inferred—see *Meteor.* 1.3.)

Now, it is true that the "static reading" of elementary essences runs into patchy waters later in the *De caelo*. For at *De caelo* 4.5, 312b3–20, Aristotle seems to suggest (a) that the proper place of water is removed from that of earth, and (b) that it would nevertheless occupy the center if all earth were to be removed. If this were right, descending water would rush on past its natural place to occupy the natural place of earth when earth is removed. Indeed, it would not stop even there if it were not for the fact that it would collide with water rushing in to the center from the opposite direction. John Sisko (2002) has argued that this is inconsistent with the static reading: since water does not come to rest in its natural place in this counterfactual situation, the role of place must simply be to define direction.

[7] This shows, by the way, that the place of rest does not wholly account for the motion of the elements—the difference between the motion of earth and that of water cannot be derived from their natural place. (See Bodnár 1997 on the need for a mixed interpretation: however, I argue below that book 4 is not on the same track as book 1.)

Sisko is certainly right to say that there is a problem here. But it is unclear that the dynamic interpretation does any better with the passage he invokes. Gill's interpretation (2009) is that the role of natural place is to define the direction of motion, and not to provide a resting spot. Sisko seems to miss what this implies for Aristotle's conception of direction: "up" and "down" are not defined in "absolute space"; instead, they are defined relative to a place. And this implies that, in the passage we are discussing, water would be moving in a contra-natural direction after it has passed its natural place. It continues to move in the same straight line, but it does not move in the same cosmological direction. So it is not clear how a separate natural place for water works even on this attenuated role for natural place.

I think it likely that Aristotle stumbled in *De caelo* 4.5.[8] But it is also possible that his treatment of heaviness and lightness in book 4 is not strictly commensurable with the treatment in book 1. I suggested a few paragraphs ago that earth is distinguished from water for reasons other than to explain motion. This dovetails in spirit (if not in detail) with the main suggestion that Robert Bolton makes in his contribution to this volume. Bolton argues that Aristotle's investigation of cosmological properties is *a priori* because of the relative inaccessibility of the phenomena, while his investigation of perceptually accessible phenomena is more empirical. It is worth noting that the dry-wet, hot-cold pairing is in terms of the "principles of perceptible body" (*De gen. et corr.* 2.2, 329a7).

Aristotle seems to have different ways of characterizing the elements in Bolton's two methodological domains. In book 4, he adds to the austere doctrine of elements found in *De caelo* book 1, which is concerned with nothing other than the three-dimensionality and finiteness of body, and the character of movement and change. The doctrine of elements

---

[8] He certainly seems to stumble when he suggests that heaviness and lightness are not symmetric (312b3–10). Because air is only relatively light, it is also heavy to some extent. Thus, air will *descend* to the place of water if the latter is removed. This raises the question: Should air not have in it more of lightness than heaviness? Why then would it descend, even if there is nothing to support it? The answer seems to be that the relative lightness of air does not have the corresponding effect: "air will not move upwards to the place of fire if the fire is removed, *except by force*." Here, Aristotle seems to allude to the action of a pipette: when water is released from it, air naturally replaces it from above, but force is required to draw water upward into it. The problem is that there is nothing in his cosmological principles to support this asymmetry. Thus, the principle seems of asymmetry to be introduced speculatively and *ad hoc* to explain an empirical observation. On this point, Bolton's (2009) "two standards of inquiry"—actually two methodologies—clash.

presented in *De caelo* 3 and 4 and elsewhere adds new premises and additional principles. These additional principles bring new distinctions, and carry less certainty (cf. Bolton 2009). They should not be read back into the doctrine of book 1.

It is possible, then, that when Aristotle talks about natural place in book 4, he means something different than what he means in book 1. My claim is that in book 1, natural place is a resting place towards which an element moves when displaced from it, provided that it is not impeded. It is possible that in book 4, he does not have natural motion in mind at all. Possibly, all that he means here is that given the weaker impetus of water towards the center, it will naturally occupy the region above that occupied by earth. This place is natural as a *consequence* of natural motion; it does not define natural motion, either as the static or as the dynamic interpretation demands. Notice, by the way, that on this reading natural place is a volume or region in book 4 (since a non-zero quantity of an element will occupy more than a point), but a point in book 1, since the center of the Totality is a point.

I do not believe that the dynamic interpretation can be sustained: at the very least, it takes some fancy footwork to reconcile it with Aristotle's notion of e-change and the conception of nature as a cause of both motion and rest.

5. *The existence of place*

The way of understanding Aristotle's theory recommended above does not as yet tell us how *place* "has a certain power." What it demonstrates is that earth possesses a tendency to move toward the center of the Totality. But this is a power that *earth* has in virtue of its being or essence. In section 1, we saw that Aristotle implicitly rejects the two most plausible accounts of how *place* might figure in the natural tendency of an element to move there. So it is unclear how the center-seeking tendency of earth translates into a power of place. Nor does the account just presented show why place exists.

Let us first address the question of existence. Suppose that a philosopher *X* is disinclined to allow that:

(1) I am in Toronto.

implies

(2) Toronto exists.

Perhaps *X* thinks that only material things exist, and that places are not material; perhaps, she thinks that place-attributions such as (1) merely relate substances one to another. On any such view, *X* would presumably be equally disinclined to think that:

> (3) I have a tendency to return to Toronto whenever I am away (provided that I am unimpeded)

implies (2). Since the mere mention of Toronto in (1) is not sufficient to force *X* to acknowledge the existence of Toronto, the mere fact that Toronto is mentioned in (3) should not be enough either.

Aristotle's argument might be taken to address thinkers like *X*. The point might be that earth's natural tendency to move to the center shows that the center has power and, hence, exists; this would differentiate (1), which ascribes no power to place, from (3). Of course, *this differentiating strategy would work only if (3) implicitly ascribes power to place.* On the face of it, however, (3) seems to ascribe power to *me*, not Toronto. It *may* be that Toronto is implicated in my tendency to return, but we still have not been given convincing reason to believe that this is so. In other words, we still do not know what it would take to get from (3) to an ascription of power to Toronto. And the same holds of the center of the Totality. So we are missing a part of the puzzle.

Now, one way of addressing the above argument is to insist that already with respect to (1) the place-sceptic *X* should back down. For it might appear that in (1), place is predicated of an underlying subject. In the *Categories*, Aristotle maintains that predicables are among the "things that are," although they depend for their existence on the individual substances of which they are predicated. It might seem to follow that, as Morison (2002, 80) says, "our practice of saying *where* something is depends on the existence of proper places." If this is what Aristotle intended, then his argument from the movement of the elements is not meant to differentiate (3) from (1) by showing that (3) implies that Toronto has the power to move me. He would simply be adducing (3) as yet another context in which we are committed to place in virtue of the ontology that he articulates in the *Categories*.

This appeal to the ontology of the *Categories* is misguided, however. The reason is that in the *Categories* place (τόπος) is not predicated of substance; rather, *where* (πoυ) is predicated of substance. The predicable in (1) is not *Toronto*, but *in Toronto* (cf. the examples of wheres given in *Categories* 4: *in* the agora, *in* the Lyceum). Being in Toronto is what characterizes me and the other denizens of this place—Toronto is not a characteristic of us. Similarly, (3) characterizes me in terms of a whither, not in terms of a place. This is not to say it makes no sense to speak of characterizing substances in terms of places; merely that it is elliptical to do so and that there is, therefore, no direct logical route from (3) to the power or existence of place.

Wheres, whences, and whithers are properties of substances: let us call them *locators*. Locators are distinct from places, and so we are back to the original problem. The tendency to move to the center is a locator predicated potentially of earth—a locator derived from the definition of earth. How can we infer from this locator the existence of place? And how does it implicate a power belonging to the center?

Now, the *Categories* does license a more indirect logical route to the reality of place. As we have seen, Aristotle characterizes the place of a body *x* as the innermost motionless surface of the thing that contains *x*. Surfaces, and hence places, fall into the non-substantial category of *quantity* in the *Categories* (6, 4b24); they, like other non-substantial items, exist because they reside in substances. So one might say that since places are surfaces, they exist. This has the somewhat surprising consequence, however, that my place does not depend on *me*. For the surface that is implicated in (1) is not a surface of me, but a surface of "the thing that contains" me. In order to validate the reality of my place, we need to look for the substance that contains me. As Aristotle says, "If there is a body outside a given body, a body which contains it, then it is in place, if there is no such body, it is not" (*Phys.* 4.5, 212a31–32).

What is this unique container? The question is important precisely because of Aristotle's strategy in the *Categories* of subordinating the reality of non-substantial things to that of the substances in which they reside. (1) locates me in Toronto. But understood as a piece of land, Toronto does not contain me. Only my feet touch this piece of land; most of my body is surrounded by air. The geometrical surface that contains me is partly air and partly earth. Do such heterogeneous surfaces exist in Aristotle's ontology? Of course, it might be that the airy surface exists, because it is a surface of a body of air—of the atmosphere, perhaps. Equally, the earthy surface exists because it is a surface of a body of earth—of the Earth, perhaps, or of Toronto, whatever exists in Aristotle's ontology. But does the mereological sum of these surfaces exist in Aristotle's ontology? It would if we could call on a single body that contains me, which partly consists of air, partly of earth. But it is dubious that Aristotle would allow us simply to count just any such assemblage as a full-fledged substance; in general, he does not allow that the sum of two individuals is an individual. So it is not as yet clear how the ontological strategy of the *Categories* could be sufficient for recognizing the reality of place.

Benjamin Morison (2002, chapter 5) has dealt with the question of the container in a convincing way. In addition to the above problem with Aristotle's theory of place, which is a problem about the *existence* of a container, Morison finds another, which has to do with its *uniqueness*. It seems axiomatic that:

(4) If y contains x, and z contains y, then z contains x.

For example, if the sphere of the Moon (y) contains the Earth (x), and the Totality (z) contains the sphere of the Moon, then the Totality contains the Earth. Thus, while there may sometimes be a smaller individual thing that contains a given substance, this smaller thing cannot be *the* thing that contains the substance, since there is always a bigger container, namely, the Totality.

Now, the Totality is a substance (*De caelo* 1.9, 278a10–15, b3–4; 3.1, 298a26–31; *Meta.* 7.2, 1028b9–13; note that the Totality survives the pruning of the list of substances at the start of 7.16). And it is, as Morison calls it, the "maximal surrounder" (2002, 138). Indeed, the Totality is the maximal surrounder for everything; for as Aristotle himself says, the "common place of all things" is carved out from the Totality, since everything is in it (*Phys.* 4.2, 209a32–33). We can solve the foregoing difficulties if we simply took the Totality to be the container that Aristotle's definition of place demands. The surface that contains me is a surface of the Totality, and derives its existence from it. By being located somewhere, I create a "hole" in the Totality. Or to put in another way, by virtue of my location in it, the Totality acquires an interior surface. I reside in this hole; I am bounded by this surface; this is my place. The inner surface created in the Totality by me is, in the ontology articulated in the *Categories*, one of the things that are. This is why (1) implies the reality of place.

Taken in this way, the reality of place is not inferred directly from our use of locator predicates. First, we have to factor in what place is, and how it determines the application of locators. In *Physics* 4, place is found to be a surface. We know from the *Categories* that surfaces, being non-substantial entities, derive their reality from substances. So we have to identify the substance in which place exists. With Morison's help, we identify it as the Totality. Only now can we conclude that places exist.

Note the surprising consequences of this indirect route to the reality of place. First, it turns out to belong to the category of *quantity*, not to the category of *where*. Secondly, the place of x is not in x and does not depend on x for its reality; although it is created in a straightforward sense by x, the surface that contains x belongs to x's container. Finally, the place of x depends for its reality on the substance that contains x. We have seen that this substance is the Totality. Thus, the reality of place depends on the reality of the Totality.

6. *The problem of power*

The indirect logical route to the existence of place finds immediate resonance in the *De caelo*.

> Every body in the form of a part is complete by definition, since it has all the dimensions. But each is made determinate by contact with what is close to it, and, hence, in a way each of these bodies is many. But the Totality of which these things are parts is necessarily complete and in every way as the name indicates, and not complete in one way but not in another.
> (1.1, 268b5–10)

In the first two sentences of this passage, Aristotle is talking about bodies that are contained within the Totality. In the context (268a6–7), we know that bodies are merely entities extended in three dimensions; they have no other individuating property, at least not *qua* body. (Though it is a three-dimensional entity, an *animal* is not individuated by its surroundings, as we shall see in a moment, but by its form. The body of an animal, however, considered in abstraction from the animal itself, i.e., merely as an agglomeration of material, is individuated by its surroundings.) Bodies, then, are "made determinate by contact with what is close" to them. In other words, these "partial bodies" are delimited by their place or the inner surface of the Totality that surrounds them.

Notice the contrast between two kinds of entity in the last sentence quoted above. Partial bodies are complete "in one way but not in another," while the Totality is necessarily complete in every way. The Totality is an independent existent; places and partial bodies derive their reality from it. Aristotle is alluding here to his belief that bodies have no intrinsic unity. "What in the world will make mathematical magnitudes one?" he asks at *Meta.* 13.2, 1077a20–23. "For things in our sensible world are one in virtue of soul, or part of soul, or of something else, reasonably enough; when these are not present, the thing is a plurality and splits up into parts." The thought he is expressing here is that a body individuated by its spatial extent has no strong individuating conditions. Suppose you added a gram of oil to 10 grams of water, with the oil floating on top. Do you have one body here or two? The problem is that there is no interesting way to answer this question. These things are individuated merely by place, and here there are two ways to define places. The place of the water is different from the place of the oil, and so we have two bodies. Look at this another way: the oil-water mixture has a single place, and so it is one body. Thus, the oil is complete in one way, incomplete in another; consequently, the entire liquid is one in a way, many in another way.

Organisms are different in this respect: they are one in virtue of functioning as a unit—mass added to an organism's body will not count as a part of it unless it is capable of participating in the unitary functioning of that body. (There is a difference between a piece of meat which I hold in my hand and a digested piece of meat: the former is clearly not a part of me, and the latter provides material that becomes a part of me.) This is what Aristotle is pointing to when he says that living things are "one in virtue of soul." A partial body cannot be considered one in the fullest sense, because "it is a plurality and splits up into parts." And at the beginning of *Meta.* 7.16, Aristotle eliminates partial bodies, "heaps" as he calls them there, from his *Meta.* 7.2 list of substances, on precisely the grounds that these are neither one nor have sources of self-motion *qua* being one. In this passage, he implicitly leaves the Totality and living organisms as the only genuine substances.

Consider now a weak unity like a clod of earth. It is one only because of its spatial continuity within certain boundaries; in other words, it is made one only "by contact with what is close to it." Its existence and unity depends on something else, its container; for to repeat our earlier conclusion, its place resides in something else. Its unity is *not* traced to something intrinsic, such as soul; the clod of earth has no soul. (Another reason for denying that clods of earth have soul is that they are totally homogeneous—homeomerous, in Aristotle's terminology—and cannot be divided into soul as agent and body as thing-acted-upon (*Phys.* 8.4, 255a15–16).

The clod has a tendency to move toward the center of the Totality—that is, it moves toward the center if it is not impeded. Now, we cannot say that the center of the Totality is a final cause for the clod of earth, a place to which it is attracted because that place constitutes its own perfection. Only things with soul are attracted to a place or to a condition because this place or this condition is good, or good-for-them. For it is only for things that have soul that there is a good. But as we have seen, weakly unified things like clods of earth lack soul; nothing is a final cause for them.[9] Nor can we say that the center of the Totality is an efficient cause of the clod's movement; earth moves toward the center because of what it is itself. Having solved the problem of the existence of place, we are back to our original puzzle concerning its power, though with deeper understanding of the difficulties that lie in the way of a solution. How does earth's tendency to move to the center translate into a power of place?

7. *The stability of a mass of earth at the center*

Machamer's own solution to these problems is flawed from the outset, though suggestive and insightful. He points to a rather opaque passage in the *Physics*, which reads as follows:

> That each thing is borne to its own place is in accordance with reason. For that which is continuous and touching without force belong to the same kind. And things that are naturally together are impassive, while things that touch one another affect one another. And, not unreasonably, everything stays in its own place. For also a part remains in its place, but what is in a place is a divisible part in relation to the whole, for example, when one moves a certain part of water or air. (4.5, 212b29–213a1)

Aristotle speaks here of the "parts" of an element, bits of the elementary mass residing in a place, for example a clod that is a part of the Earth, i.e., not displaced from it. The claim is that since the bits of earth that constitute the globe stay together without force, and without mutual interaction—they must belong together, and stay together impassively. Machamer draws from this the idea that "the power that natural place has is the power to make a natural or organic unity among those things that are alike by nature" (1978, 379), adding one page later that: "It is not that natural place is a form, but rather that an element attains its proper form when it is in a natural or organic unity with its like." Let's call this proposal the *Organic Unity Schema*.

[9] Not all commentators accept this constraint on final causation, it should be said. Richard Sorabji (1988, 187), for example, claims that natural place is a final cause for elementary motion but does not explain how being at the center is good for a clod of earth. Keimpe Algra (1995, 208) too thinks that the natural motion of the elements is "obviously teleological." He acknowledges (1995, 218n) that Aristotelian final causes are for the sake of a good, but pleads that being at the center might be good relative to some [other?] substance. It seems to me that while it may well be better for the Totality that *some* earth should be at the center, it cannot be better even from this point of view that any particular clod should be so. How then does the center constitute a cause of this clod's movement?

The *Organic Unity Schema* underestimates the role of place. Machamer seems to suggest that a bit of earth is impassively at rest simply because it is together with other bits of earth, the role of place being simply that clods of earth come together there. Aristotle is quite explicit about natural place having a role:

> The old saying that "like moves to like" is not true in all circumstances. It would not be the case, if one were to remove the Earth to where the Moon is now, that individual parts would move toward it: instead they would move to where the Earth is now. (*De caelo* 4.3, 310b1–5)

The Earth does not stick together just because it is a great homogeneous mass of stuff. Removed from its natural place, the "natural or organic unity of earth" will break down as all the little clods race to the bottom. It is only when it is in its natural place, that Earth is a natural unity. Place is part of the explanation why all the bits of earth that constitute Earth stay together; mere consanguinity cannot do the trick by itself.

This said, it must be acknowledged that Machamer's *Organic Unity Schema* touches on an important principle. To see why, let us examine another plausible (but, according to Aristotle, wrongheaded) explanation of why Earth is a continuous mass.

> *Materialist Schema* Consider a clod of earth on the surface of the globe. It has a tendency to move toward the center. Why then is it not at the center? Because it is prevented from getting there by all the earth between it and the center. If it were separated from the surface of the globe by air, it would set off toward the center, and come to rest when it contacted the surface of the Earth. Similarly if it were on the surface of the ocean it would sink. It can get below any other element, but not below other bits of earth. This capacity of earth to replace other elements below it results in an agglomeration of earth around the center.

The passage from *Physics* 4 quoted right at the beginning of the present section indicates that this *Materialist Schema*, which attaches no importance to properties of the whole globe, shifting the explanatory burden to the stuff out of which the Earth is made, is mistaken. The *Materialist Schema* suggests that only the one clod of earth at the exact center of the Totality manages to realize its form. All the other bits of earth are *prevented* from getting there. This is at odds with Aristotle's claim that the mass of earth is impassive when it is at the center; its rest is not maintained by the mutual interactions of bits of earth.[10]

---

[10] I am inclined to think that the Materialist Schema is similar to Mary Louise Gill's claim that pieces of earth stay still because they are vying with other pieces of earth for the center.

So his claim must be understood in this way:

> *Aristotelian Schema* When it is at the center (but not otherwise) the Earth constitutes a stable and impassive unity.

Parts of the contiguous mass of Earth that are not at the center should not therefore be regarded as impeded; they are not *prevented* from realizing their form; on the contrary, they do actualize their form. The *Aristotelian Schema* adds the power of place to Machamer's *Organic Unity Schema*; it adumbrates—far from transparently—how place is invoked.

The *Materialist Schema* is attractive to us, but not to Aristotle. It is attractive to us—modern materialists that we are—because it undermines the explanatory value of *wholes*. Certain large entities—populations, ecosystems, states—look as if they are designed as a whole: one is inclined to think that in order to explain certain stable properties of these large entities, there is no choice but to appeal to autonomous laws with specific application to entities of this kind. For instance, the stability of ecosystems leads some ecologists to posit laws that apply to these systems autonomously, i.e., with no reference to the organisms and environmental features that constitute it.

A common materialist strategy in opposition to such particularism is what Robert Nozick (1974, 18–22) calls the "invisible hand". (He takes the term from Adam Smith.) An invisible hand strategy attempts to show how the nature of the parts is quite sufficient to explain stable properties of the whole; for instance, it might be that these parts tend of their own accord to find themselves in certain equilibrium states regardless of where they start from—these equilibrium states turn out simply to constitute the large entities and stable properties that were the object of investigation at the outset. Nozick explains the strategy in the following terms:

> Fundamental explanations of a realm are explanations of the realm in other terms; they make no use of any of the notions of the realm. Only via such explanations can we explain and hence understand everything about a realm; the less our explanations use notions constituting what is to be explained, the more (*ceteris paribus*) we understand. (1974, 19)

The stability of the Earth at the center of the Totality tempts one to think that it is *meant* to be there. The *Materialist Schema* purports to show, from authentically Aristotelian principles concerning the nature of the elements, that from any random starting point, earth will tend to congregate at the center of the Totality. It thus eschews the explanation of the Earth's position in holistic terms. It is, in essence, an invisible hand explanation.

Aristotle emphatically rejects the strategy of the invisible hand in cases where the large-scale phenomenon to be explained is stable, *a fortiori* when this phenomenon is eternal. In *De caelo* 2.1, for example, he rejects the notion that the motion of the heavens can be achieved by an external mover: eternal motion cannot involve constraint or effort. (He makes a related point in De *motu animalium* 4: earth does not move to the center by constraint. See also *Phys.* 2.4 196a25–b4.) Motion of this type requires effort and is contingent upon the will of an external mover; eternal motion cannot be contingent since it is impossible that something that could cease has never ceased and will never cease (1.12, 283a25). States of affairs of this type cannot be "by coincidence or spontaneous"; hence, they are "for the sake of something," i.e., because they are good (*Phys.* 2.8, 199a4–5).

> If nothing that happens by chance or spontaneity could be eternal, and the heaven and its circular motion are eternal, by what cause is this motion in one direction rather than the other? Necessarily, either this is a principle or there is a principle for it. (*De caelo* 2.5, 287b25–27)

Eternal things cannot be by chance. The invisible hand makes the large-scale phenomenon a matter of coincidence or spontaneity, Aristotle believes. (Here he is in bad company: think of modern creationists and their arguments against Darwinian explanations.)

Now, just as the heavens eternally revolve, so also the Earth is eternally at the center of the Totality. Aristotle infers that its rest at the center cannot be a matter of force, constraint, or material interaction: this is why he thinks that the *Materialist Schema* is on the wrong track. To explain the Earth's position merely by reference to the downward-moving and center-resting tendencies of earth is like explaining the structure of a wall by saying that the heaviest things, the stones, fell to the bottom, with earth on above this, and the lightest components of all, wood and thatch, finding their way to the top (cf. *Phys* 2.9, 199b35–200a5)—this would be a structurally correct, but obviously inadequate invisible-hand explanation. While the principles of motion and rest that this schema employs are correct, the eternal presence of the Earth at the center implies the independence of the principle that the Earth is impassive at the center. This is just the *Aristotelian Schema* stated above. It is a *sui generis* principle with specific application to the Earth and needs to be added to those mentioned in the *Materialist Schema*. (The latter must also be modified: the Earth is not held in place by material interactions, but stays where it is impassively.)

9. *Conclusion*

One might think that the *Aristotelian Schema* renders the strategies of the *Materialist Schema* not just incomplete, but also superfluous, at least for the explanation of the position of the Earth. True, we may need to invoke the element-potentials of earth, its downward-moving and center-resting tendencies, in order to explain why a clod of earth held high falls when it is released. But this does not apply to the cosmic order. Once a domain-specific higher-level principle has been invoked to explain facts such as that the Earth is at the center, that the stars are above, and so on, do the element-potentials not cease to play any further role in explaining these facts?

The line of thought just set out misses a central procedural tenet of Aristotle's holisitic explanatory strategies. Whenever Aristotle posits a matter-form analysis of an entity, he also posits a nexus of material causes that will achieve and maintain the form (cf. Matthen 1989, Meyer 1991). When an architect designs a wall, he desires to build a strong and stable structure. So he puts the heavy materials below where they will support the mass of building above, and the lighter elements above where they will put the least burden on the foundations below. Thus, the explanation of the structure wall in terms of element-potentials, the heavy below and the light above, is not entirely misplaced: if you put the stone on top, it *would* come crashing downward through the lighter wood. As far as the cosmos goes, his point is not that its order is explainable *without* reference to element-potentials, but that any such explanation has to be subordinated to principles like the *Aristotelian Schema* above.

This is illustrated by his treatment of the rain cycle. He gives a teleological account of this process in *De gen. et corr.* 2.10: it is "the closest approximation to eternal being," he says there. But having understood it in this way, he gives an account of the cycle in terms of heating, cooling, rising, falling, and so forth both in the text just mentioned and in *Meteor.* 1. These actions cannot be understood in abstraction from their role in the cosmos, however: they merely manifest the independent principles of how the cosmos is arranged and ordered.[11]

The element-potentials mentioned in the *Materialist Schema* play the role they play *because* they serve the cosmic order—the form of the Totality. The elements have defined positions relative to the sphere that

[11] Cf. the criticism of the *Timaeus* at *De caelo* 3.2, 300b17–26.

constitutes, and defined movements relative to these positions. These higher level principles are defined in terms of place. It is important, however, to differentiate between cosmic phenomena in which place has power, and those relative to which it does not. Our analysis of e-change gives place no power with regard to the actual movement of earth. The power of place is manifested rather in the stability and impassivity of the Earth at the end of its natural movement to the center.

Bibliography

Bodnár, I. (1997) "Movers and Elemental Motions in Aristotle," *Oxford Studies in Ancient Philosophy* 16: 81–117.

Bolton, R. (2009) "Two Standards for Inquiry in Aristotle's *De caelo*," in A.C. Bowen and C. Wildberg (edd.), *New Perspectives on Aristotle's* De caelo (Leiden) 51–82.

Cohen, S. (1994) "Aristotle on Elemental Motion," *Phronesis* 39: 150–159.

Gill, M.L. (1991) "Aristotle on Self-Motion" in L. Judson (ed.), *Aristotle's* Physics: *A Collection of Essays* (Oxford) 243–265.

——, (2009) "The Theory of Elements in *De caelo* 3 and 4," in A.C. Bowen and C. Wildberg (edd.), *New Perspectives on Aristotle's* De caelo (Leiden) 139–161.

Machamer, P.K. (1978) "Aristotle on Natural Place and Motion" *Isis* 69: 377–387.

Matthen, M. (1989) "The Four Causes in Aristotle's Embryology," *Apeiron* 22.4: 159–179.

——, (2001) "Holistic Presuppositions of Aristotle's Cosmology," *Oxford Studies in Ancient Philosophy* 20: 171–199.

Matthen, M. and Hankinson, R.J. (1993) "Aristotle's Universe: Its Form and Matter," *Synthese* 96: 417–435.

Meyer, S.S. (1992) "Aristotle, Teleology, and Reduction," *Philosophical Review* 101: 791–825.

Morison, B. (2002) *On Location: Aristotle's Concept of Place* (Oxford).

Nozick, R. (1974) *Anarchy, State, and Utopia* (New York).

Sisko, J. (2002) Rev. of D. Sedley (ed.), *Oxford Studies in Ancient Philosophy* 20 (2001), *Notre Dame Philosophical Reviews* 2002.03.01.

THE THEORY OF THE ELEMENTS IN *DE CAELO* 3 AND 4

Mary Louise Gill[*]

The four simple bodies—earth, water, air, and fire—compose all perishable bodies in Aristotle's universe and are themselves perishable. Either separately or in combination, the four elements fill up the entire sublunary region, the region containing a large mass of earth (the Earth) at the center and extending outward to the sphere of the Moon. The region beyond the periphery of the sublunary region, stretching outward to the sphere of the fixed stars, is filled with a fifth simple body, aether, which is imperishable. *De caelo* 1 and 2 distinguish the superlunary element from the sublunary elements, argue for its primacy, and define the main features of the universe as a whole. *De caelo* 3 and 4 then investigate the nature and behavior of the sublunary elements.

The *De caelo* is only one of several works in which Aristotle investigates the four sublunary elements. They are central to his discussion of coming-to-be and passing-away in *De generatione et corruptione* and to his analysis of the composition and dispositional properties of compounds in *Meteor.* 4; they figure prominently in all of Aristotle's treatises on natural philosophy and regularly appear on his lists of commonly accepted substances (οὐσίαι). We will first consider the special perspective Aristotle brings to the elements in the *De caelo*, where he defines them by their weight and lightness, features associated with their natural motion. We will then examine his treatment of elemental natural motion and end with some observations about how his conclusions about natural motion reinforce his rejection of the elements as substances.

1. *The elements in* De caelo

Many of Aristotle's works discuss the behavior of the four elements, but most of his treatises investigate the interaction of the elements in the formation of compound bodies or in their own mutual transformations.

[*] Brown University.

In those contexts, Aristotle examines the elements' active and passive potencies (δυνάμεις)—the hot, the cold, the wet, and the dry (*De gen. et corr.* 2.2, 329b24–26). Each element has two such tangible differentiating features, one from the temperature spectrum hot/cold, and one from the humidity spectrum dry/wet: earth is cold and dry, water is cold and wet, air is hot and wet, and fire is hot and dry (*De gen. et corr.* 2.3, 330b1–7). The elements act on and respond to one another in virtue of those features.

The *De caelo* ignores the hot, the cold, the wet, and the dry, and concentrates instead on weight and lightness. A passage in *De gen. et corr.* 2.2 indicates why weight and lightness are discounted in that work, and by implication why they figure prominently in the *De caelo*. Aristotle lists a number of tangible features of the four elements. In addition to hot, cold, wet and dry, he lists heavy and light, hard and soft, sticky and brittle, rough and smooth. He writes as follows about heavy and light:

> Of these, heavy (βαρύ) and light (κοῦφον) are not active (ποιητικά) or passive (παθητικά): things are not called "heavy" or "light" because they do something to something else (τῷ ποιεῖν τι ἕτερον) or because they undergo something by something else (πάσχειν ὑφ' ἑτέρου). The elements, however, must be mutually active and passive, since they combine and change into one another. (*De gen. et corr.* 2.2, 329b20–24)

Since *De gen. et corr.* is interested in the ways in which the elements act on and respond to one another in the composition of compound bodies and in their own transformations, it appeals to the features involved in those changes and transformations. The basic features that explain the elements' interactions are the hot and the cold, which are active potencies, and the dry and the wet, which are passive potencies. The *De caelo*, by contrast, is interested in how the elements behave in their own right (καθ'αὑτά), independent of one another. So the *De caelo* focuses almost exclusively on weight and lightness, the features involved in the elements' natural motion to their own places.

2. *Definition of the simple bodies*

De caelo 1.1 announces that the science of nature is mainly concerned with bodies (σώματα) and magnitudes (μεγέθη), the affections (πάθη) and motions (κινήσεις) of these, and further with the principles (ἀρχαί) of that sort of substance (οὐσία) (268a1–4). Aristotle offers some definitions.

A *magnitude* is that which is infinitely divisible, either in one dimension (a line), two dimensions (a surface), or three dimensions (a body) (268a6–8). Although Aristotle gives no argument here, he denies the existence of indivisible magnitudes.[1] No magnitude is divisible in more than three dimensions (268a9–10). Bodies are complete (τέλειον) magnitudes, since they can be divided in all three dimensions (268a30–b7). Bodily parts of a larger whole, though complete in that sense, are incomplete in a different sense. Since bodily parts are determined (ὥρισται) by contact with their neighbor, Aristotle says that each is in this way "many." Only the totality (τὸ πᾶν)—the universe as a whole—of which the parts are parts, is necessarily complete (268b7–10). We shall return to the claim that the parts are somehow many, because it reflects Aristotle's judgment that the elements are not substances. For now it is enough to note that the body of the universe is a plenum, composed of three-dimensional infinitely divisible bodies.

De caelo 1.2 states that all the natural bodies and magnitudes can move locally (κατὰ τόπον) in virtue of themselves (καθ'αὑτά). The *nature* (φύσις) of a natural body is an internal principle of motion (ἀρχὴ κινήσεως) (268b14–16). All local motion is either straight or circular or a combination of the two (268b17–18). There are just two kinds of simple motion, in a straight line and in a circle, because straight lines and circular lines are the only simple magnitudes (infinitely divisible in one dimension). Rectilinear motions are, however, distinguished into two sorts, motion up and motion down. So there are three basic types of simple motion, all defined with reference to a single fixed point, the center of a sphere. *Motion up* is away from the center, *motion down* is towards the center, and *motion in a circle* is around the center (268b18–24). Because there are three simple motions, there are three simple bodies (1.3, 270b26–31).[2]

De caelo 1 bundles earth and water together as bodies that move naturally downward, and fire and air together as bodies that move naturally upward. Aristotle still speaks as though there were three elements—an indestructible element and two others—in 3.1 (298b6–8). Not until book 4 does he mark off the intermediate elements, water and air, from earth and fire.

---

[1] Contrary to the Atomists and Plato. See *De caelo* 3.4, 303a20–24, which refers us to arguments about time and motion in *Phys.* 6.1–2. Cf. *De gen. et corr.* 1.2, 315b24–317a12.

[2] On Aristotle's method of argument here and in much of the *De caelo*, see Bolton (2009) in this volume.

Bodies are simple or compound (*De caelo* 1.2). Simple bodies, such as fire and earth, have a natural principle associated with a single simple motion. Compound bodies have composite motions, though they may move in accordance with the prevailing ingredient (268b26–269a2). *De caelo* 3.3 offers this definition of a bodily element:

> Let an element of bodies be that into which other bodies are divided, which is present in them either potentially or actually (which of the two is still debatable), but is itself indivisible into things different in form.
>
> (302a15–18)

Elements are the ultimate stuffs into which compound bodies can be analyzed and into which compounds are finally destroyed.[3] The elements themselves can be broken up into smaller bits of the same stuff, but not into simpler material components which differ from them in form. The elements are the ultimate material bodies.[4]

3. *Weight and lightness*

Aristotle's universe is a finite sphere with a determinate center, which coincides with the center of the Earth (*De caelo* 2.14, 296b6–18), and a determinate extremity, which is the boundary of the sphere of the fixed stars (1.6, 273a7–13). There is only one cosmos. Though the form of the cosmos is repeatable, it is not repeated, because all the matter is contained in this one (*De caelo* 1.8–9).[5] The cosmos is filled with bodies of the three fundamental sorts we have discussed. Aristotle argues that circular motion is prior to rectilinear and that the simple body that moves in a circle, which he calls by the traditional name "aether" (*De caelo* 1.3, 270b20–25), is more divine than and prior to the elements that move in straight lines (269a18–b17).

---

[3] Since the *De caelo* is not itself concerned with the composition of compounds out of the elements, Aristotle here leaves open the question whether the elements are present in such bodies actually or potentially. *De gen. et corr.* 1.10 (esp. 327b22–31) argues that the elements are present in a compound (μίξις) only potentially. I discuss this topic in Gill (1989) ch. 5.

[4] Some scholars think that Aristotle believes in matter more ultimate than the elements, traditionally called "prime matter" (see, e.g., Solmsen 1958, Robinson 1974, Cohen 1984). This view has been challenged by a number of scholars (e.g., King 1956, Charlton 1970 and 1983, Furth 1988). In Gill (1989) ch. 2 and appendix, I side with the challengers and argue that the sublunary elements are Aristotle's ultimate matter. See also below, the section "Natural Motion in *Phys.* 8.4."

[5] On the argument in 1.9, see Matthen (2001) 174–177.

According to *De caelo* 4.1 we call things "heavy" or "light" because they have the capacity (τῷ δύνασθαι) for a certain natural motion. A body is *heavy* if it moves naturally toward the center. A body is *light* if it moves naturally away from the center. Weight and lightness are thus dispositions of bodies to move naturally toward or away from the center of the universe, and such motions spring from a body's own nature.[6] *Absolute* weight and lightness are defined by determinate places. The extremity of the universe (τὸ τοῦ παντὸς ἔσχατον) is up (308a21–22), and the center of the universe is down (308a22–29). Something is *absolutely light* which moves upward towards the extremity; something is *absolutely heavy* which moves downward towards the center (308a29–31).

The distinction between absolute weight and lightness provides a criterion for marking off earth from fire (*De caelo* 1.8, 277a12–23). Earth is absolutely heavy (and possesses no lightness), because it sinks below everything else and moves toward the determinate center. Once at the center it comes to a rest, since it can travel downward no further. Since all earth tends toward the center, a mass of earth (the Earth) rests in a ball at the center of the universe.[7] The behavior of fire is more puzzling. Initially, Aristotle simply specifies the light element as a body that moves away from the center. Later he speaks of a definite goal, the extremity of the universe. Fire cannot reach that location, however, because the mass of revolving aither impedes its passage beyond the sphere of the Moon. Aristotle says little explicitly about fire's upward motion. He does say that fire rises above air and has no weight even in its own location (4.5, 312b14–16), that the body is lightest which rises to the top (ἐπιπολάζον) of all things that move upward (1.3, 269b25–26), and that necessarily "what rises to the top of everything moves toward the extremity of the place in which the motion of these bodies is performed" (4.4, 312a4–5). Many scholars take these claims to indicate that the goal of fire is the extremity of the sublunary realm.[8] But if they are right, Aristotle's presentation is misleading. In 4.1 he refers to the extremity of the universe (τὸ τοῦ παντὸς ἔσχατον) as uppermost in position (308a21–22), and

[6] A body can move unnaturally in a direction opposed to its natural motion—e.g., a clod of earth can be thrown upward. Motions contrary to nature are brought about by an external cause (*Phys.* 8.4).

[7] For further discussion of this topic, see Matthen (2009), in this volume. Whereas Matthen argues that the Earth stays where it is impassively, I think that the Earth is held in place by many bits of earth vying for the center. See n. 10 below.

[8] Bodnár (1997) 97–98, argues for this claim; many scholars simply assume it.

specifies the absolutely light a few lines later as what moves upwards and towards the extremity (πρὸς τὸ ἔσχατον) (308a29–30). The text gives no reason to construe the simpler phrase as anything other than the extremity of the universe, which has just been mentioned. The phrase at 4.4, 312a4–5, quoted above, "the extremity of the place in which the motion of these bodies is performed" (τὸ ἔσχατον τῆς χωρας, ἐν ᾗ ποιοῦνται τὴν κίνησιν), presumably refers to the place where fire's upward progress stops—hence the actual place of fire—and not the goal of its motion.[9] Fire stops moving up when it reaches the sphere of the Moon, not because that is its natural goal, but because it can move upward no further.[10]

---

[9] The same can be said about 296b13–15: fire "moves toward the extremity of the place surrounding the center" (πρὸς τὸ ἔσχατον φέρεται τοῦ περιέχοντος τόπου τὸ μέσον). Although this phrase could refer to the extremity of the universe, it would be an unusually elaborate phrase for that purpose (Guthrie 1971, 244).

[10] On my view, even if fire could reach its goal at the extremity of the universe, where it would naturally rest, it would rest not of its own accord but because its upward motion is stopped by the extremity of the cosmos. Earth ceases to move downward when it arrives in the vicinity of the center, but not because it is programmed to stop—the first clod to reach the center would stop because it could proceed downward no further (beyond the center it would have to move upward); all other clods would be stopped by the ball of earth already there. Whereas the soul of a living organism is a principle of motion and rest—soul directs, coordinates, and limits the organism's natural motions (see *De an.* 2.4, 416a9–18)—on my view the elements possess only a principle of motion and no principle of rest: they automatically move according to their natures, unless impeded, and what needs to be explained is why they *stop* moving, not why they *start*. Aristotle's conception of being is *dynamic*, not *static*: a thing is most actual when it is *acting* according to its nature, and this includes an element's activity when confined at its own place.

My view most directly opposes Cohen (1994 and 1996, 37–45) who argues that the elements have only a principle of rest and no principle of motion. Cohen and I both attribute to Aristotle's elements what Bodnár (1997) 90, calls a one-sided nature—a principle of motion or a principle of rest (but not both). For various criticisms of my view (Gill 1989, 235–240; 1994, 31–32), see Cohen (1994) 154–155, (1996) 40–41; Bodnár (1997) 93–98, Matthen (2001) 180 and (2009), in this volume.

De caelo 1.8, 276a22–30 might be taken as counter-evidence to my position. Here Aristotle distinguishes natural motion and rest from enforced motion and rest, and the mention of natural rest could suggest that the elements have a principle of rest, as well as a principle of motion. I think that both natural and enforced elemental motions are limited by something external to the element. Rest at the center is natural for earth, because earth has arrived at its "like" (as we shall see below, *De caelo* 4.3 treats the place of earth as its form and designates that place as its "like"). Rest away from earth's own place is enforced, because earth's downward motion is impeded by something other than its "like."

Sisko (2002) defends my view against Matthen (2001), citing *De caelo* 4.5, 312b2–19, where Aristotle discusses the intermediate elements, air and water, and says that if the next lower element were removed, air would move into the place of water, and water into the place of earth. This passage suggests that the intermediate elements are not programmed to stop in the intermediate places either: they stop where they do because

Fire remains very active at the extremity of the sublunary region: unusual phenomena occur when hot dry stuff is ignited by the circular motion above, resulting in shooting stars and the like (*Meteor.* 1.3, 340b10–14; 1.4, 341b18–24).[11] Aristotle speaks in the *De caelo* of combinations of straight and circular motion (1.2, 268b18), which may characterize these occurrences.[12]

The two intermediate bodies, water and air, do not have locations determined by the structure of the universe. Aristotle has independent reasons for positing two elements between earth and fire. Since there are two active and two passive features—hot and cold, wet and dry—which combine into four pairs consisting of one active and one passive member, and since each pair characterizes one element, there are four elements (*De gen. et corr.* 2.3, 330a30–b7). Elements are observed to change into one another (e.g., water into air and air into water), and Aristotle explains such transformations by appeal to the elemental contraries.[13] The *De caelo* marks off water and air from earth and fire by distinguishing relative from absolute weight and lightness. According to *De caelo* 4.1,

their motion downward is impeded by the element below. I welcome Bodnár's (1997) 94, label for Aristotle's cosmos as I conceive it: it is *pressurized*—at least the sublunary region is.

[11] All such events are said to occur below the level of the Moon (*Meteor.* 1.4, 342a27–33).

[12] Carl Pearson has pointed out to me that Aristotle may have inconsistent views about these meteorological events: Aristotle sometimes speaks of fire at the periphery of the sublunary region being ignited by the rotation of the lunar sphere, but sometimes attributes such phenomena to dry exhalations, which are distinct from fire. Pearson has also called my attention to some puzzles about Aristotle's account which exercised the later tradition and especially Philoponus. For instance, if Aristotle takes fire's own place to be the extremity of the universe, the potency of fire to be in its place is never actualized. Below I argue that the upward motion of fire is an *activity* of fire, which is actualized as soon as fire starts to move. Fire is *actually* light (and so moves upward) as soon as it has been generated. Fire's own nature is actualized even if it is not in its own location. At the same time, fire's lightness is defined by the extremity of the universe, the goal of its upward motion, and its potentiality to be *there* is never realized. The same might be said about the bulk of earth, since its goal is the center of the universe, which is a geometrical point. I think the problem arises because the elements are intrinsically incomplete (formless), and could not be completed—fully actualized—even if they reached their own location. I say a bit more about this topic below in the sections "Elemental Natural Motion" and the "Status of the Elements." Another worry: the place of fire is also the place of aether. This may not be a serious problem, since Aristotle has said that the place of aether is *around* the center. Aether's place would still be around the center, even if fire could get past it, which it cannot.

[13] Aristotle gives his account of the various sorts of elemental transformation in *De gen. et corr.* 2.4. He tackles the issue from a different perspective in *De caelo* 3.6–7.

a body is *relatively light*—light in relation to something else (πρὸς ἄλλο κοῦφον)—if two bodies equal in bulk both have weight, and the second body moves downward with greater momentum than it (308a31–33). *De caelo* 4.4–5 argue that whereas earth is absolutely heavy (any portion of it moves downward if not impeded) and fire is absolutely light (any portion of it moves upward if not impeded), air and water are both heavy and light. Both are lighter than earth (any portion of them rises to the top of earth) and heavier than fire (any portion of them sinks below fire). But in relation to each other, air is absolutely light and water is absolutely heavy: air, whatever its quantity, rises to the top of water; and water, whatever its quantity, sinks below air.[14]

4. *Enforced motion and self-motion*

Elemental natural motion differs from both enforced motion and self-motion. To see how, we should first outline Aristotle's general theory of change in the *Physics*.

Enforced motion

In response to Parmenides, who had denied the possibility of change (*Phys.* 1.8, 191a23–33), Aristotle proposes in *Phys.* 1.7 that every change involves three principles: a pair of opposites, and an underlying subject, which is informed first by one opposite (the privation), and then by the other (the form). A change is the emergence of something new, because the form replaces the privation. The change is not a sheer emergence, however, because the underlying subject is there all along, characterized first by one opposite and then by the other. When a man comes to be healthy from having been sick, the man survives the change, but is characterized first as sick and later as healthy.

[14] Aristotle argues that water and air cannot be combinations of the other two elements. If they were, we could not account for their distinctive behavior—water always sinking below air and air always rising above water. There could be instances of a large quantity of water containing more fire than a small quantity of air, or a large quantity of air containing more earth than a small quantity of water. In the second case a large amount of air would move downward faster than a small amount of water. Since this is never observed to happen, air and water move to their distinctive places because each has a distinctive feature (*De caelo* 4.5, 312b32–313a13).

The pair of opposites must be properly opposed on some continuum, so that there is a path from one to the other. A body comes to be hot from cold, or comes to be musical from unmusical, or travels to Athens from the Piraeus.[15] Aristotle recognizes changes in three categories—quality (alteration), quantity (increase and diminution), and place (locomotion). He locates the pair of opposites in one of these three categories, and the underlying subject (e.g., living organism, an artifact, a complex material stuff, an element) in the category of substance. He also recognizes a fourth sort of change, the generation and destruction of a substance, and in this case the form and privation, as well as the continuant (matter), are located in the category of substance.[16] The continuant has a *passive* potency to be in the end-state—that is, it is of a *suitable kind* to be in that state (bricks and stones can be made into a house, whereas wool cannot). Before the change the subject's passive potency coincides with the privative state. The change replaces the privation with the positive form.

So far we have described change from the perspective of the object changed. *Enforced* changes are typically brought about by an external mover, which has the property it transmits to the patient or (in artificial production) has it in mind.[17] According to *Phys.* 8.5, "the mover is already in actuality, for instance, the hot thing heats and generally that which has the form generates" (257b9–10).[18] In artificial productions, a builder has in mind a house, a sculptor a statue, and a doctor health.[19] Aristotle calls the form, which enables a mover to cause a change, an *active potency* and defines it as "the principle of active change in another thing or ⟨in the thing itself⟩ as other" (*Meta.* 5.12, 1020a4–6; cf. 9.1, 1046a10–11). An active potency does not initiate changes but, like a code of instructions, enables its possessor to direct and coordinate motions in the moved so

---

15 An individual does not become musical from having been pale, except accidentally, if the unmusical thing happens also to be pale (*Phys.* 1.5, 188a30–b3).

16 In the *Metaphysics*, Aristotle often lists the form, the matter, and the composite as three ways to be substance (*Meta.* 7.3, 1029a30–33; 7.10, 1035a1–2; 8.1, 1042a26–31; cf. 5.8, 1017b23–26; 9.7, 1049a34–36; *De an.* 2.1, 412a6–9). Substantial generation and destruction pose special problems for the replacement model in *Phys.* 1.7, which we cannot address here. I argue elsewhere that Aristotle's considered view about substantial generation depends on his conclusion that matter survives in the product only potentially. See n. 3 above. For my current thoughts on this topic, see Gill (2008). We shall touch on elemental transformation below in connection with natural motion.

17 Bodnár (1997) 88–89, aptly calls the account the Transmission Model of causation.

18 Cf. *Phys.* 3.2, 202a9–12; *Meta.* 7.7, 1032a24–25.

19 *Meta.* 7.7, 1032b21–23.

that they lead to a particular outcome. Aristotle defines the patient's *passive potency* as "a principle of passive change by another thing or ⟨by the thing itself⟩ as other" (*Meta.* 5.12, 1019a20–23; 9.1, 1046a11–13). Mover and moved act and respond in virtue of their active and passive potencies, which are opposed to each other but aimed at the same actuality, a form to be realized in the patient. By means of the change the mover assimilates the moved to itself (*De gen. et corr.* 1.7, 324a9–11). Fire, in virtue of its heat, causes something cold to be hot; a doctor, in virtue of his knowledge of health, causes someone sick to be healthy. Aristotle defines change as the joint actuality of the mover and the moved. This actuality is incomplete, because it lasts only as long as the patient is *coming to be* in the end-state. The motion terminates in the complete actuality, the end-state realized in the patient (*Phys.* 3.1–3).[20]

Usually the patient acquires merely a *passive* state (a πάθος)—a new passive potency which corresponds to the agent's active potency. If the doctor's efforts succeed, the sick person becomes healthy, not a doctor. Health is a passive state which enables its possessor to respond in certain characteristic ways to its surroundings (*De gen. et corr.* 1.7, 324b13–18). Only in some cases does the patient acquire an active potency which transforms it into a new agent, as when man generates man or a doctor teaches a student the art of medicine.

Self-motion

As we earlier noted, *De caelo* 1.2 defines the natural bodies as bodies which can move locally (κατὰ τόπον) in virtue of themselves (καθ' αὑτά), and defines the nature (φύσις) of natural bodies as their principle of motion (ἀρχὴ κινήσεως) (268b14–16). *Phys.* 2.1 defines nature more explicitly as an *internal* principle:

> The nature is a certain principle (ἀρχή) and cause (αἰτία) of being moved and resting in the thing to which it belongs primarily in virtue of itself (ἐν ᾧ ὑπάρχει πρώτως καθ' αὑτό) and not accidentally (μὴ κατὰ συμβεβηκός). (192b20–23)

The phrase "not accidentally" excludes instances in which the mover and moved happen to be the same individual, as when a doctor cures himself. The doctor cures and the sick person is cured in virtue of opposed properties (192b23–27).

---

[20] I discuss Aristotle's definition of change in Gill (1989) ch. 6 and (2003) § 3.

In a self-mover, whose active principle is a nature (the organism's soul), the mover and the moved are more intimately related. *Metaphysics* 9.8 distinguishes a nature from an active principle located in a separate mover:

> I mean by potency (δύναμις) not only the one that has been defined, which is called an active principle of motion in another thing or ⟨in the thing itself⟩ as other, but generally every active principle of motion or rest (πάσῃ ἀρχὴ κινητικὴ ἢ στατική). For nature (φύσις) is also in the same genus as potency; for it is an active principle of motion—not, however, in another thing but in the thing itself *as itself* (ᾗ αὐτό).
>
> (1049b5–10; cf. *Meta.* 5.4, 1014b18–20; 1015a13–15.)

The nature of a thing is an active principle of motion or rest located in the thing itself *as itself*. The modification "as itself" distinguishes this case from that of the doctor who cures himself. Whereas the doctor acts on himself *as other*, a natural body acts on itself *as itself* (ᾗ αὐτό), or in the language of *Phys.* 2.1, *in virtue of itself* (καθ' αὑτό). The agent and the patient are one and the same individual—and not merely accidentally. Unlike enforced changes, the active principle of the agent and the passive principle of the patient are not opposed. They are active and passive versions of *corresponding features*, which are jointly realized in an activity which does not alter the patient's actual state but *expresses* it.

Consider the production and use of an artifact. A violin-maker constructs an instrument out of pieces of wood, glue, and other materials. As in other enforced changes, the craftsman acts on separate materials, imposing on them a form they initially lack, and gradually assimilating them to a form which corresponds to his active principle, the form of the violin. A violinist who plays the violin, on the other hand, does not change the instrument, that is, give it a form it initially lacks. Instead, the violinist enables the violin to be fully what it already is, to express its capacities as a violin. The music performed is the joint expression of the active capacities of the violinist and the passive capacities of the violin. This is an artificial case in which the mover and moved, though separate, act and respond in virtue of active and passive capacities which are not opposed but correspond. In a self-mover, the mover and the moved act and respond in virtue of corresponding features which are located in one and the same object—as though the violinist and her violin were the same individual.[21] That is Aristotle's conception of an ensouled body, a self-mover.

---

[21] Cf. *De an.* 2.1, 412b10–17, where Aristotle considers what it would be like if an ax were a natural body.

The elements are natural bodies, which move naturally upward or downward in virtue of themselves. In *De caelo* 3.1 and elsewhere Aristotle lists them as substances, along with plants and animals, their parts, and the heaven and its parts. We might therefore expect the elements to contain an active and passive principle of their own motion, like other natural things. But Aristotle distinguishes the elements' natural motion from both enforced motion and self-motion.

5. *Elemental natural motion*

De caelo 4.3 gives Aristotle's account of elemental natural motion. Before we turn to that difficult chapter, we should consider Aristotle's treatment in *Phys.* 8.4.

Natural motion in Phys. 8.4

Phys. 8.4 tries to drive a wedge between self-motion and mere natural motion. Self-motion, says Aristotle, is restricted to living things, which are able not only to move themselves but also to stop and to move locally in opposite directions. The province of fire, on the other hand, is to move only up, not down (255a5–11). Self-movers can be analyzed into a mover and moved, whereas elemental natural movers are continuous and naturally unified (συμφυῆ), and therefore admit no such analysis (255a12–18). He concludes that the elements have a principle of motion—but not of actively *causing* it (τοῦ κινεῖν, τοῦ ποιεῖν), only of *undergoing* it (τοῦ πάσχειν) (255b30–31). Since the elements lack an inner active principle, what is the active principle of their natural motion?

Phys. 8.4 considers two possibilities: what generated the element in the first place (the producer), and what removes the impediment to its natural motion (the trigger)—for instance, someone who removes a column supporting a weight, so that a heavy object falls. Since Aristotle identifies the trigger as an accidental mover (255b24–29),[22] we seem to be left with the other alternative. Is the cause of an element's generation also the cause of its natural motion?[23]

[22] Self-motion too involves such triggering causes (*Phys.* 8.2, 253a11–20; 8.6, 259b6–16). On this topic, see Furley (1978) and Gill (1991).

[23] Errol Katayama defends this thesis in "Soul and Natural Sublunary Motion in Aristotle," presented at the American Philosophical Association, Eastern Division Meetings, 2007.

At 255a30–b25 Aristotle recalls his distinction between levels of potentiality spelled out in *De anima* 2.5 (417a21–b16). A child is a potential knower in one way before she learns, because she is a human being, and human beings are the sorts of things that sometimes have knowledge. An adult is a potential knower in another way once she has learned, because she actually has knowledge and can exercise it when she wishes, if not impeded. The transition from the first level of potentiality to the second is an Aristotelian change, a transition between opposites, caused by an external mover with a relevant active potency. The knower's use of her knowledge is caused by the knower herself. The exercise of knowledge is an *activity*, not a change from opposite to opposite.

Aristotle applies this model to the elements (255b5–11). Water is potentially light in the first way, because it can be transformed into a light body. The transformation of water into air is a substantial generation and is brought about by an external mover, a producer—for instance, someone who heats the water and turns it into steam. The resulting air is potentially light in the second way. If Aristotle is true to his model, the upward motion of air is analogous to a person exercising her knowledge. Although motion from one location to another is a local change in the category of place, this local change does not alter the element's nature. Motion upwards no more alters the nature of fire than Socrates' walk from Athens to the Piraeus alters his. Socrates' walk, though itself a change from one location to another, is an expression of his nature as a biped animal. The upward motion of fire is similarly an *expression* of what fire is, a light object. That expression can of course enhance the character the element already has (as playing the violin makes the player a better violinist), but it does not replace that character.

The active cause of Socrates' walk is his own nature as a locomotive animal of a certain kind. Fire lacks such an internal active cause. In this special case, should we say that what produced the light object in the first place also causes its upward motion? In other situations Aristotle distinguishes the producer from the cause of activity. The teacher causes someone to learn, while the one who has learned is herself responsible for exercising her knowledge. A violin is produced by a violin-maker and played by a violinist, and these two individuals have different skills. A male parent generates his offspring, and his active principle controls embryonic development up to a point; but once the heart or analogous organ is formed, the active principle becomes internalized and the organism takes charge of its own further development.[24] Thereafter fetal development, though it involves many local changes, is overall an *activity*, since the organism already *is* the sort of thing it will later be more perfectly. Consider what Aristotle says about newly generated air:

> It is *already light*, and will straightaway act (ἐνεϱγήσει), if nothing interferes. The *actuality* (ἐνέϱγεια) of the light is to *be* somewhere, namely *up*, and it is prevented when it is in the opposite place. (255b10–12)

Once air has been generated, the producer has completed its work and is no longer relevant. Air is *already light* and moves upward immediately if not prevented. Like someone with newly acquired knowledge, who needs only the opportunity to exercise her capacity, air exercises its lightness in upward motion unless prevented. A trigger may need to release it, but the trigger is an accidental cause. There seems to be no adequate answer to the question: what is the active principle of elemental natural motion?

In both enforced motion and self-motion the active principle is a *form* possessed by the agent. Perhaps we can answer the question about the elements by identifying their form. Absolute lightness is the property that differentiates fire from other natural movers, and so appears to be its formal feature.[25] Although lightness is a *passive* potency, a principle of *being moved*, not of actively causing it, in that respect the lightness of fire resembles the form of many other objects. The form of a violin, too, is a passive potency and depends for its activation on an external mover. Nonetheless, the lightness of fire differs in one chief respect from the passive form of complex objects. Unlike the form of a violin, which structures various material parts into a musical instrument, the differentiating feature of fire does not structure simpler material parts. Recall that fire is a simple body—an element into which other things can be divided but which cannot itself be divided into things different in form (*De caelo* 3.3, 302a15–18). Because it is not worked up out of simpler matter, fire is not a matter-form composite like a violin. Its simplicity also explains why fire cannot be analyzed into mover and moved, as self-movers can. Fire is *pure matter*. The absolute lightness of fire, though it differentiates fire from other things, is properly speaking its *material* nature. Fire has no intrinsic form.

---

[24] Aristotle compares that moment to a son setting up house for himself, apart from his parents (*De gen. an.* 2.4, 740a1–9).

[25] Heat and dryness are also differentiating features of fire, but they specify fire in its interactions with other things.

Yet something plays the role of elemental form. Aristotle answers his question as follows:

> This is the question we were seeking to answer, why light things and heavy things are moved to their own place. The reason is that they *are naturally somewhere* (πέφυκέν ποι), and that is the *being* for the light and the heavy, the one defined by the *up* and the other by the *down*.
>
> (*Phys.* 8.4, 255b13–17)

Whereas Socrates' locomotive soul enables Socrates to express his nature when he walks from Athens to the Piraeus, and controls his mode of travel, the *goal* of fire controls and limits its upward motion. Since fire is defined with reference to its own place, that location functions as its form, its active principle, and its principle of rest.[26]

Natural motion in De caelo 4.3

De caelo 4.3 again asks: what is the cause of elemental natural motion? The chapter ends by recalling the two options that Aristotle considered in *Phys.* 8.4: the producer and the trigger. Again he opts for neither alternative, but recalls them as causes *in addition* to some other mover (311a9–12).[27] Earlier in the chapter he says:

> If what causes motion (κινητικόν) upward and downward is what makes something heavy (τὸ βαρυντικόν) and what makes something light (τὸ κουφιστικόν), and the movable (κινητόν) is what is potentially heavy or light (τὸ δυνάμει βαρὺ καὶ κοῦφον), then (δ')[28] each thing being moved to its own place (τὸν αὑτοῦ τόπον) is being moved to its own form (τὸ εἰς τὸν αὑτοῦ τόπον φέρεσθαι), and this would be the best way to understand the ancient saying that like is moved to its like. (310a31–b2)

Scholars have apparently found this passage so puzzling that they ignore its conditional structure in their translations, then construe the next 15 lines as a lengthy digression explaining how we should understand the old saying "like is moved to like," and have Aristotle return to his topic only at 310b16–19, where they finally observe the conditional structure

---

[26] In saying that the place of fire *functions* as its form, I differ from Morison (2002, 50–53) who argues that the place of fire is *part* of its form. I also disagree with Algra (1995, 216, 217–221) who takes the form of an element not to be its proper place but its *being* in its proper place.

[27] Note the καὶ ("also") in καὶ κινεῖ δὲ κτλ. at 311a9.

[28] The final clause of this sentence appears to contain an apodotic δέ at 310a33.

of the earlier passage.[29] In the later passage Aristotle says: "So to ask why fire goes up and earth down is the same as to ask why the curable if it is moved and changes *qua* curable, comes to health and not to whiteness."

The first passage is an argument, which Aristotle unpacks in the main part of the chapter. *Phys.* 8.4 sheds light on the passage. What moves fire upward? According to *Phys.* 8.4, fire moves to its own place because it *is naturally somewhere* (πέφυκέν ποι) (255b13–15): the upward *place* explains fire's natural motion upward and limits its motion if it arrives. That place functions as the *form* of fire, because it defines what the element is and how it naturally behaves (255b15–17). The first premise in the *De caelo* argument states that an element's own place (the cause of its upward or downward motion) is its form (what makes the element heavy or light).[30] The second premise focuses on elemental matter, the thing moved. According to *Phys.* 8.4, once a light object has been generated from a heavy object, it is *already* (i.e., *actually*) light (255b10). But it is also *potentially* light, because—depending on impediments—it may or may not be exercising its lightness in upward motion and is not yet *there* (in the upward place). The *actuality* (ἐνέργεια) of the light, says Aristotle, is *to be somewhere*, namely *up* (τὸ ποὺ εἶναι καὶ ἄνω) (255b11), and the light object's motion is blocked when it is in the opposite place (255b12). If the actuality of the light object is *to be* up, that upward place is its form. The light object manifests its lightness by *going* there and attains its full actuality by *being* there.[31] Aristotle's conclusion about motion relies on his definition of motion in *Phys.* 3.1–3, which I mentioned earlier: motion is the incomplete actuality of the mover and moved, which terminates in the full actuality, the end-state realized in the patient. Although I have filled in a good deal to make sense of the argument, Aristotle can

[29] See Stocks (1930) 310a n. 1; cf. Guthrie (1971), 345–349. As at 310a33, the manuscripts read (apodotic) δέ at 310b16–19, which Allan 1955 has emended to δή in his Oxford edition following Simplicius' report of Alexander (Heiberg, 1894). See Stocks (1930) 310b n. 4.

[30] Recall that *De caelo* originally defines the simple bodies with reference to the center of a sphere and later with reference to the center or the extremity of the universe. The center and the extremity, though not intrinsic features of earth and fire, function as their formal causes—earth is defined with reference to the center, and fire with reference to the extremity, the goals of their natural motions. Scholars often refer to the "natural" places of the elements, but Morison (2002) 34, points out that Aristotle nowhere speaks of "natural" places. For that reason I refer to the place of fire as its *own* or *proper* place (ὁ αὑτοῦ τόπος).

[31] On my view, the light object would retain its potency to move upward, even if it arrived at its own place. Its upward motion is limited by its place; see note 10 above.

get his conclusion from his premises, with the added assumption about motion. If what can move an element (the downward place or the upward place) is what makes it heavy and light (its form), and if the movable object (what can *go* to the downward place or upward place) is potentially heavy or light (i.e., can *be* in the downward place or upward place), and if motion is the joint actuality of the mover and moved, then a thing's being moved to its own place is its being moved to its own form.

Aristotle does not simply rely on his earlier discussion, but goes on to spell out his conclusion in the next section, where he clarifies how we should understand the old saw "like is moved to its like." He first points out that not everything moves to its like in the ordinary sense: if the Earth were transposed to where the Moon now is, bits of earth would not move toward the bulk of earth, but would still move to the center of the cosmos, where the Earth in fact is (310b2–7). This imagined scenario shows that the direction of earth's natural motion is determined by its proper *place*, not by the bulk of similar stuff, wherever that bulk happens to be.

Next he discusses the determinate places of fire and earth, and starts with his definition of place from the *Physics*:[32]

> Since place is the limit of what surrounds (τὸ τοῦ περιέχοντος πέρας), and both the extremity and the center surround all things that are moved, this ⟨limit⟩ turns out in a certain way to be the *form* of the thing surrounded, and *being moved to its own place is motion to its like.* (310b7–11)

According to the *Physics*, place is the first motionless limit of the surrounding body. The center and extremity of the cosmos do not satisfy the definition of relative place: the center is a geometrical point, not something that literally surrounds anything; and the extremity is not the limit of a surrounding body, since there is no body surrounding the extremity of the universe. Even so, Aristotle claims that the two limits of the cosmos, the center and the extremity, surround all movable bodies, which are contained within those limits. The scenario of the displaced earth shows that we should not take the ancient saying "like to like" to mean that an element is moved toward stuff like it. On the contrary, since an element's proper *place* functions as its *form*, we should understand "like to like" to characterize the motion of an element to its own place or form.

The center functions as the form of earth, and the extremity functions as the form of fire. What about the intermediate elements water and air? Aristotle continues:

---

[32] *Phys.* 4.4, 212a20–21; cf. 212a5–6.

> The things in succession are like one another, e.g., water is like air and air is like fire. We can speak in reverse (ἀνάπαλιν) in the case of the intermediates, but not the extremes. For instance, air is like water, while water is like earth. In fact the higher item is always related to the one below it: as form is to matter, so they ⟨the elements⟩ are related to one another. (310b11–15)

The universe displays a permanent cosmic order: the extremity of the universe is the determinate up and the center of the universe is the determinate down, and the hierarchy of the elements reflects that cosmic order: the higher item is always related to the lower as form to matter.[33] This ordering cannot be reversed in the case of the extreme elements, fire and earth.[34] But apparently the ordering of the intermediate elements can be reversed, since air is like fire, but also like water; and water is like air, but also like earth. Because we can speak in reverse about the intermediate elements, the consecutive element above *or* below air can function as its form: the consecutive elements on both sides play a limiting role.[35]

The next sentence, which commentators usually regard as the conclusion of the argument 15 lines earlier, should be interpreted in light of Aristotle's identification of the center and the extremity as the form of earth and fire: "So to ask why fire goes up and earth down is the same as to ask why the curable, if it is moved and changes *qua* curable, comes to health and not to whiteness" (310b16–19). The goal of an element's motion determines its direction. Fire goes up and not down, because its goal is the extremity of the universe, just as the curable goes to health and not to whiteness, because its goal is health.

The active principle of elemental natural motion lies outside the element, yet the elements *seem* to have an inner active principle. Why? Aristotle observes that someone can recover his health with only the slightest

---

[33] Cf. *De caelo* 4.4, 312a12–16.

[34] Stocks (1930) 310b n. 2, construes the claim thus: whereas water is like earth, earth is not like water, and whereas air is like fire, fire is not like air.

[35] This passage enables me to respond to Bodnár (1997) 96 n. 20, who criticizes my claim (Gill 1989, 240) that the confining element is form to the one confined. He points out that this would allow earth (the cosmically lower element) to be the form of water (the cosmically higher element), since earth confines the downward motion of water. Aristotle appears to allow for precisely this possibility here, when he says that the order of the intermediate elements can be reversed. There is a permanent cosmic hierarchy among the elements, with earth at the bottom, and fire at the top, but each of the intermediate elements is limited by the adjacent element above and below, and so both adjacent elements can function as its form.

external stimulus (310b26–29). Why do the elements seem different from that? The difference between the elements and everything else is that they are *simple*, not compounds worked up from simpler materials. Because they are simple, they have only one simple natural motion, and so need no internal active principle to direct or coordinate their motion. Aristotle explains why the elements seem so much like self-movers:

> More than those ⟨e.g., a sick person who needs only a small stimulus to recover his health⟩, the heavy and the light appear to have in themselves the principle, *because their matter is closest to substance.* (310b31–33)

Because the elements are simple, they behave like self-movers without needing an internal active principle. Once some earth has been generated, it automatically moves toward the center if unimpeded. What serves as its active principle is the place itself, which controls and terminates its motion.[36]

Aristotle says a little more to clarify his claim that the elements are closest to substance. Locomotion is a special sort of motion. It belongs to separate bodies and is the last motion in the order of generation and the first motion in the order of substance (κατὰ τὴν οὐσίαν) (310b33–311a1). Although locomotion, growth, and alteration are all non-substantial changes, locomotion takes precedence over the others and belongs to more elevated objects.[37] So the fact that the elements engage in locomotion gives them a distinctive status. He continues:

> When air comes to be from water and light from heavy, it goes up. And at the same time it *is* light, and no longer *becomes* ⟨light⟩, but *is there* (ἐκεῖ) ⟨i.e., up⟩. So it is apparent that being in potentiality going to actuality, it goes to the place and quantity and quality appropriate to the actuality. And the same fact explains why things that already exist and are earth and fire are moved to their own places, if nothing interferes. (311a1–8)

[36] Machamer (1978) points out that in *Phys.* 4.1 Aristotle says that place has a potency (δύναμις) (208b8–11) but denies that it is any sort of cause (209a18–22). There has been considerable discussion about these claims (see Matthen (2009), in this volume for more on the topic). I think that Morison (2002) 49–53, is right that place has a δύναμις because it makes a difference in the world. I agree with Sorabji (1988, 187 n. 6), and Morison (2002, 52–53) that the denial that place is a cause is part of an aporetic conclusion. As I understand Aristotle's view in *De caelo* 4.3, the center functions as the formal cause of earth's heaviness and final cause (the τέλος or goal) of its downward motion. Although I do not accept Lang's (1998) conception of Aristotelian place, I agree with her that the determinate places of earth and fire are causes. For objections to the idea that a natural place is any sort of cause, see Algra (1995) 192–221.

[37] For instance, plants have the capacity to nourish themselves and to grow, but they cannot move themselves locally. In the world of living things, local motion is reserved for higher objects, and is the only sort of motion displayed by celestial bodies.

Air is already *actually* light as soon as it has been generated. Motion to its own place does not alter the nature of air (its lightness), but expresses and enhances it. At the same time, natural locomotion is intimately connected with elemental being, more so than it is for living organisms whose lower functions (nutrition and perception) are crowned by locomotion. The elements are incomplete because what controls their natural motion lies outside them, and is still outside them, even if an element arrives at its own place. An element is constantly active, but that unflagging motion cannot compensate for the fact that an element's form is and remains external to it.

6. *Status of the elements*

The elements regularly appear on Aristotle's lists of agreed substances.[38] They seem to be substances, because they are *subjects* (ὑποκείμενα): they have properties, interact with other things, and move locally. Like complex bodies in the sublunary realm, the elements are generated and destroyed. In some respects they seem superior to complex bodies, including living organisms. In *Meta*. 9.8 Aristotle compares them to eternal things, because they never tire of their activity:

> The things that are subject to change, for instance, earth and fire, imitate the imperishable things: these too are always active, since they too have motion in themselves in virtue of themselves. (1050b28–30)

The elements have a claim to be what is most real in Aristotle's system, because they are the basic components of all compound bodies in the sublunary realm. Since they are utterly simple, their internal unity—unlike that of complex material objects—is unproblematic. If being an ultimate subject is a criterion for being a substance,[39] the elements arguably have a better claim to be substances than the complex objects they constitute.

But they are not substances. In *Meta*. 7.16, following a detailed investigation of substance, Aristotle famously says:

---

[38] *De caelo* 3.1, 298a29–b1; *Meta*. 5.8, 1017b10–14; 7.2, 1028b8–13; 8.1, 1042a6–12; cf. *Phys*. 2.1, 192b8–13.

[39] Subjecthood is the chief criterion of substantiality in the *Categories*, and I believe that it remains a criterion in the *Metaphysics*. Some scholars would disagree or reduce it to a secondary status.

> It is evident that the majority of things that seem to be substances are potencies (δυνάμεις)—the parts of animals (for none of them exists separately, and when they *are* separated, even then they all exist as matter) and earth, fire, and air. For none of them is one, but as it were a heap (σωρός), until it is worked up and some one thing comes to be from them.
> (1040b5–10)

Aristotle's ground for excluding the elements as substances in *Meta.* 7.16 is that they are heaps, not unified things.

With this assessment of the elements, let us return finally to a passage in *De caelo* 1.1 which we noticed at the outset. Aristotle claims that bodily parts of the universe are complete in one sense, incomplete in another. They are complete in that they are magnitudes divisible in three dimensions. They are incomplete in that they are determined by contact with their neighbor. In that way each is "many." Only the totality—the universe as a whole—of which the parts are parts, is necessarily complete (268b5–10). Aristotle denies that the elements have an internal active principle, a principle that would direct and limit their motion. That direction is provided by a determinate place, which also limits an element's motion if it can be reached. More often an element's upward or downward motion is limited by the adjacent body. Because the elements are intrinsically formless, they lack an inner principle that makes them unified *things*. They are like heaps—simple formless stuffs—whose boundaries are determined by surrounding bodies.[40] Aristotle's treatment of the elements in the *De caelo* supports his verdict that they are not genuine substances. Still—and this is interesting—of all the things that seem to be substance and fail, the elements come closest to succeeding. That is because "their matter is closest to substance" (*De caelo* 4.3, 310b31–33).[41]

[40] On this passage, see also Matthen (2009), in this volume. Matthen thinks that bodies which lack an inner source of unity are delimited by their *place* (the inner surface of the surrounding body). I prefer my interpretation of the passage because Aristotle mentions a thing's contact with its neighbor and not its place.

[41] I read a version of this paper at a conference on Aristotle's Natural Philosophy at Rice University and am grateful for the discussion on that occasion and to Donald Morrison for organizing the event. I especially thank Carl Pearson for his probing questions, which deserve further investigation (see n. 12 above). I also thank Mohan Matthen for helpful conversations and the editors of this volume, Alan Bowen and Christian Wildberg, for encouraging me to reconsider several topics. The paper is better because of these interventions.

Bibliography

Algra, K. (1995) *Concepts of Space in Greek Thought* (Leiden).
Allan, D.J. (1955) ed. *Aristotlelis de caelo* (Oxford).
Bodnár, I. (1997) "Movers and Elemental Motions in Aristotle," *Oxford Studies in Ancient Philosophy* 15: 81–117.
Bolton, R. (2009) "Two Standards for Inquiry in Aristotle's *De Caelo*," in A.C Bowen and C. Wildberg (edd.), *New Perspectives on Aristotle's* De caelo (Leiden) 51–82.
Charlton, W. (1970) *Aristotle's Physics: Books 1 and 2* (Oxford).
——, (1983) "Prime Matter—a Rejoinder," *Phronesis* 28: 197–211.
Cohen, S.M. (1984) "Aristotle's Doctrine of the Material Substrate," *Philosophical Review* 93: 171–194.
——, (1994) "Aristotle on Elemental Motion," *Phronesis* 39: 150–159.
——, (1996) *Aristotle on Nature and Incomplete Substance* (Cambridge).
Furley, D. (1978) "Self-Movers," in G.E.R. Lloyd and G.E.L. Owen (edd.), *Aristotle on Mind and the Senses* (Cambridge) 165–179. Repr. in D. Furley, *Cosmic Problems* (Cambridge, 1989) 121–131; and in M.L. Gill and J.G. Lennox (edd.), *Self-Motion: From Aristotle to Newton* (Princeton, 1994) 3–14.
Furth, M. (1988) *Substance, Form, and Psyche: An Aristotelean Metaphysics* (Cambridge).
Gill, M.L. (1989) *Aristotle on Substance: The Paradox of Unity* (Princeton).
——, (1991) "Aristotle on Self-Motion," in L. Judson (ed.) *Aristotle's Physics: A Collection of Essays* (Oxford) 243–265.
——, (2003) "Aristotle's Distinction between Change and Activity," *Axiomathes* 14:17–36.
——, (2008) "Form-Matter Predication in *Metaphysics* Θ.7," in M. Crubelier, A. Jaulin, D. Lefebvre, and P.M. Morel (edd.), *Dunamis. Autour de la puissance chez Aristote* (Louvain) 391–427.
Guthrie, W.K.C. (1971) *Aristotle: On the Heavens* (London/Cambridge, MA).
Heiberg, J.L. (1894) *In Aristotelis de caelo commentaria*, Commentaria in Aristotelem Graeca 7 (Berlin).
King, H.R. (1956) "Aristotle without *Prima Materia*," *Journal of the History of Ideas* 17: 370–389.
Lang, H.S. (1998) *The Order of Nature in Aristotle's Physics: Place and the Elements* (Cambridge).
Machamer, P. (1978) "Aristotle on Natural Place and Natural Motion," *Isis* 69: 377–387.
Matthen, M. (2001) "Holistic Presuppositions in Aristotle's Cosmology," *Oxford Studies in Ancient Philosophy* 20: 171–199.
——, (2009) "Why Does Earth Move to the Center? An Examination of Some Explanatory Strategies in Aristotle's Cosmology," in A.C. Bowen and C. Wildberg (edd.), *New Perspectives on Aristotle's* De caelo (Leiden) 119–138.

Morison, B. (2002) *On Location: Aristotle's Concept of Place* (Oxford).
Robinson, H.M. (1974) "Prime Matter in Aristotle," *Phronesis* 19: 168–188.
Sisko, J. (2002) Review of David Sedley (ed.), *Oxford Studies in Ancient Philosophy* 20 (2001), *Notre Dame Philosophical Reviews* 2002.03.01.
Solmsen, F. (1958.) "Aristotle and Prime Matter: A Reply to Hugh R. King," *Journal of the History of Ideas* 19: 243–252.
Sorabji, R. (1988) *Matter, Space, and Motion: Theories in Antiquity and Their Sequel* (London).
Stocks, J.L. (1930) *De caelo*, in W.D. Ross (ed.), *The Works of Aristotle Translated into English* (Oxford) vol. 2.

THE ARGUMENT FOR THE SPHERICITY OF THE UNIVERSE INARISTOTLE'S *DE CAELO*: ASTRONOMY AND PHYSICS

Pierre Pellegrin*

1. *Introduction*

Aristotle's *De caelo* has undergone a series of misinterpretations regarding either Aristotle's general intentions in the text or the major epistemological principles according to which he conducted his research. A glaring example of the former would be to consider, as has been done, the *De caelo* as an astronomical treatise or rather as an Aristotelian astronomical treatise. Nevertheless, several passages of the *De caelo* clearly explain the connections drawn by Aristotle between contemporary mathematical astronomy and his own project.[1] These passages uncover two different ways in which Aristotle—the Aristotle of the *De caelo*, to be more precise—appeals to the findings of astronomical research. On some occasions, he calls upon these findings in order to confirm the results of his own research, for example, this comment regarding the central position of the Earth:

> What the mathematicians say in astronomy also testifies to this, since the appearances, given the changing of the figures by which the arrangement of the stars is determined, happen as if the Earth were at the center.
>
> (*De caelo* 2.14, 297a2–6)

On other occasions, Aristotle cites astronomers as having established findings which are somehow adjacent to his own. Concerning the order of the wandering stars, an issue dealt with in chapter 10 of book 2, Aristotle is content to make the general remark that the closer a planet is to the sphere of the fixed stars, the slower its movement will be, without any further investigation into the ratio of their velocities. His approach to this question is *physical* in that the stars are said to move in a direction

* CNRS, Paris.

[1] It is important to remember that the term μαθηματικός can be translated as either "mathematician" or "astronomer". Only the context permits distinction between the two.

opposite to that of the sphere of the fixed stars, those which are closer to it thus being more subject to its arresting influence. Still, he is aware of the different problems which his reader might encounter when it comes to the order of the stars. Thus, he begins his chapter as follows:

> As to the order of the heavenly bodies—the way each of them is located given that some are before and others behind, i.e., the way they relate to each other from the point of view of their distances—this should be left to astronomical works which actually speak of these matters sufficiently. It can be drawn from their demonstrations that the motions of each are in proportion to their distances, some being faster, others slower. (*De caelo* 2.10, 291a29–34; see Dalimier and Pellegrin (2004) 250 for the text.)

According to Aristotle, therefore, the physical theorist may support his work with mathematical astronomy. There is a particularly noteworthy passage in this vein which merits significant commentary. Aristotle intends to demonstrate that the size of the Earth "is not huge" (297b32), a demonstration which he backs up with various arguments. The more memorable of these is how the presence of elephants in both Africa and India indicates a continuity between India and the Pillars of Hercules (it should also be remarked that Aristotle takes this argument from other authors whom he does not acknowledge, accepting it as "not too incredible"). He concludes his reasoning with the observation that "all the mathematicians who attempt to calculate the size of the circumference say that it is about 400,000 stades" (298a15), a number which is difficult to compare with the actual circumference of the Earth, due to the uncertainty surrounding the length, or lengths, of a stade. It is a number, however, which seems to be much greater than the Earth's actual circumference. In other words, Aristotle satisfies himself with an approximation which the science of his time would have permitted him—at least according to his belief—to refine.

In fact, the connections between Aristotle's speculations in the *De caelo* conducted from the point of view of a physical theorist and those of the "mathematicians" are much more apparent when considered from a general point of view. Since I do not aim to tackle this question myself, I will limit myself simply to indicating certain difficulties which might be encountered by someone intending to go far enough into the details.

With regard to the relative velocities of the planets, for example, one must first of all confront the problem of the interrelations between mathematical and physical explanations. Aristotle appears to hold, as he does in *Phys.* 2.2, that the astronomer and the physical theorist (who, on this occasion, is actually a cosmologist) study the same object, in this case, the velocity of the heavenly bodies. However, his position in the *Physica* does not correspond exactly to that in the *De caelo*. In the *Physica*, the physical theorist considers the Moon as a natural entity endowed with certain properties, whereas the mathematician regards it as a sphere. In each case, both mathematical and physical science employ their own explicative machinery; the former by calculating, for example, the surface of the lunar sphere, and the latter by demonstrating, for instance, that the Moon's elemental composition renders it incorruptible. The two modes of research are evidently not parallel in that the mathematician, for example, is not concerned with final causes. That is, both modes bear upon the same object, the Moon, yet each does so through a different approach that is characteristic of the science in question. From the perspective of the *De caelo*, however, the mathematicians observe and perhaps calculate (how and to what extent is up to scholars of ancient astronomy to say) the relative velocities of the planets, whereas the physical theorist explains the cause of this phenomenon, namely, the retarding action of the sphere of the fixed stars. It seems, therefore, that the objective is the same *stricto sensu*, and that the physical and mathematical approaches share the same relation as that between ὅτι and διότι explanations. In other words, the astronomer demonstrates that the planets farthest away from the Earth take the longest time to complete their revolution, and the physical theorist explains why. Would that imply an isomorphic yet inverse correlation between mathematics and physical theory, similar to that found in *An. post.* 1.13, where the physical theorist knows the fact of the rainbow, while the geometer, or in this case, the optical theorist, knows the cause, one science being subordinate to the other? No, because in the case of the planets, astronomy and physical theory both provide a complete scientific explanation within their own domain: physical theory invokes the arresting influence of the sphere of the fixed stars, while astronomy establishes a correlation between velocity and distance, the sort of demonstration which could generate mathematical astronomy and which is doubtless of the type that could produce mathematical optics in the case of the rainbow. In the *De caelo*, therefore, astronomy and physical theory share an original relationship in respect to the manner in which Aristotle describes elsewhere the correspondences between the sciences. It is neither a relationship of subordination nor one of complete alienation.[2]

---

[2] In the *Analytica posteriora*, Aristotle highlights two instances when more than one science deals with the same object; the case of subordinate sciences, "but, Aristotle adds, many sciences which do not fall under one another are in fact related in this

A further problem which arises from the Aristotelian explanation and which was well recognized by Simplicius (Heiberg 1894, 472.5) is that, if the velocity of astral movements is the result of two factors—the natural tendency of stars to move in a circle and the opposing influence of the sphere of the fixed stars—how can these contrasting movements both be considered natural? Furthermore, it is difficult to understand how Aristotle reconciles this account with the theory of homocentric spheres which he supposedly upheld. This provokes the formidable question of whether and to what extent he adheres to that theory in the *De caelo*. It is, thus, necessary to conclude (with great caution) that, as confirmed in the opening line of the *De caelo*, the research Aristotle conducted comes under physical theory and that mathematics (pure or applied) may have no more than an appended role. It, therefore, seems reasonable to claim that in the *De caelo* the more physical an argument is, the stronger it will be.

The misconception surrounding the epistemological principles which I intend to examine is closely linked to this physical theory-based position of *De caelo*. Paul Moraux, for example, in the introduction to his edition of the treatise, argues that, according to Aristotle, "when applied to celestial entities, investigation by the senses is exposed as being quite impotent and even marred with error" (Moraux 1965, cxvii: cf. Moraux 1951). Up to this point, it is difficult to disagree with Moraux's analysis, even if the term "impotent" is not the most appropriate. However, Moraux continues: "without doubt, it is for this reason that Aristotle usually prefers the investigative methods which he considers to be more effectual and precise: the reasoning and application of the principles of terrestrial physical theory." This affirmation is completely insupportable. In the passages where Aristotle recognises the limited nature of sensory

way—e.g., medicine to geometry: it is for the doctor to know the fact that curved wounds heal more slowly, and for the geometer to know the reason why" (*An. post.* 1.13,7 9a13–16; Barnes 1993). The problem of the relationship between mathematics and physical theory with respect to the study of celestial phenomena is recurrent throughout the history of philosophy and the sciences, most often based upon the Aristotelian distinction. Geminus' summary of Posidonius' Meteorology, known through a quotation by Alexander of Aphrodisias which is reported by Simplicius in his commentary on Aristotle's *Physica* (*In phys.* 291.21–292.31), presents a particular interest from this point of view. Until when was it upheld that it was the job of physical theory to mediate between conflicting mathematical hypotheses? Duhem believes that Adrastus of Aphrodisias was the first to break with this position. Ptolemy, in any case, would later strive hard to restructure astronomy upon a physical basis.

knowledge in celestial studies, he does not claim that non-sensory argumentation, that is, recourse to reasoning from principles, is "more effectual and precise"; it is simply a last resort in situations where nothing else is possible. This he says explicitly a few lines later: "On these matters it is right to seek also after greater knowledge, although we have very small starting points and we are at a great distance from the phenomena of this region" (*De caelo* 2.12, 292a14–17).[3] Nevertheless, Aristotle displays no trace of the "mistrust towards sensory knowledge" attributed to him by Moraux. He continues to recognise a certain primacy behind arguments based on sensory data. This is clearly confirmed in the following often quoted passage from book 3. It is a criticism of the Platonists who, due to their refusal to abandon their principles, come to uphold the idea that all elements can transform themselves into one another except for earth:

> But if [the elements] are produced by a decomposition in surfaces, first it is strange that not all of them could be generated from each other. But this is what these people are compelled to say. For it is not rational that there be one exception to this transformation, and this is not what sense perception shows us: all of them, in the same way, change into one another. Dealing with phenomena, they come to make some statements which are not in accordance with the phenomena.... Because of their commitment to their principles, they seem to fall into the attitude of people who defend a thesis in an argument. For they accept any consequence under the assumption that they possess true principles, as though some principles did not require to be judged from their consequences, and particularly from their final

---

[3] Cf. *De caelo* 2.3, 286a3–7:

> Since movement in a circle is not contrary to movement in a circle, one must examine why there are several locomotions, though we must endeavour to conduct the inquiry from far off—far off not in the sense of spatial distance, but much more by virtue of the fact that we can perceive altogether few of the characteristics of the heavens. Nonetheless let us speak of the matter. (Leggatt 1995)

and *De caelo* 2.8, 290a13–20:

> It is reasonable that they all perform the same type of movement, but alone of the stars it is the Sun that seems to do this in rising and setting, and it does do not in itself but as a result of the distance from which we see it; for our vision, in being extended a long way, skews through weakness. This is also perhaps the reason for the apparent twinkling of the fixed stars, and the non-twinkling of the planets. (Leggatt 1995)

This is put even more neatly at *Meteor.* 1.7, 344a5: "For the phenomena inaccessible to observation, we think that a satisfactory explanation according to reason (κατὰ λόγον) has been given when we have brought them back to the possible."

> result. But the final result, which in the case of a productive science is the product, in the physical science is the evidence of sensory perception, which always prevails. (*De caelo* 3.7, 306a1–17)

The definitive and decisive test of a principle's validity is, therefore, compatibility between one's results and data drawn from sensory experience. Robert Bolton (2009), drawing upon this passage as well as others in the paper which he contributes to this collection, distinguishes "two standards of evaluation" at work in the *De caelo*, one based on "physical" argumentation (φυσικῶς) and the other on "rational" argumentation (εὔλογον, κατὰ λόγον). The main aim of Bolton's text is to demonstrate that the "rational" arguments are, in fact, dialectical, a thesis which does not concern me here. Bolton also shows, in reference to the passage in *Phys.* 1.2 which is dedicated to refuting the Eleatics, that recourse to rational procedure does not only take place when sensory data is unavailable, but also in the proposing of a primary physical principle, namely, the existence of motion, when sensory data *is* available. However, the reason that makes the *De caelo* the one physical treatise in which Aristotle resorts most frequently to rational procedures is principally that our perception of the phenomena which he studies is so limited. Bolton also recognises that "Aristotle regards εὔλογος procedure as distinctly *inferior* to the alternate more scientific level of procedure" (68). Evidently, this recalls one of the most famous passages of the Aristotelian corpus, found in *De partibus animalibus* (1.5, 644b23 ff.). Here Aristotle compares physical theory, or rather biology in this particular case, to the study of "divine" celestial bodies. Our understanding of these bodies is "altogether limited," because it is only possible for us to perceive a small amount of data. This is not the case for animals and plants. Knowledge of the former brings us "more pleasure," due to the excellence of the object, just as a fleeting glance of some beloved object brings us more pleasure than a long consideration of other things. Yet our knowledge of objects down here is *better and fuller*; we have "more science" (*De part. an.* 1.5, 645a2).

To this I should add that there is nothing rational or logical which does not yield itself to the results of observation. Consider this famous passage:

> Since there are two difficulties concerning which anyone might reasonably be troubled, we must endeavour to give what we take to be the explanation, reckoning zeal to be a mark of respect rather than of over-boldness if a person, through a thirst for knowledge, is content with even tentative solutions in matters concerning which we have the greatest difficulties.

> There are many such difficulties, not the least amazing of which is why it is not the case that at their successively greater distances from the primary locomotion the several heavenly bodies undergo a greater number of motions,[4] and instead that the intermediate bodies undergo the most. For it would seem reasonable that, since the first body[5] undergoes a single locomotion, the body nearest undergo the least number of movements. … As it is, just the opposite happens, since the Sun and the Moon undergo fewer movements than some of the planets.[6] … With some of them this has even been clearly observed; for we have seen the Moon, when half-full, pass under the star of Ares, which was occulted on the Moon's dark half, and emerged on its bright, radiant half. … One may rightly be troubled by this difficulty, as well as for what reason the number of stars in the primary locomotion is so great that their whole array seems to be innumerable, whereas each of the others has a single, separate star, and two or more stars are not observed to be fixed in the same locomotion.
>
> (*De caelo* 2.12, 291b24–292a14: Leggatt 1995)

Therefore, when perception is not a possible option, only "tentative solutions" can be offered. The difficulties raised in this passage are interesting because Aristotle takes them further than he does elsewhere. Bolton correctly observes that, although the "logical" procedure was inferior, it was still useful in its own right. Here however, there is a conflict between the supposedly "logical" (rational) procedures and the observations themselves. It should be noted that Aristotle does not seem to doubt for a minute the fact that a theory founded upon "what seems to be rational" must be adapted if confronted with more abundant perceptual data.

2. *Arguments in favour of the sphericity of the universe*

I should now like to examine the strength of some arguments according to their relation to observation, all these arguments aiming at establishing one and the same thesis, that of the sphericity of the universe. Although

[4] That is to say, requiring a larger number of spheres in order to explain their movement.

[5] Here, the expression "first body" does not mean the element which is moved in a circle and makes up the celestial bodies, but rather the sphere of the fixed stars.

[6] This assertion spread confusion among the commentators because, in Aristotle's system of homocentric spheres, the Sun and the Moon have no fewer spheres than the other planets and, therefore, no less movements. In his annotated edition of the first two books of the *De caelo*, Leggatt (1995), following Dicks (1970, 204–205), suggests that Aristotle distinguishes the Sun and the Moon from the other planets by their lack of retrogradation, which means that, in a certain way, they have fewer types of movement than the other planets.

this thesis has without doubt an essential position in Aristotelian cosmology, Aristotle dedicates only a few pages to it; and the demonstration that he provides in *De caelo* 2.4 has no more than a decidedly average size. Commentators have distinguished four or five arguments in this chapter (depending on whether the remarks on figures which open the chapter can be considered as an argument), their diversity of form and status giving the chapter a certain epistemological scope. Aristotle does not in fact seem to have difficulty rubbing elbows with "demonstrations" which appeal to speculations drawn from what could be called "transcendent geometry," an analysis of spatial relations that includes geometrical, physical, and logical arguments. His arguments are certainly not all of equal strength, which is explained by what we have seen above. Such an accumulation of arguments of different strengths is by no means reserved for the examination of the sphericity of the universe but, as has been noted, appears rather often throughout the *De caelo*, creating a remarkable combination of procedures which could be regarded as belonging to physical theory with arguments which are dialectic in the wider (but nevertheless Aristotelian) sense of the term. It is not, however, my aim to analyze the extent to which Aristotle's arguments are scientific or dialectical in the chapter dedicated to the sphericity of the universe. Nevertheless, it does fall within my intentions to attempt an estimation of the *physical* foundations which could be found in Aristotle's procedure to support this thesis, a thesis which is so difficult to verify through sensory means.

Before turning to the chapter itself in which Aristotle establishes the universe's sphericity, let us remind ourselves of where this fits into his study of the "heavens." Relying on considerations which are based on anything but observation or experience, Aristotle determined that the universe was an entity which involved perfection in its very concept, on the one hand, because its body is itself perfect—it is at the peak of an ascending succession of realities—and, on the other hand, because the universe is completely perfect in that it comprises the totality of all currently existing bodies. Furthermore, this universe is complete and unique, without a beginning and without possibility of corruption. It also incorporates an eminent component, the fifth element, whose essential property is to move in a circle. Being an animated entity, the heavens have absolute directions: high and low, left and right, forwards and backwards.[7] Commentators have been much engaged in assessing the

[7] See Lennox (2009) in this volume.

epistemological position of the arguments Aristotle employs to establish the properties of the heavens, because, in a treatise which, from its very first sentence, places itself within the cadre of natural philosophy, Aristotle resorts principally to lines of reasoning which depend little upon observation and experience. By the example of Paul Moraux, we have seen the sorts of erroneous interpretations that these data could give rise to. However, we have also seen what Robert Bolton (2009) can offer by way of opposition to these mistakes.

And so, Aristotle establishes the sphericity of the universe in *De caelo* 2.4. That the spherical nature of the universe (the meaning of which we will examine in greater depth) has not been introduced until this moment, after the series of properties given in book 1, is without doubt because the acknowledgement of the universe's sphericity is a sort of prerequisite to the study of celestial movements which Aristotle begins to undertake in the next chapter. The structure of the chapter is as follows. First of all, Aristotle announces the thesis which he is about to explore, namely, that the universe is spherical (286b10–11). He then proposes (286b11–287a2) general reflections on the figures or shapes that the universe might have, reflections which have either been considered as forming arguments in themselves (some, such as Thomas Aquinas, have even regarded them as the most convincing arguments in favour of the thesis), or as being linked to the following paragraph in order to form the "first" argument. They have also been considered as a reservoir of ideas which are then utilised within various other arguments of the chapter. This is a problem which we will try to tackle later. Next come four arguments which have been neatly separated by the commentators (287a2–11; 287a11–22; 287a23–30; 287a30–b14), before some lyrical final considerations (287b14–21) in which a hint of Platonism has been noted. However, if we want to draw a general lesson from the line of argument proposed by Aristotle in chapter 2.4 by gaining a more precise idea of the form of these arguments and their epistemological position, it is necessary to define, above all things, the nature of the thesis, or theses, that these arguments are supposed to establish.

This thesis is announced in the very first sentence of the chapter.[8] It has been generally and perhaps unanimously understood as an assertion

[8] There are several ways in which this sentence can be understood. It seems to me, in fact, that since the expression "primary by nature" (τῇ φύσει πρῶτον) can only be applied to the form, the same must hold for the expression "the most appropriate with regard to its substance" (οἰκειότατον τῇ οὐσίᾳ). Therefore, the text means "It is necessary that the

that the universe as a whole has an external spherical form. Having recognised that Aristotle aligns himself with the "spherists" in opposition to those who prefer to give the world a flat form such as that of a drum, interpreters commenting on this chapter have struggled to assess how to connect and differentiate Aristotle's views and those expressed in Plato's *Timaeus* or the doctrines put forward by the Eleatics, for example. This unanimous effort on the part of the commentators is worthwhile given that in this chapter, there is indeed a demonstration of the thesis that the exterior envelope of the universe is spherical. Thus, the whole of the first part, which I have designated as general considerations upon the figures, prepares the way for the thesis of the sphericity of the world's envelope, which is established in the first lines of the first argument. The general line of this argument is as clear as its details are obscure and difficult.[9] It goes as follows: granted that the circle is primary among the plane figures (a premise which is unused at this point), and the sphere is primary among the solid figures, "since the first figure is that of the first body, and since the first body is the body located in the outermost revolution, then the body which undergoes a circular revolution will be spherical" (287a2–5). Let us call this thesis that the external envelope of the universe is spherical T_1.

The status of this argument which establishes T_1 is also fairly clear. The thesis according to which the first body possesses the first form is a teleological principle comparable to those often voiced by Aristotle ("nature does nothing in vain," "nature always does the best," and so forth). More precisely, this principle expresses what could be called internal adaptation, which determines, for example, that nature gives the appropriate organs to whoever is most capable of using them, just as one gives a flute to someone who knows how to play it (cf. *De part. an.* 4.10, 687a9); only the mode of expression is different. The idea that the hierarchy of forms corresponds to the hierarchy of bodies is evidence of nature's perfection, which is not expressed under the providentialist or intentional forms habitually used by Aristotle in biology, when, for example, he compares the nature of an artisan or the master of a house. Several commentators

Heavens have a spherical form, because this is the form both most appropriate with regard to its substance and primary by nature," whereas some translators have understood this to mean that the spherical form is the best suited to the substance of the Heavens. This is not, however, of the utmost importance for my present argument.

9 For example, the allusion to "those who order figures according to their numbers" (286b33), is not only historically opaque, but also difficult to understand in the argument which Aristotle proposes.

(see Leggatt 1995, 231) have attempted to interpret the expression "first body" as having the same meaning as it does elsewhere in the *De caelo* where it designates the fifth element moving in a circle, which the post-Aristotelian tradition would later call ether. However, Aristotle excludes this interpretation by specifying that by "first body" he means "the body located in the outermost revolution." Here, therefore, and in the passage quoted above, the expression "first body" designates the sphere of the fixed stars.[10]

Having established T_1, Aristotle then introduces another thesis which states that the universe is spherical "in depth," i.e. that it is comprised of concentric strata which are spherical both internally and externally. This spherical structure extends into the supralunary region as well as the sublunary, where, unsurprisingly, it is not quite as fully accomplished. These theories are consistent with Aristotle's customary usage of the term "heavens," which designates the universe in its totality. Let us call this thesis, which states that the universe is an interlocking network of spherical homocentric strata T_2. The establishment of T_2 leads Aristotle to introduce into his argument a distinction between three regions of the universe: the sphere of the fixed stars, the region where the planetary movements take place,[11] and the sublunary sphere.

This development raises a simple question: which is the one, or at least principal object of our chapter, T_1 or T_2? In other words, when Aristotle says at the beginning of the chapter that "the heavens must have a spherical figure," are we to understand that the sphericity which he has in mind is a sphericity in depth? It is hard to deny that T_2 is richer in content and more interesting than T_1 from both a cosmological and philosophical point of view. However, if the chapter does indeed aim to establish T_2, this seems difficult to reconcile with some of the arguments identified by the commentators. Let us examine these arguments, one after the other.

What can be regarded as the first argument establishes both T_1 and T_2. Roughly speaking, it seems to define T_2 in the following manner: if the exterior figure of the universe is spherical, according to the already established T_1, the body which fits perfectly into the interior would

[10] Simplicius (Heiberg 1894, 408.1) explains that the body the movement of which is not irregular is said to be first.

[11] There is really no reason to replace the word "planets" in *De caelo* 2.4, 287a9, as Allan (1955) and Longo (1961) have done. This is not the only occurrence of the word in *De caelo* (cf. 2.2, 285b28), where the expression "errant stars" can also be found. The region of the planets is entirely composed of the fifth element.

also have to be spherical. Gradually, therefore, we recognise that all the strata which constitute the world must also be spherical. The part of the first argument which establishes T_2 is thus dependant upon T_1 as a principle. The argument itself could be regarded as a bit of elementary geometry, once we subscribe to a fundamental principle of Aristotle's physical theory which is not evoked here, according to which there is no empty space in the universe, because whatever is contiguous[12] with something spherical must itself be spherical. Nevertheless, this argument is remarkably weak. If the part of a shell which touches the inside of a sphere is in fact spherical, this says nothing about its other side, the one which is closer to the centre of the universe. Alexander of Aphrodisias makes this objection, but addresses it to the fourth argument of our chapter (*apud* Simplicium in Heiberg 1894, 414.30). It may be agreed, certainly, that such criticism has no force with regard to the supralunary sphere because that region is entirely composed of an element whose principal property is to move in a circle. So, in this region of the universe, the presence of an externally spherical and internally polygonal shell made of the fifth element makes no sense. However, the criticism does have force with regard to the sublunary sphere. It should also be noted that T_2 inherits T_1's force of conviction as well as its faults, in that T_1 is a teleological thesis based upon geometrico-numerological speculations of a Pythagorean nature. Moreover, if it is true that T_1 is ill-founded because it is based upon neither observation nor experience, then T_2 will also lack a firm empirical foundation.

Concerning what the commentators designate as the second argument (*De caelo* 2.4, 287a11–287a22), the objective is clear. This argument intends to establish T_1, without any regard for T_2. Briefly summarised, it goes as follows: if the universe as a whole was a polyhedron, its revolution would occupy a larger space than its volume. Now on the one hand, the universe is animated by a revolving motion—"which is both observed and assumed as a principle" (287a11)—and on the other hand, it has been established that there is no empty space beyond the sphere of the fixed stars (cf. 1.9, 279a11). In the second part of the argument (287a19–22), Aristotle considers the idea that the universe has a shape with a regular curve (he provides the example of a lentil shape, or that of an egg), only to conclude that this concept brings him back to where he started, given that the rotation of a

[12] "Contiguous" as used in the text (287a6) should be taken in a looser sense than Aristotle's customary technical usage of the term.

such a universe would occasionally leave a space empty. This analysis is open to dispute and has indeed been contested. Alexander, for example, observed that a lenticular or oviform shape, as well as a cone, cylinder or rhomboid, could actually turn on itself without occupying a space any larger than its volume. Still, this argument introduces a new aspect in that T_1 is no longer founded upon what I have called geometrico-arithmological speculations. At present, T_1 has found a more physical basis, namely, the empirical observation of the rotation of the sphere of fixed stars.

In fact, one possible response to Alexander's objection is very relevant to our own intentions. The difference between a sphere and a lentil is that it is possible for a lentil to rotate without leaving any space, but only when it rotates about one of two axes perpendicular to one another. The sphere, on the other hand, can manage this if it rotates about an axis along any diameter. Therefore, if it is a matter of deciding the exterior form of the universe, then Alexander's objection is perfectly well founded: the universe could have the form of an egg or a lentil, for instance, without leaving any space empty around its rotation. Nevertheless, what is lost with the idea of an oviform or lenticular universe is the concept of the system of homocentric spheres, as adhered to by Eudoxus or Aristotle, which provides an explanation for planetary movements. We know, in fact, that in order to reconcile the apparently irregular movements of the planets to the idea of nested spheres homocentric about the Earth, Aristotle follows in the footsteps of Eudoxus and Callippus by having each of these spheres turn about an axis which is a diameter of another sphere. Taking into account this complex mechanism, Aristotle's objection applies completely: if the movement of a planet is the result of the rotational composition of several lenticular bodies, the space taken up by these rotating bodies would be greater than their volume. We can only suppose, therefore, either that Aristotle had only the external envelope of the universe in mind and simply neglected to think about the solid shapes mentioned in Alexander's objection,[13] or that, as he saw it, one of the superior aspects of the spherical shape is that it allows for the proper nesting of rotating spheres. Alexander is thus correct: this argument does not establish T_1. Nevertheless, it renders T_2 quite possible. We can even go as far as to say that in considering T_1 from a more physical point of view, that is to

[13] Xenarchus of Seleucia also reproached Aristotle for not having seen that the helix of a cylinder was a simple curve along its straight side and around the circle, "because each of its parts corresponds to any other" (Heiberg 1894, 13.22).

say, a more scientific one, the second argument lends T_2 a greater solidity because in the previous argument T_2 was established on the basis of T_1.

The third argument (287a23–30) rests upon a property which is attributed to the sphere of the fixed stars: the movement of the heavens is the fastest of motions. This property does not only feature in the *De caelo*, it is also to be found in *Meta.* 10.1, 1053a8–12. In book 8 of the *Physics*, Aristotle showed that the movement of the heavens was primary, eternal, continuous, and uniform, without mentioning its velocity. In 8.9, 265b8, this movement was declared "the measure of the other movements." When it comes to our passage of the *De caelo*, the line of argument seems to be as follows:

> The movement of the heavens is eternal, continuous, and uniform; therefore, it is a measure for all other movements. The measure is the minimum; therefore, the movement of the heavens is the least movement, that is to say the shortest and the fastest. The fastest movement takes place along the shortest trajectory. The shortest trajectory which leaves from and returns to the same point is the circle. Thus, the movement of the heavens is a circular motion. Therefore the heavens are spherical.

I am not too interested in the finer details of the argument nor in the significant textual difficulties which allow for several different readings.[14] Thus, I have no views on what Aristotle precisely means when he says that "the least of the lines which leave from and return to the same point is the circumference of the circle" (287a27–28). Simplicius resolved this problem by identifying an allusion to what geometers would later call isoperimetry: the idea that, on any given surface, the smallest perimeter which can bound this surface is a circle. This means, in stereometric terms, that the smallest surface which envelops a given volume must be a sphere. Simplicius himself indicates that this theorem had not been presented before Archimedes or Zenodorus, and that at the time of Aristotle it was only known about "in general." It is, thus, entirely possible that the thesis of isoperimetry had been put to use before being formally established. Furthermore, it does not seem particularly useful to introduce the notion of velocity here.

[14] In 287a27, Alexander's reading as reported by Simplicius (Heiberg 1894, 412.11, 413.10) ἀπὸ τοῦ αῦτοῦ ἐπὶ τὸ αὐτό seems better. Allan (1955) and Moraux (1965) chose ἀφ' αὑτοῦ ἐφ' αὑτό, which would mean "the lines which leave from themselves to return to themselves."

What interests us in this argument is, first and foremost, Aristotle's goal. It seems that this third argument, in the same way as the second argument, aims to establish T_1. The main subject of the study, namely, the part of the heavens which performs this "final revolution" is, in fact, the sphere of fixed stars. Once again, Aristotle offers a basis for T_1 which seems much more secure than the one which produced his considerations regarding the correspondence between the first figure and the first body. It is an argument which could be described as both geometric and kinetic. On the other hand, there is no trace of T_2.

The situation is quite different, however, with regard to the fourth argument. "One may also," writes Aristotle, "derive just this conviction from the bodies situated about the center of the universe" (287a30–31). This "conviction" is T_1. The approach here is actually inverse to that of the first argument. Aristotle goes on to show that the structure of the universe is a series of spherical layers and that, therefore, the outer envelope of this universe is also spherical. In other words, he demonstrates T_1 by supporting it with T_2 in a line of reasoning which appears to be heavily "physical." Aristotle in fact proposes, by way of a supportive structure, an argument based upon an empirically observed premise: when he speaks of accepting "the hypothesis that water by nature flows always into what is more hollow" (287b5), it probably indicates the "sensible obviousness" which Aristotle uses as the foundation to all his physical speculations (cf. 3.7, 306a17). The faults in this reasoning, which come under the same criticism as the one addressed in the first argument, were recognized by commentators in antiquity. Alexander brought to light the fact that, if the surface of the water is spherical and the layer of air adapts itself perfectly to the layer of water, this means that the internal surface of the air must also be spherical. Nonetheless, this proves nothing about the external surface of the air, which is in contact with the fire.

As a matter of fact, these four arguments establish T_1. This is less apparent in the first argument because T_1 seems to be established straight away: "Since the first figure is that of the first body, and since the first body is the body located in the outermost revolution, then the body which undergoes a circular revolution will be spherical" (287a2–5). This thesis, though, is deduced from the general considerations on the figures which come at the beginning of the chapter, and so I believe that these considerations should be viewed as part of the first argument. T_2 only surfaces in the first and fourth arguments. In the first, it is shown to be a consequence of T_1; and in the fourth, it represents a starting point for the establishment of T_1. We can take this a step further by pointing out that the arguments in favor of T_1 gradually increase in strength, according to the physical criteria mentioned above. The first argument establishes T_1 from speculations of a "transcendent-geometrical" nature; the second, from a combination of geometry with a physical proposition (there is no empty space); the third, from geometrical and kinetic considerations; and the fourth, by induction from sensory observations. So far as establishing T_1 goes, the arguments progress as if the "easy" direction offered by the first argument is just not sufficient, with Aristotle adding various arguments until he finds one which is truly *physical* and, therefore, appropriate to the research carried out in the *De caelo*. This physical argument relies upon T_2 and particularly upon its establishment through sensory data.

Is this sheer incoherence? Are we simply caught up in a vicious circle, where the first argument appears to establish T_2 out of T_1 and the fourth produces T_1 out of T_2? In other words, why does Aristotle not begin with the fourth argument, given that it simultaneously establishes T_1 on a physical basis whilst also constructing T_2 which is necessary for the continuation of the *De caelo*? The reason for this, it seems to me, is that T_1 retains a pre-eminence over T_2, because, to employ Aristotelian terms, it is naturally prior to T_2. In fact, it is because the supralunary region is spherical that the sublunary region is the same and not the opposite, and it is because the sphere of the fixed stars is spherical that the supralunary region is the same and not the opposite. The sphere of the fixed stars is effectively a principle for the entire universe, while the supralunary region does the same for the sublunary region, where sphericity begins to degrade and continues to do so the nearer one approaches the center of the Earth. In this way, the first argument exposes the sequence of causes according to nature. Indeed, the course that leads from T_1 to T_2 and back in the other direction illustrates the distinction made by Aristotle between deductions which start from principles, and inductions which lead towards them. In this case, it is T_1 which represents the principle itself, whereas T_2 is a property deduced from the principle.

However, the version of T_2 which appears in the first argument (T_{2a}) is not identical to that made in the fourth (T_{2b}). T_{2a} is actually a property deduced from T_1 as a principle. Alexander's argument takes little effect under these conditions because, if the outer envelope of the heavens is spherical and the world is composed of concentric spherical rings, it is not simply for the reason that whatever is contiguous with a sphere must be spherical but, more fundamentally, because the entire universe is filled with a desire to imitate the sphere of the fixed stars, which is directly moved by the Prime Mover. T_{2b}, on the other hand, is inferred from the natural properties of water, "that the surface of water is spherical, is clear for those who accept the hypothesis that water by nature flows always in what is more hollow" (287b4–6). The sphericity of water cannot, in any case, be used as a principle to explain the sphericity of the air, an observation which here gives strength to Alexander's objection.

It is also worth noting that, so far as Alexander is concerned, the fourth argument is only coherent when it is confined to the sublunary region and does not seek to establish T_1. He proposes to salvage Aristotle's argumentation in the following manner:

> Humid air, like water, runs into hollow space in such a way that its external surface becomes spherical. Therefore, the internal surface of fire which adapts itself to fit the air must also be spherical. "Its external surface is also certainly so, if the divine body [the ether] is so, as it has been shown earlier." (Heiberg 1894, 415.2–4)

It is hard to tell whether Alexander establishes this property of humid air by experience (observation) or through a direct analogy between water and humid air when regarded as two fluids. What is certain, is that this property of air is not deduced from T_1. However, in order that this empirical argument may be used to infer the sphericity of each layer in the sublunary region, the outer layer of the supralunary region must necessarily be spherical. Aristotle's argument only recovers its force when it abandons its proposed aim, which is to establish T_1 out of T_2. T_1 must consequently be established elsewhere.

The fourth argument, therefore, has a paradoxical nature. In one sense, it is the strongest epistemological argument for establishing T_1 because it is the most "physical." In another sense though, it is the weakest because Alexander's objection applies most fully. To regard the situation in a negative light, the way in which T_1 is produced in the first argument is not "physical" enough, and its establishment by induction in the fourth argument causes it to run up against the difficulties pointed out by Alexander. Consequently, it is within the second and third arguments that T_1 is most effectively established. The difficulties with T_2, on the other hand, are not quite the same. T_2 is established in two different ways: in the first argument, it is deduced from T_1 as a principle (T_{2a}); and in the fourth, it is derived empirically (T_{2b}). In the first case, T_2 is established more securely for the supralunary region; and in the second case, for the sublunary. It can, therefore, seem that the two arguments in which T_2

appears are in competition with each other to establish T_2 more firmly. It must, thus, be remembered that *De caelo* 2.4 only partially manages to establish T_1 upon a properly physical basis, but that it does succeed in establishing T_2. This could even be considered as the principal object of the chapter, even if its "official" aim is actually to establish T_1.

3. *The physical basis supporting the doctrine of the universe's sphericity*

T_2 is not completely established until 2.14, where Aristotle demonstrates the sphericity of the Earth using three arguments. The first rests upon both the notion of natural movement towards a particular place, and upon the idea that a clustering of material around a center (where the tendency of each matter particle is to move as close to the centre as possible) will form a sphere: "that, if the parts move from all directions towards a unique center in the same way, it be necessary that the bulk they constitute must be equal in all directions, this is clear. ... But such a figure is a sphere" (297a18–25).

The second argument is based upon the downward movement of two bodies which, because their trajectories are not entirely parallel, converge towards the center of the Earth. Scholars are unclear about on what experiential basis Aristotle formed such a conviction.[15] I am inclined to think that it is constructed in the imagination and not founded upon any observation; and this is, in my view, what Aristotle himself says at the introduction to his last argument: "Further support [for the sphericity of the Earth] also comes from perceptual evidence" (297b23–24), an expression which suggests that he considered the trajectory of falling bodies not as an object of observation. The "phenomena" referred to by Aristotle are the lunar eclipses, in which the line which limits the Moon is convex; and given that the eclipse takes place due to the interposition of the Earth, it may be concluded that the Earth is spherical.[16] It should

[15] Moraux (1965, cxxxi note 2) has suggested that it comes from an observation not unlike the one used by Eratosthenes in order to demonstrate the sphericity of the Earth: at a place situated under the tropic during the summer solstice, the trajectory of a stone falling into a pit strictly follows the rays of sunlight falling into the same pit. However, the trajectory of the stone creates an angle with the rays at places further away from the tropic.

[16] Alan Bowen has pointed out to me that this reasoning is flawed, given that there are an indefinite number of forms which could project a curved shadow onto a sphere.

be remarked upon that this is not simply a "crude" observation, but one which must be combined with a developed astronomical knowledge in order to know the true cause of the eclipse, the interposition of the Earth between the Sun and the Moon.

It seems that these three arguments become stronger one by one, if their force is to be measured by their degree of proximity to sensible perception. In any case, we have in *De caelo* 2.14 a solid "physical" basis for T_2, which becomes further edified the more Aristotle considers the inherent constraints of the sublunary region. At the end of the second argument, in fact, he concludes that "the Earth is therefore spherical, or, at least, its nature is to be spherical. But one should speak of each thing as being what it tends to be by nature and is actually by nature" (297b20–23). This contains an allusion to the irregularities in the Earth's surface, which do not in any case constitute a disclaimer or an exception to T_2.[17]

T_2 also bears a strong unifying factor for the universe in that this spherical structure encompasses both the sublunary and supralunary regions. I should now like to examine the consequences of this layered universe for each of these domains.

The establishment of T_2 has an enormous scope with regard to the supralunary region. It would here seem advisable to return briefly to a doctrine which we have already mentioned. We know that in order to explain the planetary movements Aristotle adopted the system of homocentric spheres invented by Eudoxus and corrected by Callippus. Simplicius even claims that Callippus had revised Eudoxus' system in collaboration with Aristotle, but it is difficult to decide whether this affirmation has anything other than chronology in its favour. There is, however, a fundamental difference between the systems of Eudoxus and Callippus and that of Aristotle, which has been well noted by the commentators (cf. Duhem 1913, 126). According to a tradition, notably that reported by Simplicius, Eudoxus was attempting to resolve a problem which had been put to him by Plato: how can the movements of the celestial bodies, and in particular the apparently erratic movements of the planets,

---

[17] In *Meteor.* 1.3, 340b33, Aristotle divides the layer of air which surrounds the Earth into two parts; one which is created by the general movement of the world and the another which is included in the circumference "which makes the Earth completely spherical," a reference to the air which makes up the space between the deepest valleys and the highest mountains. It is doubtlessly necessary to consider that the irregularities of the sublunary region also account for the fact that sometimes it is air which is in contact with earth and sometimes water.

be explained using only circular orbits, the circle being the only type of line which is compatible with the supposed perfection of the celestial region? His response, however, was purely astronomical. He conceived of a system of spheres explaining the apparent movement of each planet, but each of these systems was treated as an independent mechanism. It is probable that Eudoxus and Callippus actually believed in the existence of these spherical globes; but in any case, they were content to draw up a model which simply made the movements of each planet intelligible. For Aristotle, on the other hand, the study of the heavens was classified as physical theory. His response was to create a real model supported by the fundamental principles of his physical theory. In such a model, the different mechanisms which move the different planets could not possibly be independent of each other because this would involve them being separated by an environment which does not transmit motion, that is, by an empty space. Indeed, we know that Aristotle's universe contains no emptiness; hence the idea of compensatory spheres which are added by Aristotle to give a physical basis to the independence of each planetary mechanism.

From this, we can begin to understand what Aristotle is referring to when he distances himself from Callippus in *Meta.* 12.8:

> But it is necessary, if all spheres altogether are to explain the phenomena, that there be other spheres for each of the planets, one fewer than those supposed ⟨by Callippus⟩, which counteract the ones already mentioned, and bring back to the same position the first sphere located the furthest from the star. (*Meta.* 1073b38–1074a4)

It is, therefore, necessary that the exterior sphere of the system of spheres belonging to each planet not be affected by the motions of the spheres in the systems above it, so that the planet may retain its proper place and motion in the overall cosmic system. If, hypothetically, each planet had a number of compensatory spheres equal to its number of motive spheres—five for the Moon, the Sun, Mercury, Venus and Mars, and four for Jupiter and Saturn—then each system of spheres corresponding to a planet would be independent and all in all, immobile. Nevertheless, since all these systems share a common feature, namely, having a sphere which turns from East to West in the space of a day, we can retain this movement for all the spheres in the universe, which means adding a number of compensatory spheres equal to the number of spheres in each system minus one. Therefore, each system of spheres transmits this single diurnal movement from east to west to the one below it, meaning that the total collection of spheres will be transported according to the daily movement of the sphere of the fixed stars.[18] Furthermore, the notion that each system of spheres receives the some motion from the preceding system and transmits it to the following one fits better with Aristotle's idea of a universe without empty space, in that all the spheres are in continuity with each other. In attempting to apply a principle of economy—a test of the perfection of the overall system—it seems that Aristotle saves himself the trouble of attributing a primary sphere to the system of Saturn, the planet furthest from the Earth. The sphere of the fixed stars may, in effect, perform this role because this would conserve the diurnal movement from east to west for the entire universe. In this way, Aristotle calculates 55 spherical globes for the whole universe excluding the sphere of the fixed stars,[19] in contrast to Eudoxus' 26 and Callippus' 33.

Aristotle's astronomical system, therefore, supposes T_1; however, it could even do without it. As soon as the fixed stars are animated with a circular motion, one can imagine that the framework which carries them would not be spherical without ruining this astronomical system. Of course, this fault would be much more serious because it would be somehow metaphysical, so firmly was it accepted that the perfection of the universe would require a spherical envelope. On the other hand, the system of concentric spheres when regarded *as a physical system* absolutely requires T_2, or at least the "part" of T_2 which concerns the supralunary region. Aristotle in fact, like Eudoxus and Callippus, felt no need to resort to epicycles or eccentric spheres in order to explain planetary movements. The supralunary region is, thus, a collection of spherical rings in contact with each other and also concentric with the Earth as the center.

Let us now return to the variable strengths of Aristotle's arguments. A closer look at T_2 reveals a physical foundation for an astronomical model,

[18] Simplicius (Heiberg 1894, 498.1–504.15) reports that Sosigenes highlighted a flaw in Aristotle's theory which does not seem entirely unfounded. Let us suppose, for example, that what Aristotle says applies to Saturn. Saturn's system of spheres would transmit to the system below it, that of Jupiter, the single diurnal movement from east to west. If this is the case, then what is the point of the exterior sphere of Jupiter's system, which is already supposed to be animated by this daily movement?

[19] In *Meta.* 1074a12–14, Aristotle makes a remark which has spread much confusion among the commentators: "If one does not add to the Moon and the Sun the motions we have mentioned, all the spheres will be 47." Perhaps, as has been suggested by T.-H. Martin (1871) following pseudo-Alexander's commentary on the *Metaphysica*, Aristotle is trying to distance himself from Callippus' reform of Eudoxus' system for the Sun and the Moon.

which, by its very nature, cannot admit any such foundation. Concentric spheres like those of Eudoxus are in fact imperceptible, even indirectly through their effects: indeed, it would seem that Aristotle considers them invisible, unlike the visible stars which they carry—despite the fact that both are made up of the same material.[20] T_2, nevertheless, when coupled with the principle that the universe contains no empty space, provides an indispensable empirical basis (concerning the system of homocentric spheres) to pass over from an astronomical model which aims to "save the phenomena" in the manner of Eudoxus and Callippus to an actual, existing network of real, existing spheres. In this respect, the final argument of *De caelo* 2.14 is of particular significance. It provides the only empirical basis which allows one to ascertain the reality of Callippus' system as revised by Aristotle—i.e., with the addition of compensatory spheres—by making it a genuine object of the science of nature.

Regarding the sublunary region, T_2 is no less important. It would certainly not bring the fundamental division of the Aristotelian universe, namely, that of the supralunary and sublunary regions, into question; but, as I have already indicated, T_2 is a strong unifying factor in the universe. It is, in the first place, worth noting that our knowledge "ascends" from the sublunary to the supralunary region, that is to say, from a less noble region to a more noble one. However, after all, one of Aristotle's constant doctrines is the idea that knowledge can take two opposing directions, one which begins with something that we know better ourselves and moves towards something that is more knowable by nature, and one which moves in the reverse direction. Nevertheless, the true scientific procedure is to begin with something that is more knowable by nature, using the other procedure as a sort of preparation.[21] We cannot assert that the celestial vault is spherical simply because the layers of water, air, and fire are of this shape, but rather because the sublunary elements are more or less perfect spherical shells due to the fact that the supralunary region has a spherical structure. The sphericity of the supralunary region, and more precisely, the sphere of the fixed stars, acts as the principle because the properties of the celestial bodies and regions which Aristotle studies in the chapters following book 2 of *De caelo* depend, directly or indirectly, on the sphericity of the universe. We know, in fact, that between chapter 4, which is dedicated to the

---

[20] Aristotle does not give a reason for this difference.

[21] This idea is best summarised in *Phys.* 1.1, despite an unusual use of terminology.

spherical form of the universe, and the two final chapters of book 2, which deal with the Earth, Aristotle considers the movement of the sphere of the fixed stars, then the stars, their movements, their order, and their form.

4. *Conclusion*

The sphericity of the world's envelope (T_1) is therefore, not fully proven by "physical" arguments. This should not surprise us, given that it is a phenomenon which cannot be grasped by the senses in that it belongs to a form of reality which physical theorists must examine using alternative methods for lack of anything better. In any case, the impressions we can gather from a brief reading—that T_1 is established "from two sides," "descending" and "ascending"—seem deceptive. However, with the opportunity to demonstrate T_1, Aristotle introduces another thesis which is much more important for his cosmology, T_2, a thesis which is simultaneously deduced from T_1 as a principle and established with physical arguments, at least in the case of the sublunary region.

Bibliography

Allan, D.J. (1955) *Aristotelis* De caelo (Oxford).
Barnes, J. (1993) *Aristotle's* Posterior Analytics, 2nd edn (Oxford).
Dalimier, C. and Pellegrin, P. (2004) *Aristote. Traité du ciel* (Paris).
Dicks, D.R. (1970) *Early Greek Astronomy to Aristotle* (London).
Diels, H. (1882) *Simplicii in physicorum libros quattuor priores commentaria.* Commentaria in Aristotelem Graeca 9–10 (Berlin).
Duhem, P. (1913) *Le système du monde*, vol. 1 (Paris).
Heiberg, J.L. (1894) *Simplicii in Aristotelis de caelo commentaria*, Commentaria in Aristotelem Graeca 7 (Berlin).
Leggatt, S. (1995) *Aristotle, On The Heavens I & II* (Warminster).
Lennox, J.G. (2009) "De caelo 2.2 and Its Debt to *De incessu animalium*," in A.C. Bowen and C. Wildberg (edd.), *New Perspectives on Aristotle's De caelo* (Leiden) 187–214.
Longo, O. (1961) *Aristotele,* De caelo, *Introduzione, testo critico e note* (Florence).
Martin, T.-H. (1871) *Mémoire sur les hypothèses astronomiques d'Eudoxe, de Callippe, d'Aristote et de leur école* (Paris).
Moraux, P. (1951) "Recherches sur le *De caelo* d'Aristote. Objet et structure de l'ouvrage," *Revue thomiste* 1951: 170–196
——, (1965) *Aristote,* Du ciel (Paris).
Simplicius, *In Aristotelis de caelo commentaria.* See Heiberg 1894.

DE CAELO 2.2 AND ITS DEBT TO THE *DE INCESSU ANIMALIUM*

James G. Lennox*

1. *Introduction*

In *De caelo* 2.2, Aristotle considers the questions whether it is legitimate to apply the principles "right" and "left" to "the body of the heaven" and if so in what manner to do so. This essay examines his attempt to answer those questions. The examination has three goals:

(a) to understand Aristotle's appeals to his conclusions about such "directional" principles in the *De incessu animalium*, and to assess his grounds for doing so,
(b) to determine whether his assertion, during the course of this discussion, that "the heavens are ensouled and have a source of motion" (*De caelo* 2.2, 285a29–30) plays a significant role in his argument,
(c) to draw out some of the implications of this discussion for how we should understand the logical and explanatory structure of Aristotle's science (or philosophy) of nature.

I will contend that a significant epistemological problem lies behind the appeal to the *De incessu* in this passage, and that Aristotle was fully aware of it.

2. *Cosmic dimensions and cosmic directions*

Aristotle's *De caelo* begins by declaring that the science of nature appears most of all to be concerned with bodies, magnitudes, their affections and their changes, but also with the first principles[1] of such substantial

* History and Philosophy of Science, University of Pittsburgh.

[1] I will adopt the following policy for translating ἀρχή: where it appears clearly to refer to the starting points of a science, I will use the term "principle"; where it appears to

beings.[2] This claim rests on ontological grounds—things constituted by nature either *are* bodies and magnitudes, *have* body and magnitude, or *are the first principles of* things having body and magnitude (268a1–6). Moreover, each of the natural bodies that constitute the cosmos is said to be *complete* in so far as it has all three of the *dimensions* of magnitude (πάσας γὰρ ἔχει τάς διαστάσεις, 268b7), i.e., in so far as it is divisible in length, breadth, and depth. Here, then, διαστάσις is naturally rendered "dimension" and refers to precisely what we designate by that English word.[3]

At the beginning of the second book, after reviewing weaknesses in the arguments of those who deny that the heavens are eternal, Aristotle considers whether the right and the left are among the principles of natural bodies—an idea defended, we are there told, by the Pythagoreans.[4]

> Since there are some, such as those called Pythagoreans (for this is one of their statements), who claim there to be a certain right and left to the heavens, one should investigate whether this is so in the way they claim, or in some other way—if indeed one ought to apply these principles (ταύτας τὰς ἀρχάς) to the body of the whole cosmos. (284b6–10)

refer to the source or origin of a change of any kind, I will use the term "source." A generic translation that could vaguely cover both ideas is "starting point." Even with this policy, however, there are unavoidable problems. The notion of a scientific starting point harbors an ambiguity which one encounters in Aristotle's Greek and which renders the above distinction suspect—Aristotelian science is above all the identification of causes and the production of demonstrations from premises that identify the causes of the facts being demonstrated. So "scientific principle" may refer to the primary premises of a scientific demonstration or to the fundamental facts identified by those premises. And since some of those fundamental facts turn out to be the causes of those stated in the conclusion, the line between "principle" and "source" is not clear-cut. As the entries in *Meta.* 5.1, 2 and 4 make abundantly clear, Aristotle was fully aware of this problem.

[2] See Johansen (2009, 10–13), for the significance of this starting point in understanding the relationship between the inquiry represented by Aristotle's *De caelo* and the very different project described in Plato's *Timaeus*.

[3] This passage also includes an odd endorsement of the Pythagorean dictum that the all and everything are defined by threes, a claim that we take the number three from nature and make use of it in worship of the gods and a claim that the completeness of three is shown by the fact that the first time we say "all" (rather than "both") is when we refer to three items. I will take up later the question of why Aristotle stretches in these various ways for support of the idea that objects with these three dimensions of magnitude are complete objects, but for now it is enough to note that it is part of a set of wider epistemological questions raised by the method and style of the *De caelo* discussed in Robert Bolton's contribution to this volume.

[4] The actual Greek term used consistently for the object of this natural inquiry is the singular noun οὐρανός. But English convention uses the plural to refer to the stars and planets, while the singular has primarily religious connotations of a dwelling place for angels and disembodied souls, so I have been encouraged to use the plural expression.

Aristotle's reference to the Pythagoreans here raises two distinct questions. The more fundamental one is whether these principles are applicable to the cosmos at all, and in a moment we will see that this is a well-motivated question. Assuming the answer to this question is that it *is* legitimate to speak of a cosmic right and left, the next question to ask about the Pythagorean claim is whether the heavens have a right and left in the way they suppose or in some other way. The very next lines of the chapter respond to both questions.

> For first off, if right and left are present in something, then one should posit even prior to these the presence in it of principles that are prior (τὰς προτέρας ὑποληπτέον ὑπάρχειν ἀρχάς). Now a determination was made regarding these matters in our studies of the movements of animals, because these principles are proper (οἰκεῖα) to the nature of animals; for at least in animals it is readily apparent that *all* such parts (I mean, for example, right and left) are present in some animals, *some* of them are present in some animals, while only up and down are present in plants. (284b12–18)

Aristotle claims that these principles are proper or appropriate (οἰκεῖα) to the nature of animals, and for that reason a determination about them was made in investigations of animal movements (ἐν τοῖς περὶ τὰς τῶν ζῴων κινήσεις).[5] One might suppose this to be a reference to the *De motu animalium*; but in fact it is pretty clearly to the *De incessu*, specifically, to chapters 2, 4 and 5.[6] Chapter 9 of the *De motu* does briefly *use* the notions of right, left, up, and down; but there is no discussion of them as principles; and no "determination" is made about them at all.

In chapter 2 of the *De incessu*, however, we find a methodological discussion of first principles or "starting points" in the study of nature as

[5] It might be thought that this amounts to the claim that the application of these principles is therefore *restricted* to animals, but that cannot be directly inferred from the use of οἰκεῖα. For example, a common use of this term in the *De caelo* is with reference to the distinct natural motions of the five different elements (cf. 290a2, 290a28, 300a22, 302b5, 303b4). The motion that is shared by air and fire, for example, is their proper motion. Considered as the motion of air, upward motion is both proper to it and not restricted to it. However, it is restricted to these elements as instances of light bodies (which is how they are considered in *De caelo* 3: cf. 298b7–8: λοιπὸν δὲ περὶ τοῖν δυοῖν εἰπεῖν). What can be directly inferred, then, is that *under some description* right and left are restricted to the objects of the *De incessu*. Note, in fact, that the above quote extends the application of up and down to plants.

[6] Cf. Heiberg (1894) 384.9–20; Leggatt (1995) 224; Stocks (1922) 284b14 n. 1; Guthrie (1939) 138. It is unfortunate, however, that these commentators and editors do not reference chapter 2, which, from a philosophical viewpoint, is critical to understanding what is said here.

important as any in the corpus, and it will guide us in the interpretation of the use of these "directional" principles in the *De caelo*. But before turning to that discussion, it will be useful to look briefly at one of his criticisms of the Pythagoreans' appeal to two of these principles in *De caelo* 2.2.

The chapter's initial questions concerned the propriety and manner of applying the concepts of "right" and "left" to the heavens. Aristotle initially remains noncommittal on the question of whether they should be applied at all, but he insists that the Pythagoreans have made a *methodological* mistake. Among these "directional" principles some are prior to others, and, in particular, some are prior to "right" and "left." Thus, if the Pythagoreans are committed to attributing right and left to the heavens, they ought to speak of those prior principles first. And so he goes on:

> And if it is necessary to apply any of these to the heavens, it would be reasonable that the one present first in animals be present in the heavens. For there are three, and each is as it were a principle (οἷον ἀρχή τις ἐστίν). The three I am talking about are the above and below, the front and its opposite, and the right and the left. For it is reasonable that all these [directional] dimensions (ταύτας τὰς διαστάσεις) are present in complete bodies. And "above" is a principle of length, "right" a principle of width and "front" of depth. (284b18–24)

The διαστάσεις here are *not* the standard three dimensions of bodily magnitude that one finds in *De caelo* 1.1, but they are rather three pairs of *directions*. But from this passage it is clear that the two uses of this term are, on Aristotle's understanding, intimately related, for the *directional* attributes are said to be the *principles* or *sources* of the three bodily dimensions.[7]

Now if the *De incessu* had not just been referred to as the appropriate place to define these principles, it might be just possible to gloss over this apparent ambiguity of reference by supposing that "ταύτας τὰς διαστάσεις" here refers forward, admittedly in a sloppy and confusing

[7] Interestingly, Immanuel Kant takes the opposite stance and derives the directional distinctions from the spatial dimensions:

> Because of its dimensions, physical space can be thought of as having three planes, which all intersect each other at right angles. ... The plan upon which the length of our body stands vertically is called, with respect to ourselves, horizontal. This horizontal plane gives rise to the difference between directions that we designate by the terms *above* and *below*. On this plane it is possible for two other planes to stand vertically and also to intersect each other at right angles, so that the length of the human body is thought of as lying along the axis of intersection. One of these two vertical planes divides the body into two externally similar halves, and furnishes the ground of the difference between the *right* and *left* side. (Kant 1768/1992, 2: 375–383).

manner, to the dimensions of magnitude mentioned in the next sentence. Unfortunately for this suggestion, *De incessu* 2 introduces, as the second of three fundamental "presuppositions" of natural inquiry, the following:

> Next it is necessary to assume how many and what sorts of dimensions (διαστάσεις) of magnitudes belong to what kinds of things. For while the [directional] dimensions (διαστάσεις) are six, there are three pairs: first, above and below; second, front and back; and third, right and left.
> (704b18–22; cf. *De incessu* 4, 705a26–29)

There is no doubt, then, that Aristotle refers to both the three dimensions of bodily magnitude and the three pairs of directional orientation with the same term (διάστασις), and that he is explicitly connecting them in *De caelo* 2.2: above is the principle of length, right of breadth, and front of depth. This allows him, then, to give priority to one of the two contraries of directional orientation.

There are three important conclusions to draw from comparing Aristotle's remarks on διάστασις in the *De caelo* and *De incessu*. The first is that the two different, though intimately related, uses of the term διάστασις in the *De caelo* strengthen the case for a closer look at the *De incessu animalium*. The second conclusion is that, since Aristotle explicitly refers to the "directional" διαστάσεις as ἀρχαί of the "dimensional" διαστάσεις, these are not merely homonymous uses of the same word. In some way that we must try to understand, Aristotle sees these three pairs of "directions" as "sources" or "principles" of the three standard dimensions of magnitude. Finally, the third conclusion to draw is that Aristotle is not at all clear about whether it is the three *pairs* of directions that are the principles of dimensionality, or whether it is *one contrary of each pair*. As we saw, the last line of our *De caelo* 2.2 passage refers to "above" as a principle of length, "right" as a principle of width and "front" as a principle of depth. Three notes of caution are in order here, however. First, perhaps we need to take seriously the lack of definite articles in this claim—Aristotle does not say that above, right, and front are *the* principles of the three dimensions of magnitude, only that they are principles. Moreover, two sentences earlier the three "principles" seem to be the three directional *pairs*. So the lack of a definite article may be deliberate. Finally, assuming that what are being discussed in *De incessu* 2 are true ἀρχαί of the study of nature, then that chapter is also stating that it is the three *pairs* of *directional* predicates that are the principles of the three *dimensions*.

I should like, then, to recommend a perfectly Aristotelian explanation for speaking both of one of each directional pair, and of the pairs themselves, as principles. For Aristotle, pairs of contraries are seldom "equals."

Cold is absence of heat (*Meteor.* 4.8, 384b28; *De gen. an.* 2.6, 743a36), the female is such due to an incapacity (*De gen. an.* 4.1, 765a9–10), and being musical is both the opposite of, and better than, being ignorant of music (*Phys.* 1.7 *passim*). Among these directional pairs too Aristotle is quite clear on this point:

> For the principle (ἀρχή) is honorable, and above is more honorable than below, front than back, and right than left. (*De incessu* 5, 706b12–13)

We may conclude, then, that the three directional orientations that Aristotle names as "sources" or "principles" of length, breadth, and depth—above, right, and front—are the more honorable members of each pair, and thus "more principles" than their opposites. However, in the same way that form and privation are both said to be ἀρχαί of change in *Phys.* 1.7 (cf. 191a12–16), the pairs of directional contraries can, in a restricted sense, also be referred to as principles.

Let us now return to Aristotle's consideration of the Pythagorean idea that there is a cosmic right and left. There are two appeals to what is "reasonable" (εὔλογον) in that discussion. The first encourages us to suppose that if one of these dimensional *pairs* is prior to the others in the case of *animals*, so should it be in the case of *the heavens*. Thus, there is a second kind of priority at stake in this discussion: not only is one member of each pair prior to the other, the pairs themselves have a priority ranking. This gives us one more reason to take a look at the *De incessu*, in order to understand the grounds for prioritizing one pair over another.

The second appeal to "the reasonable" encourages us to believe that *all three pairs* will be present in the heavens, for as we were told in the very first pages of the *De caelo*, the heavens, in the sense of the whole encompassed by the outermost sphere,[8] is complete because it is the entirety of complete, i.e., three dimensional, bodies. This "good reason" thus appeals directly to the opening chapter of the *De caelo*, and provides further evidence that these two uses of διάστασις are intimately related. If directional orientations are the *principles* of the three dimensions of complete bodies and the cosmos is complete (i.e., three dimensional), then these pairs of directional concepts must be fundamental in cosmology as well.

---

[8] The three meanings of οὐρανός distinguished in *De caelo* 1.9 are (a) the substantial being of the outermost circumference of everything, (b) the body continuous with the outermost circumference that includes the sun, moon and planets, (c) the entire body surrounded by the outermost circumference (278b11–22).

One important motivation for taking a close look at the discussion in the *De incessu* to which Aristotle refers, then, is to understand better how these concepts of directional dimension which Aristotle claims are proper to the nature of animals are related to dimensions of magnitude that seem to apply to *any* three dimensional body, living or non-living. Thus, before exploring *De caelo* 2.2 further, we should know what exactly is said in the *De incessu* about these "sources" of bodily magnitude that are "principles" of natural science. And I will begin with his introduction of these concepts as scientific starting points or "principles."

> We begin the investigation [of the reason why][9] by presupposing (ἀρχὴ δὲ τῆς σκέψεως ὑποθεμένοις) those things that we are accustomed to using frequently in the study of nature, assuming (λαβόντες) that this is the way things are in all of nature's works. One of these [presuppositions] is that nature does nothing in vain, but always does the best for the being of each kind of animal given the possibilities—which is why if it is better in a certain way, that is also how things are by nature. Again, it is necessary to assume (δεῖ λαβεῖν) [as presuppositions] the dimensions of magnitudes, both how many and what sort belong in what way. There are in fact six [directional] dimensions, but three pairs, the first above and below, the second front and back, and the third right and left. And in addition to these [presuppositions it is necessary to assume] that the starting points (ἀρχαί) of movements in place are pushing and pulling. Now these are starting points of motion *per se*, while that which is transported by another is moved *per accidens*; for that which is transported by something does not seem to move itself but to be moved by another.
>
> (*De incessu* 2, 704b11–705a2)

These principles appear to be presuppositions of natural investigation in a strong sense, indemonstrable starting points, perhaps. In an earlier paper,[10] I argued that the full statement of the "nature does nothing in vain" principle, the principle identified first here, performs precisely the task of a first principle in Aristotle's studies of animals.[11] The same case

[9] The closing lines of chapter 1, to which this sentence is syntactically linked, says "For *that* (ὅτι μὲν) these facts about animal locomotion happen to be so is clear from the inquiry into nature (τῆς ἱστορίας τῆς φυσικῆς); we must now investigate *the reason why* (διότι δέ)" (*De incessu* 1, 704b9–10).

[10] Lennox (1997) 199–214; repr. Lennox (2001b) 205–224.

[11] There are two uses of the negative principle ("Nature does nothing in vain, or by chance") in the *De caelo* as well, at 271a25 (or a32 if one adopts Moraux's suggested reordering of the text) and 290a30–33. And later in this paper we will have reason to look carefully at a use of the positive version of the principle ("Nature always does the best among the possibilities") at 288a2–12, in arguing for the applicability of "front" and "back" to the heavens, an especially problematic move. As with all uses of the principle, Aristotle proves his conclusion by showing that the contrary would involve something

can be made for positing the three pairs of directional dimensions. In our *De caelo* passage, however, their use introduces another sense of "presupposition" (ὑποθέσις). Aristotle says these ἀρχαί are appropriate to the study of animals, and that is why establishing them as ἀρχαί belongs to another investigation. They are to be *used* in the *De caelo*, but they are to be properly defined and established elsewhere.

On the other hand, in the passage that we are now looking at, he refers to these as principles that we "use frequently in the study of *nature*, assuming that this is the way things are in *all* of *nature's* works." He does not *explicitly* restrict them to the study of animals. We thus are left wondering what he means in the *De caelo* when he claims that these principles are appropriate or proper (οἰκεῖα) to the nature of animals.

The first step in answering this question must be to look at the actual account of the directional orientations given in *De incessu* 4 and 5, in order to determine how the principles are used there and what is and is not appropriated from that discussion for use in *De caelo*. We will see that by far and away the most attention is directed to right and left. We should not make too much of this, however. The treatise is a causal investigation of animal locomotion; and of our three pairs of concepts *only* right and left are *restricted* to animals that move from place to place and defined in fact by reference to animal locomotion.

De incessu 4 opens by relating these concepts immediately to living things, and in fact to living functions:

> Although[12] the [directional] dimensions (διαστάσεις) by which animals are naturally bounded are six in number—above and below, front and back, and again right and left—*all* living things have the above and below part. For above and below are present not only in animals but also in plants. It [viz. above and below] is delineated by function and not by

being in vain, and then uses the principle that nature does nothing in vain to establish the conclusion. In this case, uniquely, the principle is "God and nature do nothing in vain," a form of the principle very common in Galen. On these passages, see Leunissen (2009) 217–221 in this volume.

[12] Ἐπεὶ δ'εἰσὶν ... The syntax of this and the next paragraph is complex. The use of ἐπεὶ with the indicative likely has the concessive force of "although." Aristotle wants to begin with a contrast between most animals, which are bounded by six directional dimensions, and a wider class of living things (including plants) that is captured if we limit the directional dimensions to "the up and down part" (τὸ μὲν ἄνω καὶ κάτω μόριον). The μὲν at 705a28 at first looks to be *solitarium*, adding emphasis to the dimensions that belong to all living things. But the next paragraph begins Ὅσα δὲ μὴ μόνον ζῇ ἀλλὰ καὶ ζῷά ἐστι ... I suspect the δὲ here loosely marks a contrast with the μὲν clause at a28—the contrast is *very* loose, however, since a group of living things that has directional orientations in addition to above and below is being contrasted not with a wider group of living things, but with the above and below *possessed by* that wider group.

> position in relation to the Earth and the heavens alone. For whence the distribution of nourishment and growth comes in each of these things is "above," and the extremity toward which this extends is "below." The first is a sort of source (ἀρχή τις), the second a limit; and above is a source. (705a26–b1)

There are three different ways of taking the contrast being drawn here between "functional" and "cosmic" up and down, as I shall call them.

(a) Up and down can be delineated in two distinct ways, either by reference to biological function or by reference to the center and periphery of the cosmos.
(b) A full delineation of up and down requires both a reference to function and to cosmic orientation.
(c) Different modes of delineation are appropriate in different contexts, or with reference to different kinds of things.

Aristotle's account of these directional dimensions makes most sense if we understand his claim along the lines of interpretation (b). For he goes on to note, first, that the above and below are positioned alike in plants and animals, and then to qualify this claim by saying that "relative to the whole [cosmos] they are not positioned alike, but relative to function they are (705b4–5)." That is (as Aristotle goes on to explain), according to the *functional* account of "above," the roots of plants are above.[13] Moreover, we should take this claim to be restricted to above and below. For it seems clear that the account which Aristotle provides of front, back, left and right is *entirely* by reference to function and makes no reference to cosmic orientation. Thus, he states:

> Both front and back, however, belong to such as are not only alive but are also animals. For all of these have perception and front and back are defined according to this. For those things in which perception is naturally present and from whence it derives in each of the animals is front, while their opposites are back. (705b8–13)

It is fixed doctrine for Aristotle that sense perception is the distinguishing feature of animal life (cf. *De part. an.* 2.1, 647a21–24 and 8, 653b19–27; *De an.* 2.3, 414b1–10) and here he simply asserts that front and back are

---

[13] This passage should be supplemented by *De part. an.* 2.10, 656a11–14; 4.7, 683b21–25, 4.10, 686b35–36 and *Juv.* 1, 467b13–468a12. In the passage first noted, Aristotle also observes that it is only in human beings that functional above and below are aligned with cosmic above and below. In fact he says that in the human case the parts by nature are disposed according to nature (τὰ φύσει μόρια κατὰ φύσιν ἔχει ...)!

defined by reference to the orientation and position of the sense organs. In chapter 6, however, we are given more by way of a justification and account of this claim.

The directional orientation marked by the concepts right and left is found only in animals that "act on their own so as to change in respect of place" (705b14–16). In this case he says not only that these are delineated by "a certain function" but also that they are *not* delineated by position (705b17–18). How then, is such a functional determination made?

In animals with organs of locomotion, these come in pairs; and it is in these animals with limbs that right and left are clearest. In all of these pairs, one half of the pair of organs must be the source of motion; and from 705b31–706a10 Aristotle provides evidence for the claim that this side is the same for all such animals and is what is designated as "the right."

There are two ways of taking this claim. Aristotle could be asserting that the term "right" is to be applied, in any kind of animal, to the side of the body that is seen to originate motion in that kind. On this view, you could study giraffes and salamanders oriented perceptually in the same direction (which determines "front") to see which side moves first—and the term "right" will refer to that side, *even if it is the opposite side in giraffes and salamanders.* On the other hand, Aristotle might be making the strong empirical claim that if you make such observations, it will turn out that all animals originate motion from the same side. The concluding summary, which follows his presentation of evidence, makes it clear that it is the strong empirical claim he is making.

> And due to the same cause the right sides of all animals are the same; for whence the source of motion is in every case the same and is positioned in the same place according to nature; and the right is whence the source of motion. (706a10–12)

The argument here is straightforward:

> The right side is the side of origination of locomotion. *In all animals* the side of origination of locomotion is by nature positioned alike. Therefore, in all animals the right sides are the same.

If Aristotle were simply recommending labeling "right" whichever side originates motion, this argument would not be necessary. He must, therefore, be asserting the stronger claim that, for any two animals, if they are oriented in the same direction, they will originate locomotion from the same side. It is, then, a claim that can be refuted if some members of the same kind originate motion on one side and some on the other, or if different kinds with the same perceptual orientation originate motion on different sides.

Aristotle concludes chapter 4 by arguing that right and left are most distinct in human beings "because of all the animals humans have things arranged most in accordance with nature; and the right is both better than the left and separated from it. And that is why the right in human beings is most right" (706a18–21). He then generalizes the argument to all of the directional orientations, concluding that the other sources, the above and front, are *also* most in accordance with nature in mankind (706a24–26).[14] The opening of our chapter 5 continues this argument by applying these abstract remarks about "above" and "front" to animals differentiated by reference to how many feet they have.

> So, then, those animals in which above and front are distinct (διώρισται), as they are in human beings and birds, are bipeds (of the four points,[15] two of them are wings in the one case and hands and arms in the other). But those that have the front and above in the same [orientation] are either four-footed, many-footed, or footless. (706a26–31)

The rest of the chapter elaborates these claims and provides more content to Aristotle's axiology of direction, as we might call it. He notes that in bipeds, and especially in man (birds have a pelvic articulation that tips them forward slightly), the up and down are "in relation to the whole cosmos" (πρὸς τὸ τοῦ ὅλου ἄνω). However, since those with four or many feet, as well as those with no feet, are not upright, their "above" is on the same axis of orientation as their "front", and thus front and above are not distinct (cf. 706b2–9). He concedes that, since the above, front, and right are more honorable than the below, back and left, the locomotive parts are arranged in a reasonable way (εὐλόγως) (706b11–13). However, he concludes by suggesting that this may have the priorities reversed:

[14] On the idea that the better of two related principles should be separate from the worse, recall *De gen. an.* 2.1, 732a3–8:

> Since the nature of the primary cause of motion, to which the *logos* and the form belong, is better and more divine than the matter, it is also better that the superior be separated from the inferior. On this account, in those cases where it is possible and to the extent it is possible, the male has been separated from the female ....

[15] Aristotle is explaining that, though blooded and therefore in possession of four limbs, two of the limbs do not qualify as "feet." The very next lines explain that by "foot" he means a part which has a "point of contact" (σημεῖον) with the ground for movement in place.

> But it is also right to say (καλῶς δ' ἔχει … λέγειν) the converse (τὸ ἀνάπαλιν) about these matters, that it is because the sources are in these places that these parts are more honorable than the opposite parts. (706b14–16)

Two distinct options are on the table here, but Aristotle does not claim that one is correct and the other is wrong. The first option treats the greater honor of the above, the front, and the right over their opposites as an unexplained premise, from which you might conclude that that is where the sources of change will be. The second option derives the greater honor of the parts in these locations from the fact that the sources of change are located there. The premise common to the two options is that sources (ἀρχαί) are honorable.[16] Aristotle does not reject the first option, I believe, because he does not think it is wrong to attribute greater honor to one of each of these oppositions in direction than to the other. The only issue is whether such attributions are to be made *a priori*, or whether they follow from an empirical determination of which of the two is the source of change. It is likely that Aristotle sees Pythagoreans, and perhaps Platonists of a Pythagorean hue, as adopting the former position. He is arguing for the latter.

Chapter 6–7 of the *De incessu* (706b17–707a22) argue that there must be a common source of control over locomotion continuous with the locomotive parts, and that these parts will be distinguished by being on the right or the left and (when there are at least four parts) by being above or below. He rejects as irrelevant to locomotion the distinction between front and back on grounds that there is no natural backward motion (706b28–32).

3. *Lessons from the* De incessu animalium

What can we take with us from this excursus into the *De incessu animalium* relative to understanding the application of the right and left to the heavens in the *De caelo*?

[16] For an extended discussion of this passage and a comparison of it with *De part. an.* 3.3, 665a9–21, Theophrastus *Meta.* 11a8–13 and Plato *Tim.* 45a–b, see Lennox (1985) 143–164, repr. in Lennox (2001b) 266–272. I am still in substantial agreement with that discussion, and review critical reactions to its original publication in Lennox 2001b, 226–228.

(a) First and foremost, Aristotle means it when he states in *De caelo* 2.2 that matters about these directional dimensions had been determined in the discussions of animal locomotion *because* these dimensions are proper to the natures of animals. That is, indeed, where the fullest discussion of these principles takes place, including the presentation of the argument *that they are principles* and arguments over priority both *among* the pairs and *between* the pairs. And since they are there defined by references to organic functions, it is absolutely correct for Aristotle to insist that these principles are οἰκεῖα to that investigation.
(b) Only in the case of above and below does Aristotle allow that these terms can be applied both by reference to organic function *and* by reference to position relative to Earth and the heavens.
(c) In the case of up and down, it is only with respect to bipeds, and most completely with respect to humans, that there is agreement between the results of applying cosmic and functional concepts of "above" and "below." In all other locomotive organisms *functional* "above" and "below" are oriented on the opposing axis to *cosmic* "above" and "below."
(d) Right and left are concepts applicable *only* in the case of locomotive self-movers.
(e) Right and left are *not* delineated by position but *only* by function.
(f) The distinction of front and back is not relevant for discussions of animal locomotion, since all natural locomotion is forward.
(g) In the case of all three pairs, Aristotle argues that one member of the pair is more of a "source" or "principle" and more honorable than the other. In the case of right and left, the source of movement and, thus, the more honorable member of the pair, is the right.

Aristotle's remark in the *De caelo* that the appropriate place to discuss these principles is in the *De incessu* should be taken seriously. In fact, Aristotle appeals to, or relies on, every one of the above seven "lessons" during the course of *De caelo* 2.2–5. Before returning to that discussion, we also should be reminded of the context of Aristotle's appeal to the *De incessu* for support.

The reference to the *De incessu* is made when it is still an open question whether "right" and "left" have any legitimate use is cosmology. Whether the answer be yes or no, an appropriate first step is to appeal to a worked out theory of these directional principles, to see what, if anything, can be applied here. This is a legitimate move *even if the ultimate answer is "Nothing."*

The reference to the *De incessu* is also made with respect to a specific criticism of the Pythagoreans in mind—they seem to introduce right and left independently from the other two pairs of directional principles. Aristotle appeals to the *De incessu* to argue that there are principles *prior* to these; and indeed the argument there makes the case that above and below are prior on three separate grounds, all of which are defended in the *De incessu*.[17] First, they apply to *all* living things. Second, they can be defined either by reference to Earth and the heavens *or* by reference to nutritive function. Third, length is prior to breadth, and above is the principle of length. He thus goes on:

> For this reason one might well wonder that the Pythagoreans spoke of only two of these principles, the right and left, leaving aside four that are no less important; for above and front are no less differentiated from below and back than right is from left in all animals. I say "in all animals" because in some cases these directional principles differ by functional capacity only, while in other cases they also differ by their configuration; and while above and below are present alike in all the ensouled animals and plants, right and left are not present in plants. And again, as length is prior to breadth, if up is the principle of length and right of breadth, and the principle of what is prior is prior, up would be prior to right—prior in generation, that is, since "prior" is said in many ways. (285a12–22)

Aristotle follows the *De incessu* discussion carefully here in every detail. He might have mentioned, also with reference to the distinction between above and below, that human beings are organized most according to nature, but it will be noted that this fact about humans is not used to justify the priority claim in either work.

There are two methodological points upon which I want to insist in considering the background in the *De incessu* of the argument in *De caelo* 2. First, the fact that Aristotle appeals to the *De incessu* theory does not commit him to the claim that the *entire* theory of the *De incessu* is applicable here. Just as a person studying optics need not be committed to the claim that *all* of geometry is applicable to optical phenomena, so Aristotle need not be committed to the claim that all of his functional theory of directional principles is applicable in cosmology. Second, the claim about the discussion of these principles' being appropriate to the works on animal locomotion should not blind us to the fact that there is an important dependence that runs in the other direction. Throughout the discussion

[17] Simplicius (Heiberg 1894: 385.15–22) rehearses Aristotle's general views about priority, but it seems clear that here he is relying directly on the arguments of the *De incessu*.

in *De incessu* 2, 4, 5–6, the orientation of animals with respect to Earth and the heavens is a constant reference point. In chapter 5, for example, organisms are compared by reference to "three places, above, middle, and below" (706b2–3):

> ... bipeds have their above in relation to the above of the whole, many-footed and footless animals in relation to the middle [of the whole], and plants [have their above] in relation to the below [of the whole].
> (706b3–5)

After he makes these comparisons, moreover, Aristotle goes on to give the causes for these differences relative to cosmic orientation. The connection to cosmic dimensionality is, however, even more subtle than this suggests. For what makes humans special is not just that our upright posture aligns us with positional above and below—that alignment also accounts for the fact that *our* three pairs of dimensions are distinct, whereas they are confused to a greater or lesser extent in all other creatures. Thus, there is a constant reminder in the *De incessu* that above and below are *prior* principles to right and left, and that the most *natural* arrangement for *functionally* defined above and below is in alignment with *cosmically* defined above and below. So while *De caelo* 2.2 appeals to the discussion of directional dimensionality in *De incessu* 2–6, that discussion also depends on a concept of cosmic above and below that derives from the *De caelo* account of movement toward and away from the center and periphery of the whole.

4. *Returning to* De caelo 2.2

Recall Aristotle's assertion that it is reasonable[18] to say that all these dimensional principles belong to "complete" bodies, where "complete" means, as we are told in *De caelo* 1.1, three-dimensional.[19] He then treats one member of each directional pair as a source relative to one of the dimensions of magnitude: above to length, right to breadth, and

[18] For a thorough discussion of this and other appeals to what is εὔλογον in the *De caelo*, see Bolton (2009) 51–82 in this volume.

[19] *Pace* Leggatt (1995) 225 who rules out this meaning because it would make the above claim "tautological". Certainly Aristotle does say that to call a body complete in this sense is to say that it is extended in width, depth, and breadth, so that to say "complete body" is, since bodies are three-dimensional, strictly redundant. But here I think it serves to remind the reader of the reason why Aristotle calls bodies complete.

front to depth. Up to this point there is no need to seek support from the *De incessu*, since none of these ideas derives from the study of animal locomotion. But he then notes, and this is where the relevance of referring to the *De incessu* arises, that these are also principles in respect of motions—"principles whence the movements, in those having them, first begin (284b26–27)."[20] And here it is worth quoting his words:

> Growth is from above, movement in place is from the right, and movement in accordance with perception is from the front—for I call front that toward which perceptions are directed. (284b27–30)

For this reason, Aristotle tells us, we should only appeal to these principles in bodies which "being ensouled, have an inherent source of motion; for in none of the things without soul do we see whence comes the source of motion (284b32–34)."[21] It is then a straightforward matter to argue for the use of these suppositional starting points in the investigation of the motion of the heavens.

> Since we determined previously[22] that such powers are present in those bodies which have a source of motion, and that the heavens are ensouled and have a source of motion, it is clear that they also have up and down and right and left. (285a27–31)

It is notable that front and back are not mentioned in this conclusion. One might suppose that this silence is due to the fact that the heavens lack movement in the direction of the objects of perception; but on those grounds one might think that above and below, which in the *De incessu* are defined functionally in terms of the intake of nutrition

[20] Note that this is the precise formula which Aristotle uses for one of his causes, the misleadingly labeled "efficient" cause.

[21] This is yet another example of the difficulty widely noted that, by occasionally listing the four sublunary elements along with animals and plants as things with a source of motion within themselves (e.g., *De caelo* 1.2, 268b16, *Phys.* 2.1, 192b9–12), Aristotle appears to contradict statements such as this in *one* way; while by denying that even animals can move themselves without an external mover (*Phys.* 8.2, 253a11–21, 6, 259b1–16), he seems to contradict himself in a quite *different* way! For a variety of discussions of this problem, see the papers by Furley, Gill, Freeland and Sauvé Meyer in Gill and Lennox (1994). Simplicius (Heiberg 1894, 387.5–24) provides us with Alexander's view on this issue (that the elements have only a passive but not an active source of motion), a view with which Simplicius, with his Platonic sympathies, takes issue.

[22] If this is a reference to the *De caelo*, the only plausible candidate would be 1.9, 279a17–28, on which see the discussions of Kosman and Judson in Gill and Lennox (1994) chs. 7–8. But the current passage is the first place in the *De caelo* where the heavens are *explicitly* claimed to be ensouled, so the reference may well be to another work.

and the direction of growth, would be equally lacking. With respect to above and below, however, recall that Aristotle has an entirely *non-biological* account, which we saw him rely on regularly *even in biological contexts*—above is away from the center of the cosmos and below is toward it. As we have seen, this grounds Aristotle's claims that functional above and below in human beings are in line with cosmological above and below, and that plants have their functional above below and their functional below above![23] And indeed Aristotle relies on this cosmological understanding of above and below in the argument currently under consideration (cf. 284b34–285a1, 285b8 ff.).

In fact the absence of "front" and "back" here is to be explained by the immediate purpose of the argument at this point. Eventually, and problematically, Aristotle does make an argument at *De caelo* 2.5, 288a3–12 for the applicability of "front" and "back" to the cosmos, an argument which we will consider shortly. Here, however, Aristotle is in the midst of criticizing the Pythagoreans for attributing right and left to the cosmos while ignoring "prior" principles. Among the three pairs, the pair that is *prior* to right and left is up and down; front and back are, according to the *De incessu*, posterior to right and left. Moreover, as we learned from *De incessu* 4, the distinction between front and back is not relevant to discussions of locomotion, since there is no natural backward motion. Thus, following the *De incessu* closely, he needs only to show that right and left are posterior to above and below.

So what, then, is his positive argument for the attribution of the right and left to the heaven? He begins with the claim that "it is not necessary to be puzzled" (οὐ δεῖ ἀπορεῖν); because the shape of the whole cosmos is spherical, all its parts are identical, and it moves eternally (285a32–285b8)! Given that, unlike with above and below, the *De incessu* is quite clear that right and left must be given a functional account, why should we *not* be puzzled? For the *De incessu* account depends on being able to distinguish two sides (easily done with most locomotive animals) and being able to determine which of the two sides initiates motion.

> ... it is necessary to conceive it as though something in which the right parts differ in relation to the left also by configuration were then surrounded by a sphere; *for it will have the difference in functional capacity*, but will seem not to have it on account of the uniformity of its configuration. And it is necessary to think in the same manner about the source

[23] Cf. *De incessu* 4, 705b1–7, 706a24–26; *De part. an.* 2.10, 656a11–13; 4.10, 686a27–687a2.

> of movement: *for even if it never began to move, nonetheless it must have a source whence it would have begun if it started moving and if it were to stop it would start to move again.* (285b1–7)

I do not want to claim that I understand this argument; in so far as I think I do understand it, I find it weak.[24] What I find interesting, however, is the struggle to apply the zoological account of right and left, an account that, it was insisted, is based on function *and not on position* (705b17–18). In this passage I take the counterfactual, thought-experimental tone seriously. Aristotle is asking us to imagine a thing that will still have right and left *differentiated by functional capacity*, but in which, due to its spherical shape, this will not be apparent. Moreover, that locomotive capacity is there, present even in a thing that is eternally in motion—in material substances every actuality is the actuality of a capacity, even if that capacity is always activated. It is by virtue of that capacity that it *would have* begun to move had it needed to, and *would begin again* if it were to ever cease moving.[25]

The argument is, famously, that cosmic length extends from pole to pole along the axis of revolution and, therefore, that the dimension around the equator is breadth. In that case, right and left of the heaven will be the motive principles associated with movement of the sphere as indicated by the passage of the stars from rising to setting; and, therefore, the pole that is invisible to those living in the northern hemisphere is in fact above. On the other hand, since the spheres of the planets rotate in the opposite direction, *their* right—i.e., *their* source of motion—is on the opposite side and, thus, the opposite pole would be up in relation to their motion (285b8–286a1).

In his note on this passage, Stuart Leggatt (1995, 226, 285b28n) comments that "in conceding this point, Aristotle may be indicating the relative unimportance of the issue for him." I draw a very different conclusion from this passage. By indicating that right and left will be defined relative to the source of motion as determined by the direction of rotation, he is following the principles laid down in the *De incessu* once more. Moreover,

[24] As indeed may have Aristotle; see the discussion of this passage in Bolton (2009) xxx in this volume.

[25] There seems to be a premise presupposed by Aristotle's counterfactual here that is shared by the argument in *Meta.* 12. 6 1071b13–32: that material objects, even those that are eternally active, have a principle of potentiality. In *Meta.* 12, this motivates the argument for a prior, immaterial actuality; here it provides Aristotle with the resources to imagine the heavens beginning to move from an inactive, resting state.

in this passage, we see the way in which Aristotle grounds the dimensions of magnitude in the directional dimensions—source and direction of motion is primary for him, and he thinks of this as "generating" magnitudes.

5. *The structure of natural philosophy*

The interaction between these two discussions of directional orientation and its relation to dimensions of magnitude raises some interesting questions about explanatory methodology, especially since in *De caelo* 2.2 Aristotle treats the three *directional* pairs (or the more honorable member of each) as principles in some sense of the *bodily dimensions*. For chapters 7–13 of *An. post.* 1 lay down some rather strict guidelines regarding using principles appropriate to one domain in proofs in a distinct domain. Chapter 7 opens, for example, by noting that since all the relationships between terms in a demonstration are *per se* relationships, "it is not possible to prove by passing from another kind, for example, to prove that which is geometrical by that which is arithmetical" (75a38–39). Such proofs require that the kind be the same "either without qualification or in some way" (75b8–9), one way being that found in sciences subordinated one to the other as optics is to geometry or harmonics is to arithmetic (75b15–17).

Chapter 9 picks up on this theme, opening with the comment:

> Since it is apparent that it is not possible to demonstrate each thing other than from its own starting points if what is being proved belongs *qua* that thing, it is not possible to understand something by proving it from things which are [merely] true, indemonstrable, and immediate. (75b37–41)

Once again, he gives as examples[26] of exceptions harmonic proofs using principles of arithmetic, but provides a little more insight into how the exceptions work. In this case, the facts investigated by harmonics are studied by a different science because the underlying kind (τὸ ὑποκείμενον γένος) is different from that of arithmetic. Arithmetic, on the other hand, knows the reason why the facts are as they are, because the attributes studied by harmonics—ratios like 4:3, for example—are actually *per se* attributes (καθ' αὑτὰ τὰ πάθη) of number. This leads him to conclude that *unqualified* demonstration of anything can only

[26] But only as examples—he is careful on every occasion to indicate that what he says about these examples will be true of any cases that abide by the same restrictions.

come about through principles proper to it—and I suggest that the use of "unqualified" (ἁπλῶς) here picks up on the idea expressed at 75b8–9 that the kind being investigated must be the same "either without qualification or in a way." That is, the demonstrations of harmonic facts by reference to arithmetical reasons why are *qualified* demonstrations based on a *qualified* sameness in kind. The chapter concludes by returning to its opening thought—one of the most difficult aspects of science is, Aristotle opines, determining whether we have proven something from principles appropriate to it, for

> we think we have understanding if we have proofs from true and primary premises. But this is not so—it is necessary for the primary premises to be within the kind (συγγενῆ). (76a27–30)

Aristotle never explicitly discusses such issues using examples from the natural sciences. And what is interesting about the examples he *does* use to illustrate the use of principles from one science to account for facts in another is that they unite to form Aristotelian *intermediate* sciences such as astronomy, optics, or harmonics. The premises in question are mathematical; the facts being explained are attributes of kinds investigated by the natural scientist. Supposing, however, that both investigative domains—say animal locomotion and the cosmos—are, broadly speaking, within the same theoretical science; how do his strictures apply in such cases? Aristotle speaks directly to this question when the theoretical science is mathematics. You cannot, for instance, prove something geometrical from arithmetic starting points (75a38–39). However there *are* starting points that are common to more than one science and can thus be used in both.

> Of the items used in the demonstrative sciences some are proper to each science and others common—but common by analogy, since they are only useful in so far as they bear on the kind under the science. Proper: e.g., that a line is such-and-such, and straight so-and-so. Common: e.g., that if equals are removed from equals, the remainders are equal. (76a38–41)

Aristotle is also well aware of the subtle ways in which certain principles in the mathematical sciences can play roles at both the common and specific levels.

> Why do proportions alternate? The cause in the cases of lines and of numbers is different—and also the same: as lines it is different, as having such-and-such a ratio it is the same. And so in all cases. ... Items which are the same by analogy will have their middle terms the same by analogy. (99a8–11; 99a17)

In all of these passages, though the examples are predominantly drawn from mathematics, Aristotle goes out of his way to tell his readers that he is speaking quite generally about "all cases." The point made in the last passage above, for example, is illustrated with a biological example in *An. post.* 2.14, 98a20–22.

Well then, if our three pairs of directional dimensions are starting points appropriate to the investigation of animals, how is Aristotle's appeal to them in cosmology to be sanctioned? I can think of at least four answers that are plausible *prima facie*:

(a) Unlike mathematics, where principles in one branch are inappropriate for another, natural science is more homogenous. The principles discussed in *De incessu* 2 are appropriate to the γένος of bodies having a principle of change in themselves generally, and thus to the subject matter of the *De caelo*.[27]
(b) Aristotle has changed his mind about "kind-crossing demonstrations": the usage of these principles here violates the *Analytics*' proscriptions, but Aristotle is no longer worried about it.[28]
(c) The subject of the *De caelo* and the subject of the *De incessu* are related in some way analogous to the relationship between geometry and optics, or number theory and harmonics. Aristotle accepts the *Posterior Analytics*' restrictions, but these two investigations are related in ways that allows for qualified "kind-crossing."
(d) Aristotle thinks of the heavens as a living thing and, thus, that the principles appropriate to animals are *directly* applicable to the heavens. Kind-crossing is not an issue, but for a very different reason than is provided by option (a).

Among these options, (a) seems to be the most problematic. Regardless of whether the study of the cosmos as a whole and the study of animal movement are, in some very general sense, part of the science of nature, this is not a sense of "sameness" that is helpful. The "directional" principles defined and articulated through three chapters of the *De incessu*

[27] As Andrea Falcon argues (2005, 14–15), it seems more plausible to say that the kind consisting of natural bodies is *less* unified than Aristotle's paradigmatic examples of kinds, or at least exemplifies a different kind of unity.

[28] This was argued by Martha Nussbaum (1978, Essay 2) regarding the use of cosmological principles in *De motu*, as part of her case for its being written by Aristotle, since one argument against its authenticity had been its failure to adhere to the *Analytics*' strictures against explanatory principles crossing from one science to another.

are, both there and in the *De caelo*, said only to be applicable where certain biological functions—nutrition, growth, self-initiated motion, and perception—are in play. In that case, there will be parts of the investigation of nature where they are appropriate, and parts where they are not.[29] Aristotle's often refers to such a "sub-domain" as a μέθοδος of this or that kind of thing.[30] So even if, as with mathematics conceived generally, there is a single, general science of nature, there are still distinct natural investigations with their own appropriate, indemonstrable principles. As with geometry and arithmetic, so with animal locomotion and cosmic motions—the borrowing by one investigation of principles said to be appropriate or "at home" in another is not sanctioned simply because there is some "universal" science to which they both belong.

I will also dismiss from consideration (b), not because I have a proof that it is false, but rather because I am adhering to an interpretive heuristic that searches for coherence between the *Analytics*' program for science and the scientific treatises themselves. It is, of course, possible that such a search will fail in a particular case, but it has been successful often enough, and born sufficient fruit, that it would be a mistake to reach that conclusion when other plausible avenues are available.

Of the two remaining options (d) seems, at first glance, the obvious choice.[31] Aristotle says clearly in *De caelo* 2.2 that the outer sphere is ensouled and thus has its own source of motion; and since the primary directional orientations are principles appropriate to such things and should only be sought in such things (284b32–33), they are straightforwardly suppositions of this study. Moreover there are a number of passages that allude to the movement of the heavens as that of a living, if immortal, body.[32]

But there are problems here too, and these problems lead me to an interpretation that combines elements of solutions (c) and (d). As we

[29] One might argue that when Aristotle defends the directional orientations as principles in *De incessu* 2, he refers to them as principles we often make use of in our study of nature. But that need not mean that they are applicable to all of nature; and at any rate the very passage in the *De caelo* that we are trying to understand appeals to them precisely on the grounds that they are principles appropriate to the investigation of animals.

[30] For examples of the term used in this way, compare *Meteor.* 1.1, 338a26, *Long.* 6, 467b9, *De part. an.* 1.4, 644b16.

[31] Simplicius (Heiberg 1894, 387.32–388.4) takes it as obvious.

[32] Cf. 1.9, 279a18–30; 2.3, 286a7–12; 2.6, 288a28–b8; 2.12, 292a15–28. Admittedly, these are the very passages, discussed by Bolton (2009) in this volume, where Aristotle acknowledges the difficulties of the questions and the relative lack of empirical support for the needed answers; e.g., II. 3, 286a3–8; 2.5, 287b29–288a2; 2.12, 291b24–28.

have seen, Aristotle self-consciously *struggles* to apply the results of the discussion of the *De incessu* to this case, and some of that work is done by way of the assumption that the heavens are ensouled. However the heavens[33] *do not perceive, nourish or regenerate themselves, or initiate motion from a state of rest; nor do they have any parts suitable for such functions*. That is, if the heavens are ensouled, they are not ensouled in a way that fits easily with the account of the soul that is familiar from the *De anima*. In that case one might fairly ask *why* he does work so hard to rely on the principles of directionality found in the *De incessu*?

In order to answer this question, I want to close by looking at methodologically important passages in *De caelo* 2. They reveal, I believe, the peculiar epistemological dilemma in which Aristotle finds himself in cosmology. It is this dilemma, combined with certain methodological commitments from the *Analytics*, that lead him to defend cosmic directionality by appeal to his account of animal locomotion in the *De incessu*.[34]

De caelo 2.5 faces the question—forced on him by the fact that the planetary spheres, though made of the same natural substance as the outermost sphere, move in the opposing direction—of why these spheres revolve in one direction rather than the other. With eternal things it cannot be a matter of chance, nor is it voluntary. Before answering, he reflects on whether it is appropriate even to attempt an answer.

> Perhaps, then, the attempt to make some statement about certain things, or even about all things, passing nothing by, might indeed seem to be the mark of great simplemindedness or of great zeal. Yet it is by no means right to censure all people alike; rather one ought to consider the reason why they are speaking, as well as the credibility of what they say, whether "human" credibility (ἀνθρωπίνως)[35] or something more secure. So then, when someone happens upon the more accurate necessities, it is necessary to be grateful to those who made the discoveries, but at the moment we must state what appears to be the case (νῦν δὲ τὸ φαινόμενον ῥητέον). (287b29–288a2)

De caelo 2.12 opens (291b24–31) with a similar "disclaimer", noting that over-zealous attempts to state what appears to be the case (λέγειν τὸ φαινόμενον), if done out of philosophical thirst to make even a

---

[33] It seems that some of the arguments in *De caelo* 2 demand that "the heavens" must be used in its "second" sense, i.e., to refer not only to the outermost sphere but also to the spheres of the planets.

[34] And I here find my conclusions are largely confirmed by Robert Bolton's study of the underlying epistemology of the *De caelo* (2009) 51–82 in this volume.

[35] On the specific force ἀνθρωπίνως of here, see Bolton (2009) 68–70.

slight advance about subjects where we have great puzzlement, should be greeted with affection. Slightly later in that same chapter, Aristotle is clearer about *why* he is being so tentative. He has just raised a puzzle about the different apparent motions associated with the spheres of the fixed stars, the sun, the moon and the planets. He pauses:

> About these things, then, it is well to seek ever greater understanding, even though we have too few resources and are at such a great distance from their attributes; nevertheless studying that about which we are now puzzling from such resources would not seem to be unreasonable. But we contemplate such things as if studying their bodies alone, and as though they are units possessed of order yet completely without soul. But it is necessary to suppose that they participate in action and life; for in this way the attribute we want to understand will seem to be nothing unreasonable. (292a14–22)

For Aristotle, understanding of perceptual phenomena is achieved once we know the causes that produce them, and that knowledge derives from experience of the phenomena (*An. pr.* 1.30, 46a17–27). Alas, cosmology is problematic in this respect:

> Since circular movement is not opposed to circular movement, we must investigate on what account there are many motions, even if we are attempting to make the inquiry from a great distance—distant not in respect of location, but much more in respect of having perceptual awareness of far too few of the attributes that belong to these things. (*De caelo* 2.3, 286a3–8)

All these passages indicate that Aristotle, a committed empiricist, recognizes the thin empirical basis for his causal theorizing in cosmology. Yet they also indicate that he felt compelled, to the point of begging indulgence, to pursue such theorizing. One may reasonably ask, why? Why not simply resist temptation and admit an inability to reach any well-grounded conclusions on this question, since we lack the data necessary for reasonable causal inferences?

There are, I believe, two plausible answers, and they reinforce one another. The first is stressed repeatedly in the *De caelo*, but its relevance to the epistemological dilemma we are discussing is most clearly displayed in a passage in *De part. an.* 1.5.

> Among the substantial beings constituted by nature, some are ungenerated and imperishable through all eternity, while others partake of generation and perishing. Yet it has turned out that our studies of the former, though they are valuable and divine, are fewer (for as regards both those things on the basis of which one would examine them and those things about them which we long to know, the perceptual phenomena are altogether few).

> We are, however, much better provided in relation to knowledge about the perishable plants and animals because we live among them. For anyone wishing to labor sufficiently can grasp many things about each kind. Each study has its attractions. Even if our contact with eternal beings is slight, nonetheless because of its surpassing value this knowledge is a greater pleasure than our knowledge of everything around us, even as a chance, a brief glimpse of the ones we love is a greater pleasure than seeing accurately many other great things. (644b22–645a1)

Here the relative lack of perceptual acquaintance is stressed, in this case in contrast to the ready availability of information about animals and plants. But Aristotle also provides all the motivation one would need for pursuing knowledge of the eternal beings of the heavens—the surpassing value of even a little knowledge of the eternal things compared to the plentiful knowledge of the perishable things around us. The extent of one's willingness to draw causal inferences on the basis of a slender data basis is proportional to the value of the prize at the end of the inferential quest.

Yet the *De caelo* provides us with evidence for another motive behind Aristotle's willingness, with apologies, to skate inferentially on thin evidential ice. The evidence is the constant attempts to counter the arguments of the *Timaeus* and of unnamed Pythagoreans, even when Aristotle agrees in some general way with their conclusions. Granted, in many cases observation provides scant grounds for reaching conclusions—but for Aristotle, that must always be the grounds from which we begin. Thus, for example, though the Pythagoreans turn out to be correct in applying the concepts of right and left to the cosmos, the *a priori* ways in which they arrive at this application render it baseless.

And thus Aristotle starts from a more secure base. The directional orientations have been defined by reference to "the animals and plants around us." As best we can, we must attempt to extend these concepts, formed on that secure empirical basis, to the heavens. As one final example of this method, I point to the concluding lines of *De caelo* 2.5 in which Aristotle appeals to the *first* "hypothetical principle used in the study of nature" mentioned in *De incessu* 2—that nature always does what is best—as a premise in an argument to show that the concepts "front" and "back" are applicable to the heavens.

> For if nature always does the best among the possibilities, and [if] just as the motion toward the upper region is, among rectilinear motions, more honorable (since the upper region is more divine than the lower), in the same way too forward motion is more honorable than backward motion,

> [then] it has,[36] if both right and left as stated previously (and the difficulty just stated testifies that it has), front and back. In fact this explanation resolves the difficulty. For if it is the best it can be, this would be an explanation of what was said. For it is best to be moved in a simple and ceaseless motion, and this in the more honorable direction.[37]
>
> (*De caelo* 2.5, 288a3–12)

Aristotle is providing an object lesson in empirical cosmology, countering the approach found in Plato's *Timaeus* and in Pythagorean doctrine. His appeal to the discussion of above, below, right, left, front and back

---

[36] I follow the text as printed in Bekker here (ἔχει, εἴπερ καὶ) rather than the emendation of Prantl (ἔχει δή, εἴπερ) adopted by Allan, Stocks, and Guthrie.

[37] István Bodnár has suggested (personal communication) that this passage does not introduce the third pair of directional contraries, front/back to the other two pairs, but rather that the use of τὸ πρότερον καὶ ὕστερον at 288a8 simply indicates that if right and left are present, so too will be the prior/posterior relation between them, i.e., the priority of the right to the left. His primary motivation for this suggestion is that he cannot see how introducing the distinction between front and back contributes to Aristotle's argument in this chapter.

I will respond to that worry momentarily but my first concern with Bodnár's suggestion is that I do not see how the immediate argument supports it. Stripped to its bare bones, the logic of our passage is this: If nature always does best, and forward motion is (just as upward) better than backward, then the heavens, if they have both right and left, also have τὸ πρότερον καὶ ὕστερον. On Bodnár's reading, it is a puzzle why the axiology of forward and backward motion is mentioned in the antecedent of this conditional argument at all. But if I am correct, not only does that make sense, but so does the claim that *if* both right and left are present, *then* front and back are, since the *De incessu* argues that right and left can only be determined once forward and backward are. Recall that the primary error of the Pythagoreans, according to chapter 2, is to have introduced right and left without introducing prior principles, principles on which right and left depend (284b10–13). More globally, if the cosmos has the three spatial dimensions and is a complete body, then the more honorable member of *all three* directional pairs must be present as a principle of each of those three dimensions, front being the principle of depth (284b25).

That takes us directly back to Bodnár's motivation for his alternative. He correctly sees that the wider argument has to do with the question of why the sphere of the fixed stars revolves one way rather than the other (and of course the wider context for that argument is that the inner spheres revolve in the opposite direction). But as I understand the argument of this chapter, to acknowledge that issue is already to explain the relevance of this closing argument to the wider context. For in order to distinguish these two revolutions one needs to distinguish the direction of the locomotion of these motions. It may seem natural to us to see this as a question about whether the motion is "to the left or to the right." But from the perspective of the *De incessu*, the prior pair of contraries for answering questions about direction of motion is front/back. Once *that* is determined, one can then raise the question of the *initiation* of motion, which for Aristotle is when the distinction of right and left comes in to play. The very strangeness of the argument in chapter 2 stems from the fact that Aristotle is explicit about applying all three of the directional dimensions to the heavens, and realizes that doing so when the object under consideration is a nested set of eternally revolving spheres strains credulity.

in the *De incessu* is an integral part of this strategy. Where concepts have been formed and defined on the basis of a rich empirical study, and those concepts are being applied *a priori* by the Pythagoreans to the heaven, investigation should begin by drawing on the results of that study. There may be difficulties in making this application, but given the value of the knowledge sought we should appreciate the effort and try to enrich the evidence and go further. Even if he is mistaken on specific points, he is indicating the proper method for dealing with such questions.

Acknowledgements

I take this opportunity to thank István Bodnár, Alan Bowen, Devin Henry, and Mariska Leunissen for valuable written comments on earlier drafts of this paper, and Kelly Jones for assisting in the final preparation of the manuscript.

Bibliography

Allan, D.J. (1955) *Aristotelis de caelo* (Oxford).

Bolton, R. (2009) "Two Standards for Inquiry in Aristotle's *De Caelo*," in A.C. Bowen and C. Wildberg (edd.), *New Perspectives on Aristotle's* De caelo (Leiden) 51–82.

Falcon, A. (2005) *Aristotle and the Science of Nature: Unity without Uniformity.* (Cambridge).

Freeland, C. (1994) "Aristotle on Perception, Appetition, and Self-Motion," in Gill and Lennox (1994) 35–64.

Furley, D. (1994) "Self-Movers," in Gill and Lennox (1994) 3–14.

Gill, M.L. (1994) "Aristotle on Self-Motion," in Gill and Lennox (1994) 15–34.

Gill, M.L. and Lennox, J.G. (1994) edd., *Self-Motion: From Aristotle to Newton* (Princeton).

Guthrie, W.K.C. (1939) *Aristotle: On the Heavens* (Cambridge, MA).

Heiberg, J.L. (1894) *Simplicii in Aristotelis de caelo commentaria.* Commentaria in Aristotelem graeca 7 (Berlin).

Kosman, A. (1994) "Aristotle's Prime Mover," in Gill and Lennox (1994) 135–153.

Leggatt, S. (1995) *Aristotle: On the Heavens I and II* (Warminster).

Lennox, J.G. (1985) "Theophrastus on the Limits of Teleology," in W.W. Fortenbaugh, P.M. Huby, and A.A. Long (edd.), *Theophrastus of Eresus: On His Life and Work* (New Brunswick, NJ) 143–164 (repr. in Lennox 2001b, 266–272.)

——, (1997) "Nature Does Nothing in Vain …," in H.-C. Günther and A. Rengakos (edd.), *Beiträge zur antiken Philosophie: Festschrift für Wolfgang Kullmann* (Stuttgart) 199–214 (repr. in Lennox 2001b, 205–224.)

——, (2001a) *Aristotle: On the Parts of Animals* (Oxford).

——, (2001b) *Aristotle's Philosophy of Biology: Studies in the Origins of Life Science.* (Cambridge).

Leunissen, M.E.M.P.J. (2009) "Why Stars have no Feet: Explanation and Teleology in Aristotle's Cosmology", in A.C. Bowen and C. Wildberg (edd.), *New Perspectives on Aristotle's* De caelo (Leiden), 215–237.

Nussbaum, M.C. (1978) *Aristotle's De motu animalium: Text with Translation, Commentary, and Interpretive Essays* (Princeton).

Sauvé Meyer, S. (1994) "Self-Movement and External Causation," in Gill and Lennox (1994) 65–80.

Stocks, J.L. (1922) *Aristotle: De Caelo*, in W.D. Ross (ed.), *The Works of Aristotle Translated into English* (Oxford).

WHY STARS HAVE NO FEET: EXPLANATION AND TELEOLOGY IN ARISTOTLE'S COSMOLOGY

MARISKA E.M.P.J. LEUNISSEN*

1. *Introduction: The scarcity of teleological explanations in the* De caelo

The most central feature of Aristotle's conception of natural science is his theory of natural teleology: everything that exists or comes to be "by nature" comes to be or changes, unless prevented, for a purpose and towards an end, and is present for the sake of that purpose or end. In the *De caelo*, which contains in the first two books Aristotle's problem-oriented exposition of his cosmology, traces of this teleological worldview are abundant. The nature of the elements is claimed to be such that it provides them with an immanent capacity to exercise their specific motions to reach their natural places. Left to their own devices, the four sublunary elements would naturally move to their natural places and thus constitute four separate, concentrically arranged spheres.[1] Teleology also permeates the heavenly domain of the stars and planets, as all celestial motions are said to be trying to reach "the most divine principle" as a final cause (*De caelo* 2.12, 292b20–25).

Although teleology as a natural tendency is thus without doubt an important part of the makeup of Aristotle's cosmology and celestial physics, his general reliance on teleology to *explain* the different motions and features of the heavenly bodies seems to be limited in comparison with the other physical treatises. For the whole of the *De caelo* contains only seven instances of explicit teleological explanations of cosmological phenomena, six of which are in the second book (there is only one

---

* Washington University in St Louis.

[1] Bodnár and Pellegrin (2006) 282. For the teleological motion of the element earth and for the notion of natural place as end, see the contribution by Matthen (2009) in this volume (especially sections 2 and 5); for Aristotle's account of the elements in *De caelo* 2.1 and 4, see Gill (2009), especially section XX on elemental natural motion.

instance of teleological explanation in book one, there are none in books three and four).[2] Moreover, with one exception (*De caelo* 2.3, 286a8–9), none of these explanations refer directly to final causes. Instead, they all proceed through the supposition of teleological principles, such as "nature does nothing in vain," which in biology are only applied in very specific explanatory contexts, namely, in those cases where the discovery of final causes is relatively difficult.[3] This suggests that teleology is not readily discernible in the case of the heavens. Aristotle's use of teleological principles is all the more remarkable, because the teleological explanations are the only fully-fledged physical explanations that Aristotle offers in this treatise. By this I mean that the teleological explanations are the only explanations in *De caelo* 1–2 that address the nature and causes of the various features and motions of the heavens and that build upon some evidence from observation. Aristotle's cosmological treatise consists for the most part of statements of fact and of arguments building upon mathematical or numerical principles, which mainly address the number, shape, and possible motions of the heavenly bodies.[4]

[2] For the teleological explanations and the principles used, see (a) *De caelo* 1.4, 271a22–33: Why is there no motion contrary to that in a circle? (*teleological principle*: nature does nothing in vain); (b) 2.3, 286a7–9: Why is there a plurality of motions of the heavens? (*teleological principle*: everything that has a function is for the sake of that function); (c) 2.5, 288a2–12: Why do the heavens move in the direction they do? (*teleological principle*: nature always does what is best among the possibilities); (d) 2.8, 290a29–35: Why do stars not move on their own or why have stars no organs for motion? (*teleological principle*: nature does nothing in vain); (e) 2.9, 291a23–25: Why do stars not move on their own or why is there no harmony of the spheres? (*teleological principle*: nature does nothing in vain); (f) 2.11, 291b10–15: Why do stars not move on their own or why stars do not have a shape fit for locomotion? (*teleological principle*: nature does nothing in vain); (g) 2.12, 292a15–b25: Why is there a difference in the complexity of the motions of the different heavenly bodies? (*teleological principle*: actions are for the sake of something and the analogy with motions of sublunary beings).

[3] Teleological principles in Aristotle's biology are generalizations pertaining to the "observed" actions of actual and particular formal natures of living beings, indicating what "nature" always or never does in generation. It is my contention that propositional principles of explanation, such as Aristotle's teleological principles, function as the framework within which the explanation needs to take place (they both limit the amount and kinds of explanations possible, and license the explanations actually given) and are only used where references to final causes are not immediately possible. Outside the framework set up by such principles, explanations lose their explanatory force and fail to make sense altogether.

[4] Bolton (2009) calls this second type of arguments *dialectical*; see in particular sections 1 and 2.1 of Bolton's contribution to this volume. According to Bolton, the dialectical arguments in the *De caelo* rely on mathematical, numerological, or even mythological starting points rather than on perceptual phenomena, and they establish only what is reasonable (εὐλόγως) rather than what is necessary and true. While I agree

The purpose of the present paper is to shed light on the specific nature of the teleological explanations in Aristotle's cosmology and on the problems related to their application within this particular branch of the science of nature.[5] In particular, I shall argue that the way in which Aristotle uses teleological principles to explain heavenly phenomena builds upon their very successful usage in biology,[6] and is thus consistent with his treatment of cosmology as a natural science. In section 2, I shall say more about the scientific status of cosmology. Next, in sections 3 and 4, I shall discuss two representative examples of teleological explanations from the second book of the *De caelo*.

2. *Cosmology as science of nature*

The approach to the study of the heavens taken by Aristotle's predecessors and contemporaries had predominantly been mathematical in nature. (In the *De caelo*, Aristotle refers to them as "mathematicians concerned with

---

with Bolton's analysis of the nature of the bulk of the arguments used in the *De caelo*, I do not believe that *all* arguments in it are dialectical. I submit that a minority of the arguments, i.e., the teleological ones, are not dialectical, but *scientific* in nature. The teleological arguments lay out explanations that are proper to the natural sciences and make use of principles that are firmly grounded in the empirical evidence of the sublunary domain. The fact that the heavenly domain is empirically underdetermined limits the explanatory force of these explanations (hence Aristotle's "warning" that the explanations which he will offer are at most reasonable), but this does not mean that the explanations themselves are not scientific (φυσικῶς): rather, they generate the best causal accounts of the features and motions of the heavenly bodies that Aristotle can offer.

[5] The issues that I should like to discuss in this paper have received relatively little attention in the scholarly literature on Aristotle. Scholars who have studied teleology in Aristotle's cosmology have focused almost exclusively on the role of the Prime Mover as a final cause in Aristotle's *Physica* and *Metaphysica*. See, in particular, Kahn (1985). Other studies on cosmology have either left out the question of teleology completely (Falcon 2005), or have subsumed it under the "normal" use of teleology (Johnson 2005). On the other hand, Leggatt, in his commentary on *De caelo*, claims that Aristotle consciously played down the role of teleology in his cosmological treatise, because of his alleged dissatisfaction with the type of intentional and psychological teleological explanations deployed by Plato in the *Timaeus*: see Leggatt (1995) 18, 36–37, 207. Hence, Leggatt offers little analysis of the teleological explanations actually provided in this treatise, because he believes them to be of little importance.

[6] The possible relative chronology of Aristotle's works (according to which the *De caelo* is an early work and the biological works are late) does not affect my claim: since none of these treatises were published during Aristotle's life time, he may well have adjusted and revised them continuously in the light of new discoveries or conceptual distinctions made. For a defense of this view (based on a pedagogical interpretation of the cross-references in Aristotle), see Burnyeat (2004) 21–22.

ἀστρολογία"—where ἀστρολογία is best rendered by "astronomy"—or simply as "mathematicians".)[7] The theory reportedly put forward by Eudoxus and revised by Callippus represented the apparent motions of the stars and planets as outcomes of systems of concentric rotating spheres. This theory as reported did not explain the physical mechanics and causes underlying those motions, perhaps because neither Eudoxus nor Callippus was concerned with those issues. In *Phys.* 2.2, 193b22–194a12, Aristotle distinguishes this theoretical manner of studying the heavens (ἀστρολογία) from the proper study of nature by pointing out that astronomers—like mathematicians—do not study the properties of bodies *qua* properties of those bodies, but *qua* separable from them.[8]

For Aristotle, however, just as for Plato, the study of the heavens is part of the investigation of nature,[9] and thus the heavenly bodies and their features will have to be studied in a manner that takes their natures fully into account—nature both in the sense of form and of matter. This physical approach to the study of the heavens is evidenced, for instance, in Aristotle's claim that each of the spheres in his system is corporeal, and thus not simply a mathematical construct (*De caelo* 2.12, 293a7–8):

> ἑκάστη δὲ σφαῖρα σῶμά τι τυγχάνει ὄν.
>
> For each sphere is some kind of body.

Therefore, if for Aristotle cosmology is part of the science of nature, and if scientific knowledge involves the knowledge of all four causes,[10] a merely mathematical approach (such as favored by the astronomers and by Aristotle himself in many of the arguments in the *De caelo*) will not be sufficient to generate complete knowledge concerning the heavens, for the following reasons. By its very nature, mathematical reasoning cannot yield understanding of final causes (there are no final causes

---

[7] See *De caelo* 2.14, 297a2–4 (Μαρτυρεῖ δὲ τούτοις καὶ τὰ παρὰ τῶν μαθηματικῶν λεγόμενα περὶ τὴν ἀστρολογίαν; "what the mathematicians say in *astrologia* also testifies to this"), 2.10, 291a29–b9; 2.14, 298a15.

[8] Cf. *Meta.* 1.8, 989b33–990a15; 3.2, 997b16–998a1 and 13.2, 1076b39–1077a4 where Aristotle describes ἀστρολογία as not dealing with perceptible magnitudes or with the heavens above. Cf. Simplicius *In phys.* 293.7–10, 290.20–24 (on the Greek conception of ἀστρολογία as being part of mathematics, not physics); Mueller (2006) 179–181.

[9] Aristotle emphatically introduces his study of the heavens as a part of the study of nature: see for instance *De caelo* 1.1, 268a1 (Ἡ περὶ φύσεως ἐπιστήμη), 3.1, 298b2–3 (τῆς περὶ φύσεως ἱστορίας), and *Meteor.* 1.1, 338a20–25.

[10] See e.g. *Phys.* 2.2, 194a10–16, 2.3, 194b17–23; *An. post.* 1.2, 71b9–13 and 2.11, 94a20–27: cf. also Falcon (2005) 15.

in mathematics, because there is no change or good in that domain).[11] Hence, astronomy only yields understanding of the shape and size of the heavenly bodies, and of their distances from each other and from the earth. This gives important information about the quantitative properties of the heavenly bodies and their motions, especially if combined with arguments drawing from principles of physics. However, as a natural philosopher, Aristotle is also interested in the nature of the heavenly bodies, in their material composition, and in the causes of their motions (i.e., in their material, formal, efficient, and final causes). The opening words of the *De caelo* are significant:

> Ἡ περὶ φύσεως ἐπιστήμη σχεδὸν ἡ πλείστη φαίνεται περί τε σώματα καὶ μεγέθη καὶ τὰ τούτων οὖσα πάθη καὶ τὰς κινήσεις, ἔτι δὲ περὶ τὰς ἀρχάς, ὅσαι τῆς τοιαύτης οὐσίας εἰσίν·
>
> The science of nature is patently concerned for the most part with bodies and magnitudes, the affections and motions of these, and further, with all the principles that belong to this kind of substance. (1.1, 268a1–4)

Because the natural sciences are concerned with all four types of causes, and especially since the understanding of final causes is crucial (because the natural sciences are concerned with things that undergo change), Aristotle needs an additional strategy to extend scientific knowledge as he understands it to the domain of the heavens. This strategy involves the application of teleological principles of the sort he employs in his biology precisely as a heuristic for finding final causes where they are not immediately observable. In short, Aristotle uses teleological principles to discover purposes and functions among the heavenly phenomena, and thereby tries to turn his study of the heavens into a proper natural science.[12]

[11] *Meta.* 3.2, 996a21–b1.

[12] Aristotle's treatment of cosmology as part of the study of nature also explains why the teleological explanations are mainly found in the second book of the *De caelo*. For it is this book that deals most specifically with the heavenly bodies *qua* subjects of motion, that is, with the plurality, direction, and complexity of their motions, the physical mechanisms underlying those motions, and the shape of the heavenly bodies required to perform those motions. On the other hand, we find no teleological accounts regarding Aristotle's views on the nature of the heavens as a whole (for instance, for such features as the heavens' size, uniqueness, or eternity—topics that are dealt with primarily in book I), or regarding the motions and features of the four terrestrial elements (dealt with in books 3–4), which are not part of cosmology properly speaking. Aristotle's use of causal language in the *De caelo* also reveals that the second book is more concerned with Aristotle's *own* attempts to provide physical explanations than any of the other books: of the 28 occurrences of the term *αἴτιον* in the whole of the *De caelo*, 10 can

For Aristotle, scientific research comprises two stages of enquiry: the first stage consists in the systematic collection of observations of the phenomena, and the second one in the attempt to detect correlations and to give causal explanations of the phenomena. However, as Aristotle makes clear several times in the *De caelo* (see his introductions to teleological explanations discussed below in section 3), it is not at all an easy undertaking to give physical explanations of cosmological phenomena. The central problem is the limitedness—or even lack—of empirical evidence: the observations of the heavens we have are too few, and the objects of observation are too far away to offer any certain evidence.[13] The only observation that seems to be rock solid is that of the rotation of the heavens (*De caelo* 1.5, 272a5–6 τὸν δ' οὐρανὸν ὁρῶμεν κύκλῳ στρεφόμενον: "we see the heavens turning about in a circle"). Notwithstanding the many methodological caveats we find in *De caelo* (I shall discuss them in section 3 below), Aristotle remains confident that it is still possible to give explanations of cosmological phenomena that go beyond the mere fact that heavens rotate, and also beyond the conclusions (mathematical or numerical) that reasoning yields about numbers, sizes, shapes, and distances, for instance.

My contention is that Aristotle's use of teleological principles, by analogy with their use in the biological domain,[14] forms an important part of his strategy to increase the possibility of gaining scientific knowledge of the heavens. Thus, when Aristotle does proceed to give "physical" or "scientific" explanations, he is unremittingly teleological in his approach. The explanations thus presented will not qualify as demonstrations in a strict sense (i.e., not as demonstrations as described in *Analytica posteriora* or *De partibus animalium*),[15] because they do not set out to demonstrate the truth but merely the reasonableness of certain causal scenarios. However, they go a long way in taking away some of the puzzlement

---

be found in book 2 (as opposed to 4 in book 1, 3 in book 3, and 11 in book 4), and of the 28 occurrences of the term *αἰτία*, 21 can be found in the second book (as opposed to none in the first and the third book, and 7 in the fourth), while the references to causes and explanations in the fourth book are often (i.e., in 4.1, 308a25; 4.2, 309a5, 309a10, 309a28 and 310a2; 4.6, 313a22)—although not exclusively—used in descriptions of views entertained by Aristotle's predecessors.

[13] Cf. *Meteor.* 1.7 and *An. pr.* 1.30.

[14] *Pace* Falcon (2005) 101, who argues that "Aristotle is reluctant to extend the results achieved in the study of plants and animals to the imperishable creatures populating the celestial world."

[15] Cf. Lloyd (1996) 182.

pertaining to the heavenly realm and thus in making sense of the heavenly phenomena. And as Aristotle has indicated elsewhere,[16] "making sense" in such difficult circumstances entails giving an account of the heavens that is free of impossibilities.

Let us finally turn to some examples of the actual teleological explanations Aristotle provides in the *De caelo*. Broadly speaking, Aristotle gives two kinds of teleological explanations. The first kind consists of explanations that stand on their own (that is, they do not form part of an interrelated sequence of arguments), and set out to explain the *presence* of certain features and motions of the heavens.[17] In these cases (i.e., 2.3, 286a7–9; 2.5, 288a2–12 and 2.12, 292a15–b25), Aristotle explains the presence of some observed fact by reference to the good it serves within the heavens as a whole. The basic teleological idea is that whatever can be seen to be present, must be there to serve some good.

The second kind consists of those teleological explanations that explain the *absence* of heavenly features (this kind is used in 1.4, 271a22–33; 2.8, 290a29–35; 2.9, 291a23–25 and 2.11, 291b10–15). They usually follow after a series of mathematical or numerical arguments following the style of the astronomers. While the latter point out that it is, for instance, mathematically impossible for some motion or feature to be present, the teleological explanation is set up to provide a counterfactual argument claiming that those motions or features in reality could not exist in the heavenly realm, because if they did, they would be in vain. The teleological principle invoked in all of these cases is that nature does nothing in vain.

In the next two sections, I shall discuss an example from each group.

3. *Explaining why there is a plurality of motions of the heavens (Example 1)*

The first example of a teleological explanation that I should like to discuss pertains to the plurality of the heavenly motions: different heavenly bodies are observed to move in different directions—why is it that they do

[16] *Meteor.* 1.7, 344a5–7: "We consider a satisfactory explanation of phenomena inaccessible to observation to have been given when we reduce them to what is possible."

[17] Evidently there are no teleological explanations of the generation of the heavenly bodies as they are eternal and not generated. Cosmological teleological explanations are thus naturally restricted to the explanation of the features and motions belonging (or not belonging) to the eternal heavenly bodies.

not all move in the same direction? Aristotle introduces his explanation as follows (*De caelo* 2.3, 286a3–7):

> Ἐπεὶ δ' οὐκ ἔστιν ἐναντία κίνησις ἡ κύκλῳ τῇ κύκλῳ, σκεπτέον διὰ τί πλείους εἰσὶ φοραί, καίπερ πόρρωθεν πειρωμένοις ποιεῖσθαι τὴν ζήτησιν, πόρρω δ' οὐχ οὕτω τῷ τόπῳ, πολὺ δὲ μᾶλλον τῷ τῶν συμβεβηκότων αὐτοῖς περὶ πάμπαν ὀλίγων ἔχειν αἴσθησιν. ὅμως δὲ λέγωμεν. ἡ δ' αἰτία περὶ αὐτῶν ἐνθένδε ληπτέα.
>
> Since there is no motion in a circle contrary to motion in a circle, we must examine why there are several locomotions, though we must try to conduct the inquiry from far off—far off not so much in their location, but rather by virtue of the fact that we have perception of very few of the attributes that belong to them [i.e., the motions]. Nonetheless, let us speak of the matter. The explanation concerning these things must be grasped from the following [considerations].

This text shows that Aristotle is very well aware of the fact that it is problematic and difficult to offer explanations of what is present in the heavens, given the lack of empirical evidence: we are simply too far removed from the objects of inquiry in distance.[18]

It is significant that Aristotle is nevertheless confident that there is a way of answering this particular question, and that this answer follows from a *teleological* consideration. As we will see shortly, the considerations "from which the explanation must be grasped" is the supposition of the teleological principle that everything that has a function must exist for the sake of that function. By positing a teleological principle and, hence, by setting a framework within which one can search for the possible functions of those very features that have been

[18] Cf. Burnyeat (2004) 15–16, who observes that "*De caelo* 1 contains an unusually high number of occurrences of words like εἰκοτῶς and εὔλογον which express epistemic modesty: this or that is a *reasonable* thing to believe." I should like to add to this observation that words of "epistemic modesty" are even more abundant in book 2 where the explanation of the presence and absence of heavenly features properly speaking is at stake (I counted only two occurrences of the word εὔλογον and none of the word εἰκοτῶς in *De caelo* 1; in *De caelo* 2, I counted 15 occurrences of the word εὔλογον and two of the word εἰκοτῶς). On the significance of words like εἰκοτῶς and cognates, see also the contribution by Bolton (2009) in this volume. Note also that Bolton reads "disclaimers" like the one quoted above as marking the inferior dialectical arguments Aristotle provides throughout *De caelo* (see in particular section 2.1). In my reading, these introductions indicate that Aristotle, after having offered dialectical or astronomy-style arguments earlier, now intends to offer genuine physical or scientific explanations, but that they will necessarily be of a weaker kind and, hence, only amount to what is reasonable.

observed, one might be able to find the explanation of why those features are present. On the other hand, the implication also seems to be that this kind of knowledge cannot be gained by other means: observation is certainly ruled out (observation in this case will only yield knowledge of the fact that there are several motions, not of the reason why), but also mathematical arguments are not what is called for in these situations, again because they cannot yield the reason why there are several motions.

Interestingly, the other two teleological explanations that stand on their own and explain the presence of heavenly phenomena are also immediately preceded by a discussion of the methodological problems related to this very enterprise of providing explanations in the strong sense for phenomena at such a remove (see *De caelo* 2.5, 287b29–288a2; 2.12, 291b24–28 and 292a14–18). In all these methodological introductions, Aristotle expresses his conviction that, even though the empirical evidence is scanty, it is still possible to state the phenomena; and that given all the limitations, the explanations offered are the best ones possible.[19] The explanations that follow these introductions are all teleological in nature, which shows that Aristotle has strong confidence in the explanatory force of teleology also in these difficult cases.[20]

Returning to our example from *De caelo* 2.3, the teleological principle from which "the explanation of why there are several locomotions of the heavens must be grasped", is the following (*De caelo* 2.3, 286a8):

> Ἕκαστόν ἐστιν, ὧν ἐστιν ἔργον, ἕνεκα τοῦ ἔργου.
>
> Each thing that has a function is for the sake of that function.

[19] *Pace* Guthrie (1939) 165.

[20] This point is also made by Lloyd (1996) 171, with regard to the explanations in *De caelo* 2.5 and 2.12: "Thus it is surely significant that both on the problem of why the heavens revolve in one direction rather than in the other—in 2.5—and on the difficulty of the complexities of the motions of the non-fixed stars—in 2.12—his positive speculations invoke teleology." I disagree, however, with Lloyd's interpretation of the significance of this connection between Aristotle's methodological disclaimers on the one hand and his use of teleology on the other. According to Lloyd (1996) 161, 173, 175, 180, Aristotle's main interest in cosmology follows from his concern to establish his teleology, and especially the orderliness of the heavens. However, I do not believe that Aristotle's epistemological hesitations are not genuine here, or that Aristotle's concern for the establishment of teleology is all that prominent in the *De caelo*. On the contrary, I believe that Aristotle uses his teleology, already firmly established on the basis of the abundance of empirical evidence discussed in his biological works, to extend—where possible—his knowledge of the heavens.

This is a common principle in Aristotle's biology (see, e.g., *De part. an.* 1.5, 645b15–18), where it is claimed that each part of the body is for the sake of the performance of some function. By stating it here, Aristotle makes explicit that in his view, teleology extends to the heavenly domain and, hence, that some of the puzzling cosmological phenomena can be explained by reference to teleology. Aristotle also *must* refer to teleology here, since material causes alone cannot account for the differences in locomotions in the heavens (for all spheres are made from the same material, which is aether).[21] The assumption that everything that has a function is present for the sake of that function allows a series of inferences that ultimately yield the explanation of why there are several motions of the heavens: if this principle applies, then each of the motions must serve its own function.

Aristotle continues his explanation by identifying the function of the first motion, in the following way (*De caelo* 2.3, 286a8–11):

> Θεοῦ δ' ἐνέργεια ἀθανασία· τοῦτο δ' ἐστὶ ζωὴ ἀΐδιος. ὥστ' ἀνάγκη τῷ θειῷ[22] κίνησιν ἀΐδιον ὑπάρχειν. Ἐπεὶ δ' ὁ οὐρανὸς τοιοῦτος (σῶμα γάρ τι θεῖον), διὰ τοῦτο ἔχει τὸ ἐγκύκλιον σῶμα, ὃ φύσει κινεῖται κύκλῳ ἀεί.
>
> The activity of god is immortality, and that is everlasting life. In consequence, it is necessary that eternal motion belongs to the divine. Since the heavens are of this sort (for this body is a divine thing), for that reason the heavens have a circular body that moves naturally in a circle forever.

The reasoning is that, if the function of the divine is immortality, and if the heavens are divine, then the function of heaven is immortality. Furthermore, if being immortal is the defining function of heaven, then it is a necessary prerequisite for it to possess an eternal motion. That is, for heaven as a whole to be able to perform its defining function or its activity of being immortal, it has to perform at least one kind of eternal motion. And the only kind of motion capable of uniform eternal continuity is motion in a circle.

This explanation, curious as it may sound, resembles a particular type of explanation that Aristotle frequently offers in his biological works.

---

[21] Cf. Simplicius (quoting Alexander), *In de caelo* 396.6–9: "it is not possible to make either natural or material necessity responsible for these things, since both spheres have the same matter, but it is necessary to give an account of the difference in terms of some divine governance or ordering." For an analysis and defense of Aristotle's arguments for the existence of aether, see Hankinson (2009) in this volume.

[22] Here I follow Leggatt in reading θειῷ instead of θεῷ with most manuscripts: see Leggatt (1995) 227. I believe Aristotle's argument to be that the celestial sphere is *like* a divine being in the sense that both partake in eternal motion, not that it itself is a god.

Consider the following example taken from *De partibus animalium*, where Aristotle provides an explanation of why birds have wings (4.12, 693b6–14):[23]

> τῶν γὰρ ἐναίμων ἡ τοῦ ὄρνιθος οὐσία, ἅμα δὲ καὶ πτερυγωτός. … τῷ δ' ὄρνιθι ἐν τῇ οὐσίᾳ τὸ πτητικόν ἐστιν.
>
> For the substantial being of the bird is that of the blooded animals, but at the same time it is also a winged animal, … and the ability to fly is in the substantial being of the bird.

Aristotle takes the essence, or the substantial being of the animal, as a starting point, and derives from this essence the necessary prerequisites of something being what it is. Just as birds must have wings because they are essentially flyers (and the only way for birds to perform their defining function is by using their wings), so too the heavens must have a spherical body and move eternally in a circle because they are essentially immortal. According to this argumentation, eternal motion in a circle is the proper attribute of an immortal body such as the heavens.

However, Aristotle has not yet explained why there are *several* motions of the heavenly bodies. The second part of the explanation of why there are several motions consists of a complicated chain of arguments, based on a total of six assumptions. The starting point of this chain is the conclusion of the first part of the explanation, which is the necessity of there being an eternal motion of the outer sphere in order for the heaven to be immortal. The reasoning Aristotle employs is deductive, but the type of necessity to which Aristotle refers is sometimes that of a necessary consequence, while at other times it is the necessity of something having to be present first if something else is to be present (the latter is what Aristotle calls conditional necessity).[24] Let me give a summary of the chain of arguments (*De caelo* 2.3, 286a13–286b2):

(a) If there is to be a body that moves in a circle eternally, it must have a center that remains at rest.
(b) For there to be a fixed center, the existence of the element earth is a necessary condition (i.e., since whatever is made of aether cannot

[23] Cf. *De part. an.* 4.13, 697b1–13 and 3.6, 669b8–12. For my analysis of the structure of this type of teleological explanation, see Leunissen (2007) 170–171.

[24] The formula "ἀνάγκη … εἶναι" is repeated six times: in *De caelo* 2.3, 286a13; 286a20, and 286b2 (see (a), (b), and (f) above), the necessity is conditional; in 286a22, 286a28, and 286a32 (see (c), (d), and (e) above), the necessity indicates a necessary consequence.

remain at rest, there must be a second element next to aether, the natural motion of which is to move towards the center and then to remain at rest in the centre).

(c) If there is to be earth, then it is a necessary consequence that there is also fire (for earth and fire are contraries, and if the one exists, so does the other).

(d) If there is to be fire and earth, then it is a necessary consequence that the two other elements exist (for water and air are in a relation of contrariety to each of the other two elements).

(e) From the existence of the four elements, it necessarily follows that there must be generation (for none of the four sublunary elements can be everlasting).

(f) If there must be generation, then it is necessary that there exists some other motion.

According to this account, generation is a necessary consequence of there being sublunary elements, whose existence is a necessary condition for there to be an eternal, cyclical motion of the outermost sphere of the heavens carrying the fixed stars. However, having established that it is necessary for there to be generation (as a consequence of there being the four sublunary elements), Aristotle turns the argument around, and reasons that if there is to be generation, then it is conditionally necessary for there to be other motions, because the motions of the outermost sphere alone cannot cause generation. Accordingly, generation is that for the sake of which all the other motions (namely, the motions of the planets) take place. This is how Aristotle summarizes his explanation (*De caelo* 2.3, 286b6–9):

> Νῦν δὲ τοσοῦτόν ἐστι δῆλον, διὰ τίνα αἰτίαν πλείω τὰ ἐγκύκλιά ἐστι σώματα, ὅτι ἀνάγκη γένεσιν εἶναι, γένεσιν δ', εἴπερ καὶ πῦρ, τοῦτο δὲ καὶ τἆλλα, εἴπερ καὶ γῆν· ταύτην δ' ὅτι ἀνάγκη μένειν τι ἀεί, εἴπερ καὶ κινεῖσθαί τι ἀεί.
>
> For the moment so much is clear, for what reason there are several bodies moving in circles, namely, because it is necessary that there is generation; and (because) generation (is necessary) if there also has to be fire; and (because) that one and the others (are necessary) if there also has to be earth; and (because) that one because it is necessary that something always remains at rest, if there has to be something that is for ever in motion.

The complete explanation of why there are several motions of the heavens is thus that there are several functions for the sake of which these motions are present. There is one eternal motion in a circle (performed by the outer sphere carrying the fixed stars) for the sake of realizing the immortality of the heavens, and there are other motions (performed by the inner spheres carrying the planets) for the sake of generation.

Here, the use of the teleological principle allows Aristotle to draw an organic picture of the cosmological system in which all the observed motions can be explained by the purpose they serve. Arguably, this picture and the type of reasoning behind it are not without problems,[25] but at least Aristotle is able to give some rationale for some phenomena the astronomers did not explain. The plurality of the motions of the heavenly bodies makes sense in the light of the need for the heaven as a whole to perform an eternal motion, if it is to be truly immortal, and as a corollary—of the need for there to be generation, if this eternal motion is to be at all. In sum, the lack of empirical evidence makes it hard to provide fully-fledged physical explanations in cases like these, but through the use of teleological principles that are well established in his biology Aristotle at least succeeds in mitigating some of the perplexities pertaining to some heavenly phenomena.

4. *Explaining why stars have no feet (Example 2)*

I shall now turn to an example of the second type of teleological explanations, where Aristotle uses some form of the teleological principle that nature does nothing in vain in order to explain the *absence* of heavenly features, usually after a series of mathematical arguments or discussions of the available empirical evidence.

The *explanandum* to be discussed concerns the question whether or not the stars and planets possess a motion independently of the motion of the spheres. One explanation is given in chapter 2.8. Ultimately, the purpose of the chapter is to show that the heavenly bodies (most likely) do not possess a motion of their own, but are carried around fixed in concentric spheres.[26] First, Aristotle sketches three possible scenarios concerning the motions of the heavenly bodies and their spheres (*De caelo* 2.8, 289b1–3). As both are observed to move as a whole, it is

[25] For instance, it does not establish an explanation for *each* of the individual motions of the planets, or for the need for there to be generation. This latter point is well brought out (along with others worth noting) by Hankinson (2002–2003) 31–32.

[26] The basic idea is that if the stars possessed their own motion, they would be self-movers; and this would make them in principle able to stop their motions, which would threaten the eternity of the heavens and life as we know it.

necessary that the change of position takes place with both the heavens and the stars being at rest, or with both moving, or with the one moving and the other at rest. Aristotle then refers to empirical evidence (*De caelo* 2.8, 289b5 οὐ γὰρ ἂν ἐγίγνετο τὰ φαινόμενα, 289b10 φαίνεται) and gives mathematical arguments (2.8, 289b27–28 τὰ αὐτὰ καὶ ὁμοίως ἔσται ἄλογα) in order to show that the scenario in which the spheres move while the heavenly bodies are at rest is "the least absurd" (2.8, 289b34–35 μόνως γὰρ οὕτως οὐθὲν ἄλογον συμβαίνει). In addition to this, Aristotle offers a final teleological argument in favor of this theory, arguing for the unlikelihood of the stars and planets possessing a motion on their own. The argument runs as follows (*De caelo* 2.8, 290a29–35):

> Πρὸς δὲ τούτοις ἄλογον τὸ μηθὲν ὄργανον αὐτοῖς ἀποδοῦναι τὴν φύσιν πρὸς τὴν κίνησιν (οὐθὲν γὰρ ὡς ἔτυχε ποιεῖ ἡ φύσις), οὐδὲ τῶν μὲν ζῴων φροντίσαι, τῶν δ' οὕτω τιμίων ὑπεριδεῖν, ἀλλ' ἔοικεν ὥσπερ ἐπίτηδες ἀφελεῖν πάντα δι' ὧν ἐνεδέχετο προϊέναι καθ' αὑτά, καὶ ὅτι πλεῖστον ἀποστῆσαι τῶν ἐχόντων ὄργανα πρὸς κίνησιν.
>
> In addition to these arguments, it would be absurd that nature gave them no organ for motion (since nature does nothing as a matter of chance), and that she should care for animals, but disregard such honorable beings; rather, it seems that nature, as though deliberately, has taken away everything by means of which they might possibly in themselves have effected forward motion, and that she set them at the greatest distance from those things that possess organs for motion.

In a way, Aristotle's explanation here parallels that of Plato in the *Timaeus* (33d–34a). In this dialogue, Timaeus explains that the divine craftsman did not think it necessary to equip the heaven—self-sufficient and perfect as it is—with hands or feet for walking:

> ἡγήσατο γὰρ αὐτὸ ὁ συνθεὶς αὔταρκες ὂν ἄμεινον ἔσεσθαι μᾶλλον ἢ προσδεὲς ἄλλων. χειρῶν δέ, αἷς οὔτε λαβεῖν οὔτε αὖ τινα ἀμύνασθαι χρεία τις ἦν, μάτην οὐκ ᾤετο δεῖν αὐτῷ προσάπτειν, οὐδὲ ποδῶν οὐδὲ ὅλως τῆς περὶ τὴν βάσιν ὑπηρεσίας. κίνησιν γὰρ ἀπένειμεν αὐτῷ τὴν τοῦ σώματος οἰκείαν, τῶν ἑπτὰ τὴν περὶ νοῦν καὶ φρόνησιν μάλιστα οὖσαν· διὸ δὴ κατὰ ταὐτὰ ἐν τῷ αὐτῷ καὶ ἐν ἑαυτῷ περιαγαγὼν αὐτὸ ἐποίησε κύκλῳ κινεῖσθαι στρεφόμενον, τὰς δὲ ἓξ ἁπάσας κινήσεις ἀφεῖλεν καὶ ἀπλανὲς ἀπηργάσατο ἐκείνων. ἐπὶ δὲ τὴν περίοδον ταύτην ἅτ' οὐδὲν ποδῶν δέον ἀσκελὲς καὶ ἄπουν αὐτὸ ἐγέννησεν.
>
> For he thought that a being which is self-sufficient would be much better than one which is in need of other things. And he did not think it was necessary to attach hands to it to no purpose—hands for which there would be no need either to grasp or to defend itself against anything; nor had it any need of feet, nor of the whole apparatus of walking. For he assigned to it the motion that is most suited to its body, [the motion] which, of the

> seven, is the most appropriate to mind and to thinking. And, therefore, he caused it to move in the same manner and on the same spot and revolving in a circle within its own limits. All other six [motions] he took away and it was made not to partake in their deviations. And as this circular motion required no feet, the universe was created without legs and without feet.

In this passage, Plato describes how the heaven was created and was given its circular motion, which is most appropriate for its spherical shape. The other types of motion—forwards/backwards, to the left/to the right, and up/down—were taken away from it (*Tim.* 34a5 ἀφεῖλεν). However, instead of this mythological account for why the stars have no feet, Aristotle opts for a naturalistic explanation.

The structure of Aristotle's argument is quite complex. In short, it consists of a *reductio ad absurdum* followed by an alternative account proclaiming the purposiveness or, perhaps even the providence, of nature. The first part of the argument builds upon the implicit counterfactual assumption that if the stars "were intended by nature" to be moving on their own, it would be absurd for nature not to have given them organs for motion, given the fact that nature did provide such organs to "lesser" beings. I take the expression that "nature does nothing as a matter of chance" to be equivalent to the principle that nature does nothing in vain: living beings always have the parts that they need, and if the heavenly bodies lack organs for motion, that lack must be for the sake of something; or, to put it the other way around, if the organs for motion are absent in heavenly bodies it must be because their presence would have been in vain (they would have had no function to fulfill in this particular kind of being). The reference to the honorable status of heavenly beings implies that Aristotle takes the teleology of nature to apply even *more* to them than to the sublunary beings.[27]

---

[27] Aristotle repeatedly offers the *a fortiori* argument that if one agrees that animals and plants neither come to be nor exist by spontaneity (but for the sake of something), then the claim that spontaneity is the cause of the heavens—which is most divine and exhibits the greatest order—must be absurd, and that one has to conclude that final causality pertains to the heavenly realm as well. See *Phys.* 2.4, 196a24–b5; 2.6 198a1–13 and *De part. an.* 1.1, 641b16–23:

> This is why it is more likely that the heavens have been brought into being by such a cause—if it has come to be—and is due to such a cause, than that the mortal animals have been. Certainly the ordered and definite are far more apparent in the heavens than around us, while the fluctuating and random are more apparent in the mortal sphere. Yet some people say that each of the animals is and came to be by nature, while the heavens, in which there is not the slightest appearance of chance and disorder, were constituted in that way by chance and the spontaneous.

The *implicit* underlying teleological principle here is that each capacity (in this case the capacity to locomote) requires an organ[28] and that thus locomotion of the stars is possible if and only if they have organs for locomotion. The absurdity lies in the fact that nature *did* provide less honorable beings with organs for motion, and that we would have to conclude, were we to accept this account as true, that nature purposely neglected more honorable beings such as the stars. Since this account is of course unacceptable within Aristotle's view of the way nature operates, the opposite scenario, set out in the second part of the argument, must be the case: nature has *taken away* every means of locomotion, and thereby set a distance between the heavenly bodies and the sublunary beings equipped with organs for motion.[29] As Aristotle explains, spherical bodies are least fit to effect forward motion on their own, because they lack "points of motion" (*De caelo* 2.8, 290b5–8):

> πρὸς δὲ τὴν εἰς τὸ πρόσθεν ἀχρηστότατον· ἥκιστα γὰρ ὅμοιον τοῖς δι' αὑτῶν κινητικοῖς· οὐδὲν γὰρ ἀπηρτημένον ἔχει οὐδὲ προέχον, ὥσπερ τὸ εὐθύγραμμον, ἀλλὰ πλεῖστον ἀφέστηκε τῷ σχήματι τῶν πορευτικῶν σωμάτων.
>
> For forward motion it is least fit, since it is least like to those things that produce motion on their own; for it does not have any appendage or projection, as does a rectilinear figure, but stands most apart in shape from those bodies equipped for locomotion.

The core of this teleological argument for why the heavenly bodies do not have a motion of their own and, hence, must be fixed in spheres, is thus the assumption (presented as a fact) that heavenly bodies do not have feet or any other organs for locomotion. For, if nature—for the most part—does nothing in vain and the heavenly bodies have no feet, then the conclusion is reasonable that nature must have "designed" the heavenly bodies not to be able to move on their own.

The teleological argument that Aristotle offers here is again in many ways similar to explanations we find in the biological works. In biology, Aristotle holds that all animals that are capable of locomotion must have organs for motion[30] and that animals without organs for motion are not

---

[28] *De gen. an.* 1.2, 716a24–25.

[29] Aristotle considers it to be better for the superior to be separated from the inferior; cf. *De gen. an.* 2.1, 732a6–8, where Aristotle explains that it is better for the male and the female to be separated, for "it is better that the superior principle should be separated from the inferior."

[30] *De part. an.* 4.10, 686a35–b1: "all (animals) that walk must have two hind feet"; *De incessu* 3, 705a19–22: "That which moves always makes its change of place by the

capable of locomotion.[31] These two "laws" are exhaustive with regard to all blooded land walkers. The one and only exception to this rule is formed by the footless snake,[32] which obviously does not have organs specifically designed for locomotion, but moves forward by bending itself. Just as in our example concerning the heavenly bodies, Aristotle explains the absence of feet by invoking the principle that nature does nothing in vain (*De incessu* 8, 708a9–20):

> τοῖς δ' ὄφεσιν αἴτιον τῆς ἀποδίας τό τε τὴν φύσιν μηθὲν ποιεῖν μάτην, ἀλλὰ πάντα πρὸς τὸ ἄριστον ἀποβλέπουσαν ἑκάστῳ ⟨ἐκ⟩ τῶν ἐνδεχομένων, διασῴζουσαν ἑκάστου τὴν ἰδίαν οὐσίαν καὶ τὸ τί ἦν αὐτῷ εἶναι· ἔτι δὲ καὶ τὸ πρότερον ἡμῖν εἰρημένον, τὸ τῶν ἐναίμων μηθὲν οἷόν τ' εἶναι πλείοσι κινεῖσθαι σημείοις ἢ τέτταρσιν. ἐκ τούτων γὰρ φανερὸν ὅτι τῶν ἐναίμων ὅσα κατὰ τὸ μῆκος ἀσύμμετρά ἐστι πρὸς τὴν ἄλλην τοῦ σώματος φύσιν, καθάπερ οἱ ὄφεις, οὐθὲν αὐτῶν οἷόν θ' ὑπόπουν εἶναι. πλείους μὲν γὰρ τεττάρων οὐχ οἷόν τε αὐτὰ πόδας ἔχειν (ἄναιμα γὰρ ἂν ἦν), ἔχοντα δὲ δύο πόδας ἢ τέτταρας σχεδὸν ἦν ἂν ἀκίνητα πάμπαν· οὕτω βραδεῖαν ἀναγκαῖον εἶναι καὶ ἀνωφελῆ τὴν κίνησιν.
>
> The reason why snakes are footless is both that nature does nothing in vain, but always, from among the possibilities, picks what is best for each thing, preserving the proper substantial being of each, and its essence; and further, and as we have stated previously, none of the blooded animals can move by means of more than four points. For from these [two premises] it is apparent that none of the blooded animals that are disproportionately long relative to the rest of their bodily nature, as are the snakes, can be footed. For, on the one hand, they cannot have more than four feet (since in that case they would be bloodless); and, on the other hand, having two or four feet they would be pretty much completely immobile—so equipped, their motion would necessarily be slow and useless.

In short, starting from this principle, Aristotle offers the counterfactual argument that if nature had equipped snakes with feet, snakes would move very badly and the feet would have been next to useless. Given the principle that nature does nothing in vain (*and* that nature cannot give snakes more than four feet, since, in that case, the snake would not be blooded), snakes do not have feet.[33]

---

employment of at least two organic parts, one as it were compressing and the other being compressed."

[31] *De incessu* 3, 705a23–25: "And so nothing that is without parts can move in this manner; for it does not contain in itself the distinction between what is to be passive and what is to be active".

[32] *De part. an.* 4.11, 690b14–18; *De incessu* 4, 705b25–29.

[33] However, the fact that snakes do not have organs for motion does not mean that they do not locomote: they move forward by bending themselves (see *De incessu* 7, 707b6–

There is, however, an important difference between the explanatory force of the use of this principle in biology as opposed to its use in cosmology, and this difference derives directly from the lack of observational evidence in the latter domain. For, in the biological domain *observation* determines the possibilities of what nature does and does not produce.[34] In the case of the footless snake, observation shows that all other blooded animals that live on land have feet; blooded land dwellers share to a certain extent the same formal nature, which explains the occurrence of certain coextensive features like the possession of a maximum of four feet. The snake also possesses all the properties that belong to blooded land dwellers, except for one. It is therefore rational to ask why this particular property is absent in snakes. It is not rational to ask why snakes lack wings, telescopic eyes, or any other part that cannot be observed to belong to the wider class to which snakes belong. As there is a virtual infinity of properties that any animal does not have, it only makes sense for a natural scientist to explain the absence of those properties that belong to the "natural possibilities" of that animal; and what those natural possibilities are, can be established inductively, on the basis of observation and through comparison with related beings.

In the cosmological domain, on the contrary, the range of possible ways in which a certain feature or motion can be present is only partly determined by observation. What cannot be observed might still be present, and what can be observed might be the result of a visual illusion. Aristotle often struggles with this question of how much credence we must attribute to our observations of the heavens, and of which observations we should explain and which we should explain away. His general strategy is to explain the phenomena and hence to save them; but on occasion, especially when there are contradictory observations, the observations that conflict with the thesis that the stars move around fixed in concentric spheres are rather explained away. This is exactly what happens in the paragraphs leading up to the explanation of why stars have no feet in *De caelo* 2.8, 290a12–29. Before giving his teleological argu-

---

31; 8, 709a25–b4; 10, 709b27–28). This may point to a problem for Aristotle's argument concerning the heavenly bodies: for the absence of *organs* for locomotion as such does not provide conclusive evidence that the stars in fact do not locomote. Of course, as the remainder of *De caelo* 2.8 points out, Aristotle is actually committed to the stronger claim that spherical bodies do not only lack organs for motion, but also "points of motion", which (at least given Aristotle's laws of sublunary mechanics) rules out any possible way of locomotion.

[34] Lennox (2001) 214–215.

ment demonstrating the likelihood of the absence of feet in stars, Aristotle argued that if the heavenly bodies were to move on their own, they would either roll or rotate, but that neither of these motions is observed to take place. The impression that the Sun rotates in rising and setting[35] is then explained away: according to Aristotle, the rotation is merely a visual illusion, caused by the weakness and unsteadiness of our vision.

What this makes clear is, first of all, that while in biology observations clearly show that snakes lack feet, observational evidence of the heavens gives much less certainty about the absence of feet in the heavenly bodies. For all we can tell, the heavenly bodies might be too far away for us to see their organs of motion. Secondly, observations of the heavens will not tell us whether there are any *natural* limitations to the possible ways in which nature could have "crafted" stars in order to make them able to move on their own. The absence of feet in the heavenly bodies in itself seems hardly enough to establish the reasonableness of the alternative theory that they do not effect any forward motion at all.

This difference between the reliability and applicability of observational evidence in biology and cosmology is important, because Aristotle's explanation in the case of the heavenly bodies is not *prompted* by the observation that they do not have organs for motion, as it is in the case of snakes.[36] There are no observations of the heavens that would reasonably lead to the expectation of heavenly bodies having feet in the first place. (One might object, however, that in this case the philosophical tradition within which Aristotle is working prompts this question).[37] Rather, Aristotle works the other way around: because he wants to make the theory that the stars do not move on their own as reasonable as possible, he uses the teleological principle that nature does

[35] Xenophanes might have observed the same phenomenon, and gives it a similar explanation: see Diels and Kranz (1951) 21A41a δοκεῖν δὲ κυκλεῖσθαι διὰ τὴν ἀπόστασιν; "[the Sun] seems to turn in a circle due to its distance."

[36] In *De part. an.*, the principle is commonly used in the context of a discussion of why animals have certain parts, where it is "discovered" that a certain kind of animal lacks that part while other members of its larger family or otherwise related animals have it. The question thus arises through comparative and empirical research. For examples, see *De part. an.* 4.12, 694a13–20 and 694a16–18; 4.13, 696a10–15 and 696a12.

[37] On this tradition, see Cornford (1937) 55–56; besides Plato, Empedocles also seems to have argued for the footlessness of the celestial sphere (Diels and Kranz, 1951) 31B29: τὴν Φιλίαν διὰ τῆς ἑνώσεως τὸν Σφαῖρον ποιοῦσαν, ὃν καὶ θεὸν ὀνομάζει, καὶ οὐδετέρως ποτὲ καλεῖ '*σφαῖρον ἔην*'. οὐ γὰρ ἀπὸ νώτοιο δύο κλάδοι ἀίσσονται, οὐ πόδες, οὐ θοὰ γοῦν(α), οὐ μήδεα γεννήεντα, ἀλλὰ σφαῖρος ἔην καὶ ⟨πάντοθεν⟩ ἶσος ἑαυτῶι). ■ non-matching parenthesis

nothing in vain to argue for the likelihood of the absence of the organs of motion in the heavenly bodies.[38]

In sum, it seems that in this example Aristotle goes out of his way to establish the reasonableness of the assumption that the heavenly bodies do not have a motion of their own and, hence, must be carried around while being fixed in concentric spheres. In the biological realm, the observation of what happens always or for the most part in nature is what allows us to draw inferences about cases in which the goal-directedness is less evident. In a domain such as cosmology, which is empirically underdetermined, such inferences are necessarily of a conjectural nature. However, if teleology extends to the heavenly realm, and Aristotle assumes that it does, then the use of teleological principles allows Aristotle to make the most sense of the phenomena, and to provide explanations appropriate to the science of nature, rather than merely astronomical or mathematical ones.

5. *Conclusion*

To a modern audience, Aristotle's teleological explanations of heavenly phenomena may sound rather unusual, but what I hope to have made clear in this chapter is that they make perfect sense within Aristotle's conception of natural science. If the heavens are part of nature, then we need at least to attempt to state all four causes for every heavenly phenomenon, even if the investigation has been made difficult because of the scarcity of empirical data. Aristotle's use of teleological principles thus follows from his treatment of the study of the heavens as part of the study of nature; and we have seen that this approach is especially prominent in the second book of the *De caelo*, where Aristotle searches for explanations of the features and motions of the heavens as a whole and of the heavenly bodies. The scientific investigation of an empirically undetermined domain such as the heavens is difficult, as his methodological reflections show, Aristotle is mostly well aware of all the problems involved. However, if one wants to gain knowledge of the heavens and its bodies, one has to try to give explanations that at least make the phenomena—both in terms of what can and of what cannot be observed—seem as reasonable as possible.[39]

---

[38] Aristotle repeats this argument in a reversed form in *De caelo* 2.11, 291b11–17.

[39] Cf. Irwin (1988) 34.

The strategy that Aristotle employs to give plausible accounts is to posit teleological principles as a way of finding final causes in difficult cases. The principles used are not *a priori* axioms, but suppositions derived from empirical evidence. They are generalizations over the actions of the formal nature of beings, based on numerous observations made in the biological domain. Just as the use of these principles helped Aristotle to find final causes in cases where these were not immediately observable in biology, in the same way Aristotle hopes to find explanations for natural phenomena in the cosmological realm. This gives a very central role to Aristotle's scientific practice in biology: one could say that where Aristotle's philosophy of science as described in the *Analytica posteriora* offers the student of nature his scientific toolbox, the accessible and rich domain of biology is the student's main workplace. The experience and knowledge acquired in studying biological phenomena may then—of course with suitable adaptations and refinements—be applied to other, less accessible domains of nature, such as that of the heavens.

The application of teleological principles to the cosmological domain is itself based on the assumption that the heavens are no less teleological —and, perhaps even more teleological—than the sublunary realm is. However, as I have pointed out, the lack of empirical evidence in the cosmological realm also weakens to some extent the inferences that Aristotle draws within this teleological framework:[40] the explanations are plausible, but not as "conclusive" as the ones we can find in the biological works.

For the *De caelo* this means that Aristotle argues as much from as towards teleology: starting from the assumption that the heavens as a whole are goal-directed, Aristotle tries to give a coherent, plausible, and reasonable picture of the heavens in which things are present or absent for a reason. This is Aristotle's main goal in the *De caelo*: even if it is not possible to give deductions that demonstrate why the heavens and the heavenly bodies have the features they have, one can still offer plausible physical accounts or inferences to the best explanation that take away some of the puzzlement concerning the heavens.

---

[40] Cf. Falcon (2005) ix: "there are features of the celestial world that outrun the explanatory resources developed by Aristotle for the study of the sublunary world."

Acknowledgments

For fruitful discussions of preliminary drafts of this paper I thank Frans de Haas, Jim Hankinson, and Ineke Sluiter. This paper also owes much to the questions asked and comments made by the audiences at the Dutch Graduate School for Classics (OIKOS); the American Philological Association, 137th Annual Meeting, Montreal; SUNY Albany; University of Tennessee, Knoxville; Wesleyan University, Middletown; and Washington University in St Louis.

Bibliography

Bogen, J., and McGuire, J.E. (1986–1987) "Aristotle's Great Clock: Necessity, Possibility and the Motion of the Cosmos in *De Caelo* 1.12," *Philosophy Research Archives* 12: 387–448.
Bodnár, I., and Pellegrin, P. (2006) "Aristotle's Physics and Cosmology," in M.L. Gill and P. Pellegrin (edd.), *Blackwell Companions to Philosophy, A Companion to Ancient Philosophy* (Malden and Oxford) 270–291.
Burnyeat, M.F. (2004) "Introduction: Aristotle on the Foundations of Sublunary Physics," in F.A.J. de Haas and J. Mansfeld (edd.), *Aristotle's On Generation and Corruption I Book 1*, Symposium Aristotelicum (Oxford) 7–24.
Cornford, F.M. (1975) *Plato's Cosmology* (Indianapolis/Cambridge).
Diels, H. (1882–1885) *Simplicii in Aristotelis physica commentaria*, Commentaria in Aristotelem Graeca 9–10 (Berlin).
Diels, H. and Kranz, W. (1951) *Die Fragmente der Vorsokratiker*, Vol. 1–3. 6th edn (Berlin).
Falcon, A. (2005) *Aristotle and the Science of Nature: Unity without Uniformity* (Cambridge).
Gill, M.L. (2009), "The Theory of the Elements in *De caelo* 3 and 4," in A.C. Bowen and C. Wildberg (edd.), *New Perspectives on Aristotle's De caelo* (Leiden) 139–161.
Guthrie, W.K.C. (1939) *Aristotle: On the Heavens* (London).
Hankinson, R.J. (2002–2003) "Xenarchus, Alexander, and Simplicius on Simple Motions, Bodies and Magnitudes," *Bulletin of the Institute of Classical Studies* 46:19–42.
Heath, T.L. (1932) *Greek Astronomy* (New York).
Heiberg, J.L. (1894) *Simplicii in Aristotelis de caelo commentaria*, Commentaria in Aristotelem Graeca 7 (Berlin).
Irwin, T. (1988) *Aristotle's First Principles* (Oxford).
Johnson, M.R. (2005) *Aristotle on Teleology* (Oxford).
Kahn, C.H. (1985) "The Place of the Prime Mover in Aristotle's Teleology," in A. Gotthelf (ed.), *Aristotle on Nature and Living Things* (Pittsburg) 183–205.
Leggatt, S. (1995) *Aristotle On the Heavens I and II* (Warminster).

Lennox, J.G. (2001) *Aristotle's Philosophy of Biology: Studies in the Origins of Life Science* (Cambridge).
Leunissen, M.E.M.P.J. (2007) "The Structure of Teleological Explanations in Aristotle: Theory and Practice," *Oxford Studies in Ancient Philosophy* 33: 145–178.
Lloyd, G.E.R. (1996) "Heavenly Aberrations," in G.E.R. Lloyd (ed.), *Aristotelian Explorations* (Cambridge) 160–183.
Moraux, P. (1961) "La méthode d'Aristote dans l'étude du ciel, *De caelo* 1.1–2.12." in S. Mansion (ed.) *Aristote et les problèmes de méthode* (Louvain) 173–195.
Mueller, I. (2006) "Physics and Astronomy: Aristotle's *Physics* 2.2, 193b22–194a12," *Arabic Sciences and Philosophy* 16:175–206.
Simplicius, *In Aristotelis de caelo commentaria*. See Heiberg 1894.
——, *In Aristotelis physica commentaria*. See Diels 1882–1885.
Solmsen, F. (1960) *Aristotle's System of the Physical World: A Comparison with his Predecessors* (Ithaca).

THE ASTROLOGIZATION OF THE ARISTOTELIAN COSMOS: CELESTIAL INFLUENCES ON THE SUBLUNARY WORLD IN ARISTOTLE, ALEXANDER OF APHRODISIAS, AND AVERROES

Gad Freudenthal*

1. *The changing role of the celestial influences on the sublunary world from Aristotle to Averroes*

One of the radical and long-lived innovations of Aristotle's physical world-picture was the introduction of the fifth substance. At an early phase of his development (reflected in his lost dialogue *De philosophia*), Aristotle still followed the Presocratics in believing in the essential material unity of the cosmos, and specifically held that the heavens were fiery. However, in his esoteric writings, those that are the core of what we construe as the mature Aristotle, the Stagirite introduced a decisive rupture of the universe by positing what later generations were to call "the fifth body" or "ether." This newly fathomed celestial matter was quality-less and impassive and, hence, in principle, a "stranger to generation and destruction" (Solmsen 1960, 289). Considered from the vantage point of his theory of matter alone, the sublunary world was now a closed system whose functioning depended solely on the laws of physics bearing on the bodies constituted of the four sublunary elements.[1]

Yet, Aristotle was well aware that the sublunary world was not entirely self-sufficient, for obviously the Sun warmed it. As is well known, Aristotle explained the Sun's heat through two *ad hoc* "mechanical" hypotheses: one states that the Sun warms by virtue of the friction between it and the air underneath; the other affirms that the heat is generated because the Sun's movement scatters the elemental fire which is in contact with the celestial sphere. Aristotle offered two versions of this account (*De caelo*

* CNRS, Paris.

[1] I described this development in Freudenthal (1995) 104, where further literature is indicated.

2.7; *Meteor.* 1.3, 341a12–36); but, as one scholar has justly commented, they are "both almost equally lame."[2]

Aristotle posits one major effect of the superlunary world on the sublunary one: the efficient cause of all generation and corruption in the sublunary world is identified as the Sun's double motion along the ecliptic (*De gen. et corr.* 2.10–11), since this motion brings about the seasons and keeps the four elements from settling into four stable, changeless, concentric spherical layers (cf. *De caelo* 2.3, 286b1–9). Aristotle, therefore, considers the heavenly bodies to be the ultimate source of all motion and change in the sublunary world (*Meteor.* 1.2, 339a21–24) and holds that "in dependence on it [= the heavens as a whole] all other things have their existence and their life, some more directly, others more obscurely" (*De caelo* 1.9, 279a28–30; trans. Guthrie 1971).[3] Aristotle occasionally alludes to specific phenomena instantiating this principle. On the level of non-animate nature, we may mention the change of land into sea and *vice versa*, which is an "orderly" phenomenon produced by the "Sun's course" (*Meteor.* 1.14, esp. 351a32–34); and the two exhalations and their various effects such as the phenomena which take place in the upper atmosphere (*Meteor.* 1.4, esp. 342a29) as well as winds, rainfall, and so forth, which also go back to the Sun's warming effect (*Meteor.* 2.4). On the level of animate phenomena, there is most famously the fact that "a human being is produced by a human being and the Sun."[4] "Spontaneous" generation too is held to be brought about by the Sun's heat and the duration of gestation and of the length of the life of animal species is also taken to depend on celestial motions.[5] Some natural phenomena (e.g., the time of menses) Aristotle similarly describes as depending specifically on the Moon's motion.[6] All these natural phenomena are said to have the orderly

---

[2] Longrigg (1975) 214; see also Moraux (1963) coll. 1204–1205.

[3] Quoted approvingly by Alexander, *De providentia*, (Thillet 2003, 21.20–21; Fazzo and Zonta 1999, 158). For a penetrating analysis of Aristotle's views of the heavenly bodies as exerting σωτηρία over the existents in the sublunary world, see Rashed (2002).

[4] *Phys.* 2.2, 194b13; *Meta.* 12.5, 1071a15; *De gen. an.* 4.10, 777b35. I quote the elegant (and politically correct) translation in Sharples (1994b) 170. Oehler (1969) has convincingly argued that the purpose of this statement is to remind the reader of the Sun's crucial role in continually mixing up the four sublunary elements, thereby producing the continued coming-to-be and passing-away; it has nothing to do with any ideas on the vivifying effect of the celestial bodies or of their heat. See also Freudenthal (1995) 38–39.

[5] On spontaneous generation, see Freudenthal (1995) 115 (and index *s.v.*); on length of life, see *De gen. an.* 4.10, 777b18 ff. For further references and discussion, see Happ (1971) 505–512.

[6] E.g., *De gen. an.* 2.4, 738a18 ff.; see also Préaux (1973), notably 88–90, 125.

heavenly revolutions, notably, the heat and cold which they produce, for their moving cause.[7]

In Aristotle's scheme, then, the paramount phenomenon of the Sun's heat and its various effects in the sublunary world were recognized, although their compatibility with the separation of the cosmos into two separate realms was problematic. The tension, not to say incompatibility, between the notion of the "fifth substance" and the posited effects of the Sun on the world of generation and corruption was well recognized by Aristotle's medieval students. For instance, in the 1320s, the Jewish philosopher-scientist Gersonides (Levi ben Gershom, 1288–1344) praised Aristotle in the following words:

> Experience is the point of departure [lit. premise] for inquiry, not inquiry the point of departure for experience ... If we find through [experience] something that is incompatible with reason we should not reject experience on this account ... Indeed, the Philosopher [= Aristotle] has done just this when he found through experience that the Sun heats the things [down] here, although it was evident through inquiry that the Sun is not warm and that anything that brings something from potentiality to actuality must of necessity somehow possess in actuality that which the other has in potentiality. Well, the Philosopher has not cast off this experience for the sake of these premises, but rather inquired how this [fact] is possible without contradicting the theoretical premises.[8]

On the basis of theory, Gersonides argues, the Sun should not have had any effect on the sublunary world. For him, it is to Aristotle's credit that he did not sweep its heating effect under the carpet.

Against this background, consider now the following two passages from Averroes (Abū 'l-Walīd Muḥammad Ibn Rushd, 1126–1298), the foremost and most faithful medieval student of Aristotle. In his *Epitome* (freely abridged summary) of Aristotle's *Metaphysics*, the Commentator writes:

> [74] We should now examine the subject of the providence exerted over things down here, i.e., below the sphere of the Moon. We proceed from previously [established] principles. We say that the existence of terrestrial things and their persistence are the object of providence [lit. are preserved] with respect to their species. This is indeed so by necessity and cannot

[7] Aristotle's famous statement that a living being's connate *pneuma*, or its soul-heat, is "analogous" to "the element which belongs to the stars" (*De gen. an.* 2.3, 736b30 ff.) should also be recalled in this context. It echoes the earlier views concerning the vivifying effects of the celestial realm upon the lower world.

[8] Gersonides, *Milḥamot ha-Shem* (= *The Wars of the Lord*) 5.1.43; text and translation in Mancha and Freudenthal (2005) 159.

be due to chance, as many of the Ancients had have claimed. It becomes evident if one considers how the motions of the heavenly bodies suit the existence and the preservation of each and every thing coming to be in this [lower] world. This can be observed first and foremost with respect to the Sun and also with respect to the Moon. Concerning the Sun it has been established that if its body were greater than it is, or if it were closer [to us], then it would have destroyed the species of plants and animals through an excess of heat. Again, if it were smaller than it is or farther away, then [the plants and animals] would have perished through cold. This is attested by the fact that the Sun produces heat through its motion and the reflection of its rays, and the fact that there are places which are uninhabited because of excessive heat or excessive cold.

[75] Similarly, providence is clearly manifest also in the Sun's inclined sphere [i.e., the ecliptic]. For if the Sun did not have an inclined sphere, then there would be no summer, no winter, no spring, no autumn. Yet it is evident that these seasons are necessary for the existence of the species of plants and animals. Providence is also very clearly observable in the [Sun's] daily motion, for without it there would have been be no night or day; but rather there would have been be a half-a-year day and a half-a-year night, and the existents existing things would have perished perish either by day due to an excess of heat or at night for due to an excess of cold.

[76] Similarly, the effect [produced by] the Moon is evident in the production of rain and the ripening of fruits. And it is clear that if the Moon were larger than it is or smaller or farther away or closer, or if it did not borrow its light from the Sun, it would not produce that effect. Again, if it [did not have] an inclined sphere, it would not accomplish different operations at different moments in time ...

[77] Clearly, what we have said with respect to the Sun and the Moon holds [lit. is to be affirmed] also of the other planets [lit. stars], their spheres, and their motions, which are in harmony [*mu'tadilah*] and at well-defined distances from the Sun. This is why Aristotle said that the motions of the planets are like those of the Sun, and he said so because it can be seen that the planets seek to conform to the Sun. Now although owing to insufficient empirical evidence we are unable to assign many effects to [specific] motions [of the planets], to their centers' being eccentric, or to their forward or backward [retrograde] motions, still we can affirm confidently that all this is for the sake of [exerting] providence over what is down here. However, it is difficult for us to acquire this [knowledge], for it requires long experience [*tajribah*], for which man's life is not sufficiently [long].[9]

---

[9] Averroes, *Epitome* of the *Metaphysics*, in Quirós Rodríguez and Puig Montada [1919] (1998) 166–168 (§§74–77); German translations in Van der Bergh (1924) 141–143 and Horten (1912) 201–203.

Averroes' comments on Aristotle's account in *De gen. et corr.* 2.10–11 (generation and corruption result from the Sun's motion along the ecliptic) are also telling. In his *Epitome* of this text he writes:

> The movement of the Sun in the inclined circle is the foremost [efficient] cause of the generation of what comes to be and of the corruption of what passes away. For when the Sun approaches, it is the cause of the existence of most of the things that come to be; and when it retreats, it is the cause of corruption of most existing things. Furthermore, the efficient cause of the four seasons, namely, spring, summer, autumn and winter, is this motion. The efficient cause of the continuity of generation and corruption, according to Aristotle, is the first continuous motion, while the efficient cause of generation and corruption is the motion of the Sun in the inclined circle. This latter movement is not limited to the Sun alone, but is also that of the Moon and of all of the planets, although the effect of that of the Sun is more apparent. For the affect of the Sun (in its course along its inclined circle) on the alteration of the four seasons is precisely that of each planet [lit. star] in its course along its specific circle. In fact, although we know not the specific affects produced by each and every planet on existing things down here, still, through a generalization, it becomes clear that all of them are involved in generation and corruption. So much so that if we were to imagine the disappearance of a single movement or planet among them, then either no coming-to-be whatsoever or that of some beings would not take place. It is indeed clear that some existing things are specifically affected by a specific planet. This is why we find that those who have observed the stars in the past have divided existing things in accordance with them, and posited that existing thing A is of the nature of a star X and existing thing B of the nature of a star Y. Generally speaking, it is clear that these planets appear to be affected by the motion of the Sun, for the differences in their effects are mostly dependent on their proximity or distance from the Sun.[10]

Comparing these passages by Averroes (many similar ones can be adduced) with the original Aristotelian outlook, one is doubly startled.

In the first place, in Aristotle's system, the Sun's effect on the sublunary world was in fact an embarrassing anomaly, not really compatible with the five-element scheme. Seeing that Aristotle's followers systematized the Master's thought, one might have expected them to downplay it the best they could. Instead, it has now been "generalized" into a paradigmatic phenomenon: not only the Sun but the other planets too act on sublunary things: "what Aristotle has affirmed with respect to the sphere of the Sun should be understood as applying to the other oblique spheres

---

[10] Puig Montada (1992) 49.5–50.11, Kurland (1958a) 121.67–122:78 and (1958b) 133. Cf. Freudenthal (2000) 343.

too," Averroes writes in his *Middle Commentary* on *De gen. et corr.* 2.10.[11] Moreover, although all the planets supposedly consist of the very same impassive matter, they are each said to act in a *specific* way; and Averroes goes as far as to allude to the specific "nature of a star X."

Second, in Averroes' outlook, divine providence over the (species of) existing things in the sublunary world as exerted through the heavenly bodies plays a major role. Averroes insists that the heavenly influences are not accidental: "it is clear that the operations [on the sublunary world] of the planets, and specifically of the Sun, are due to their *very essence* [*fī'l dhātī*]."[12] This means that the effects of the heavenly bodies are part and parcel of a providential scheme. The idea that the good in this lower world depends on the heavenly bodies is not altogether absent from Aristotle's thought, but it is only very discreetly alluded to through the notion of σωτηρία, as Marwan Rashed (2002) has perceptively pointed out. Nor does Aristotle claim explicitly that the Sun's motion along the ecliptic, the ultimate cause of generation and corruption, is a part of a providential scheme.

For Averroes, then, the sublunary world is *not* a closed system: it could not have existed without the sustenance of the superlunary one. Now Averroes was not the only medieval Aristotelian philosopher to accommodate ideas running against the postulate of the strict partition of the cosmos. On the contrary, his position was an essential component of all medieval philosophy of nature: "There is a consensus of all philosophers to the effect that the governance [*tadbīr*] of this lower world is perfected by means of the forces emanating to it from the [heavenly] sphere," observed Moses Maimonides, Averroes' exact contemporary and one-time countryman.[13] Moreover, the medieval version of Aristotelianism posits that, in Maimonides' formulation, the planets "act at some particular distances—I refer to their nearness to or remoteness from the cen-

[11] Alaoui (1995) 125. 8–9, Kurland (1958a) 90.47–48 and (1958b) 102.

[12] Alaoui (1995) 30.6, Puig Montada (1992) 54.4, Kurland (1958a) 123.4 and (1958b) 135. The same position is taken in *Tahāfut al tahāfut*; see Van den Bergh (1954) 299–300.

[13] *Guide* 2.5, Qafah (1978) 283; Pines (1963) 260. Elsewhere Maimonides writes in greater detail:

> It is known and generally recognized in all the books of the philosophers speaking of governance that the governance of this lower world—I mean the world of generation and corruption—is said to be brought about through the forces overflowing from the spheres ... [Y]ou will find likewise that the Sages say: "There is not a single herb below that has not a *mazzal* in the firmament that beats upon it and tells it to

ter, or to their relation to one another."[14] This assumption is consequential: through it, as Maimonides (a staunch opponent of astrology[15]) comments, "*astrology comes in.*"[16] This statement carries a twofold message:

(a) Distinctively astrological motifs (e.g., that the planets "act at some particular distances" and according to "their relation to one another") are present in contemporary Aristotelian natural philosophy.
(b) Consequently, the demarcation line between the latter and astrology is somewhat blurred: Maimonides, who was a particularly good connoisseur of astrological literature, must have had in mind the fact that, as a result of this rapprochement between Aristotelian natural philosophy and astrology, upholders of the latter could make a good case for the claim that their art had a rationale in Aristotelian natural philosophy.

Maimonides himself, as also other thinkers in the tradition of the *falsafah* like Ibn Sina (Avicenna) and al-Fārābī, indeed had a hard time when they tried to refute astrology while yet recognizing that the existence of celestial influences on the sublunary world seemed indubitable.[17]

It would be a very worthy enterprise to describe in detail the evolution from the original Aristotelian viewpoint to that described by Averroes and Maimonides and which threatened to allow astrology to "come in." This cannot be done here, so I will limit myself to one, albeit in my view crucial, phase of this development, the work of Alexander of Aphrodisias. I will endeavor to point out how and why Alexander modified the original Aristotelian framework, making the boundaries between the celestial and the sublunary realms permeable.

> grow ... Now they also call a star *mazzal*. ... By means of this dictum they have made clear that even individuals subject to generation have forces of the stars that are especially assigned to them. And though all the forces of the sphere pervade all existents, yet there also exists a force specially assigned to a certain species."
> (*Guide* 2.10, Qafah (1978) 292–293; Pines (1963) 270)

Maimonides' quotation comes from *Genesis Rabbah* 10.6. He says the same in his "Letter on Astrology"; see Freudenthal (1993) for references.

[14] Maimonides, *Guide* 2.12; Qafah (1978) 304; Pines (1963) 280.

[15] See Langermann (1991), Freudenthal (1993), Stern (1997).

[16] See n. 14 above. This comment was reiterated some four or five decades later by Judah ben Salomon ha-Cohen, an enthusiastic supporter of astrology, in his encyclopedic work *Midrash ha-ḥokhmah*; see Freudenthal (2000) 358.

[17] Druart (1978, 1979, 1981), Michot (2006) 47*–54*, Freudenthal (1993); see also p. 274 below.

2. *Alexander of Aphrodisias: Providence and celestial influences*

Writing in the closing years of the second century CE, Alexander of Aphrodisias faced a world in which astrology and various star cults had become central cultural forces. "Chaldean" astrology had been a mere craft, a purportedly empirical know-how; but within Greek culture (which it began to penetrate in or just after the third century BCE), it was theorized in a series of works culminating in Ptolemy's *Tetrabiblos*, which is contemporaneous with Alexander.[18] As a discipline whose claims had manifold philosophical implications, astrology elicited reactions from philosophers of all contemporary schools, some of whom rejected it while others approved of it.[19] In Alexander's time, astrology had thus become a presence which a philosopher could not ignore. Astrology was associated with a large number of "popular" doctrines that posited a nexus of causalities linking the heaven to the sublunary world, making material processes in the latter depend on the former, such as, notably, astrological botany and medicine and occasionally alchemy as well.[20] Under the Empire, the sway of astrology was next to universal,[21] all the more so as it coalesced with religious beliefs that were in harmony with it. Franz Cumont, whose book of 1912 is still the most authoritative work on the subject, writes that the oriental religious cults which penetrated the Empire were all "influenced in different degrees by astrology and star-worship," and that we are

> struck with the power of this sidereal theology, founded on ancient beliefs of Chaldean astrologers, transformed in the Hellenistic age under the twofold influence of astronomic discoveries and Stoic thought, and promoted, after becoming a pantheistic Sun-worship, to the rank of official religion of the Roman Empire [in 274 CE]. Preached on the one hand by men of letters and by men of science in centres of culture, diffused on the other hand among the bulk of the people by the servitors of Semitic, Persian or Egyptian gods, it is finally partronised by the emperors, who find in

[18] The classic work is still Bouché-Leclerc (1899). For a history of astrology, see Gundel and Gundel (1966) esp. 202ff. (from the middle of the second century CE onward). See also the insightful Long (1982) and the overviews in Boll, Bezold, and Gundel [1917] (1931); Tester (1987); Barton (1994).

[19] For a systematic review of philosophers' attitudes to astrology in the first centuries CE see Lawrence (2006). Despite its shortcomings, this is a very helpful account and I am not aware of a comparable systematic overview in print.

[20] Festugière [1950] (1986) esp. 89ff.

[21] Cumont (1909) chap. 7, notably 240–269.

> it at once a form of worship suitable for all their subjects and a justification of their autocratic pretensions.
>
> (1912, 90; cf. 99, 133 and Gressmann 1925)

These developments coalesced into a cultural ambiance which made commonplace the doctrine that all the heavenly bodies play an essential role in the workings of nature and especially in the governance of the lower world by the supernal one.[22] Within philosophy, the question whether human destiny depends on the planets was directly linked to the question of providence. From a quite marginal topic in Aristotelian philosophy,[23] it became a central one as a consequence mainly of the challenge posed by Stoicism, specifically, by the idea that God, identified with the pervasive material πνεῦμα, governs the universe down to the most minute details.[24]

We do not know with which of these developments Alexander was familiar. In particular, the question of whether he was acquainted with Ptolemy's *Tetrabiblos* or more generally with the specific postulates of astrology has received some attention but remains undecided, although the odds are that he was not.[25] It nonetheless stands to reason that Alexander was aware that around him the idea that the heavenly bodies govern the sublunary world was almost universally accepted as indubitable. Now, generally speaking, Alexander's philosophical project can be characterized as an attempt to safeguard the Aristotelian doctrine in the face of critique by competing systems of thought.[26] His exegesis of Aristotle's texts creatively elicits from them solutions *more aristotelico* to scientific and philosophical problems that had hardly crossed the

---

[22] For an admirable synthesis, see Boll (1922–1923).

[23] For the history, see Moraux (1970); Thillet (2003) Introduction, 17–45. In Alexander's time, Aristotle was understood to have held that the objects of divine providence were the (eternal) heavenly bodies, not the ephemeral ones of the sublunary world; see Moraux (1970) 54ff.; Sharples (1982) 198–199 and (1983a) introduction, 25.

[24] Alexander defended his Aristotelian view of providence also against Platonist attacks; see Sharples (1987) 1216 and (1990) 91.

[25] Fazzo (1988) has suggested that Alexander knew Ptolemy's *Tetrabiblos* and accepted some of its postulates. But in his *De fato*, Alexander tellingly avoids any allusion to astrology; see Sharples (1983), introduction, 18–19, 128 and (2001) 522–523, 527, 529. Like his contemporary Galen (Toomer 1985), e.g., Alexander could have upheld the general view concerning the effects of the heavenly bodies on the lower world and yet oppose astrology.

[26] For a superb overview of Alexander, see Sharples, (1987); it is presupposed throughout what follows, even in the absence of specific references. For a comprehensive listing of Alexander's writings, see also Goulet and Aouad (1989) and Fazzo (2003).

Stagirite's mind, but that were on contemporary agenda.[27] This holds also for the issues with which we are concerned here. We will now see how Alexander responded to the theoretical challenge concerning the celestial influences on the sublunary world.

We broach the issue by considering Alexander's theory of providence, albeit only on the level of physics. The problem with which we are concerned grew out of an entrenched idea of Greek natural philosophy that goes back to Anaximander: any compounded substance, owing to the very fact that it consists of opposites, has a built-in or intrinsic tendency to decay (see Freudenthal 1986). In Aristotle's words: "never are [substances] eternal when they contain contrary qualities" (*De long. et brev. vit.* 3, 465b29 f.: see Freudenthal (1995) 11–17). *Aristotelians* were thus confronted with the following double theoretical challenge.

First, according to Aristotelian doctrine, the universe is eternal because none of the four elements exceeds the others in quantity and, therefore, in power; and because it is encompassed by the impassive and changeless fifth substance (Freudenthal 1995, 14–15, 101–105). But this argument did not appear convincing to everyone. The pseudo-Aristotelian treatise *De mundo*, for one, with which Alexander was apparently familiar,[28] posits the Aristotelian aether but still asks, "How it is that the universe, if it be composed of contrary principles—namely, dry and moist, hot and cold—has not long ago perished and been destroyed" (*De mundo* 5, 396a32; trans. Forster 1950)?[29] The Stoic doctrine of ἐκπύρωσις too (although on very different grounds) affirmed the natural perishability of the universe (or rather its temporary transformation to its supreme, god-like, fiery state). Similar positions were expressed in certain interpretations of the *Timaeus*, to which Alexander responded by rejecting attempts to "save" the world by invoking a divine will.[30] The problem of the physical persistence of the universe as a whole was thus high on the contemporary agenda, and Alexander was intent on showing that the universe was eternal by its own nature.

Next, with respect to the material persistence of individual sublunary substances, the Stoics affirmed that the ubiquitous πνεῦμα exerts provi-

[27] Thillet (1984) lxxix–lxxx.

[28] Moraux (1967) 160 n. 2; Sharples (1983) 24 In Alexander's time this work circulated widely, see Festugière [1949] (1983) 478.

[29] It answers this question in terms of a power "extending through all" and "forcing the most contrary natures in it to live in agreement with one another" (*De mundo* 5, 396b31).

[30] See *Quaest.* 1.18 and 2.19 (being "eternal by its own nature," the world as a whole needs no being to exercise providence on it); cf. Sharples (1983c) 62–63.

dence throughout matter and, specifically, that it maintains the cohesion over time of all substances by virtue of its different states of τόνος (tension). Stoic physics thus very easily accounted for the fact that matter is informed and that the material substances persist despite the disintegrating potentialities of the opposing elements or qualities within each of them. *De mundo* 6, 397b32–398a6, for its part, argues similarly that the power originating in the heavens penetrates even the most distant things, thereby vouchsafing their persistence. The Peripatetic school was slower to come forward with an answer. Aristotle had been aware that the persistence of sublunary material substances composed of the four opposing sublunary elements calls for an explanation. The problem looms both in his *De philosophia* and in his esoteric writings. Elsewhere I have tried to show that he worked his way toward a solution in terms of vital heat and connate πνεῦμα but without bringing it to completion, thus leaving the problem essentially unresolved.[31] In Alexander's time, this problem became more pressing because it was at the very heart of Stoic physics, which Alexander took seriously enough to devote to its criticism an entire work (*De mixtione*).

Alexander rose to this twofold challenge by elaborating a theoretical account of how the sublunary world is governed by the Unmoved Mover. It is germane to our inquiry inasmuch as it accords the heavenly bodies a central role in exerting providence over the sublunary world, and thus constitutes a decisive first step in the development of what I will call "the astrologization of the Aristotelian world-view" (p. 262 below), a development that allowed astrology to "come in." In this context, Alexander introduced the notion of a "divine power" originating in the heavenly bodies which he held to bring about and preserve form and order in the sublunary realm. Steering middle course between a position denying all providence in the sublunary realm and the Stoic deterministic view, Alexander construed divine providence as the preservation of the species (only), and affirmed that it was guaranteed through the continuous coming-to-be and passing-away brought about by the orderly motions of the heavenly bodies. Alexander's ideas on this subject straddle many disciplines, including metaphysics, psychology, physics, theory of matter, and cosmology; and my account will necessarily be schematic and tailored to the purpose at hand.

[31] Freudenthal (1995). The problem is also identified by Furth (1978, 1988), whose approach is quite different from mine.

We begin with Alexander's theory of the active intellect. On Alexander's account, the primary function of the active intellect (an enigmatic notion going back to Aristotle's *De anima* 3.5, 430a10–15) is noetic, viz. to activate forms in man's material intellect.[32] It is introduced on the general, metaphysical, principle that forms can be activated only by something possessing them most fully, viz. pure form. Alexander thus famously compares the active intellect to light, which, being supereminently visible, makes material objects visible (Schroeder 1981). But Alexander also ascribes to the originally noetic or psychological notion of the active intellect a further, cosmological, function: inasmuch as the active intellect is pure form separated from matter and impassive, Alexander identifies it with Aristotle's Unmoved Mover of *Metaphysica* 12,[33] which he construes as the ultimate cause of all things.[34] The gist of this theory is that the cause of the intelligibility of forms (and only forms are apprehended by the intellect) cannot but be the same Being that is the cause of their existence in the first place (see Moraux 1942, 92). Therefore, as Davidson (1972, 1992) has pointed out, in Alexander's system, the active intellect holds two *prima facie* distinct and unrelated roles, one physical and the other psychological: it is at once the source of forms in *material sublunary reality*[35] and the source of the forms in the human *intellect*. Henceforth, the forms in the *world* and in the human *mind* were taken to derive from one and the same origin.[36]

To understand why Alexander assigned to the Unmoved Mover functions that Aristotle had not dreamt of attributing to it, we turn to his cosmology. We begin with some metaphysical considerations expounded by Alexander notably in his treatise preserved in Arabic under the title *Mabādī' al kull* (*The Principles of the Cosmos*), the authenticity of which

---

[32] Sharples (1987) 1206. It has been pointed out (e.g., in Sharples 1987, 1207) that on this account the role of abstraction is problematic, but this needs not concern us here.

[33] Moraux (1942) 93, Sharples (1987) 1206. Alexander may follow an earlier tradition; see Moraux (1942) 162–164 and (2001) 350–351, Sharples (1987) 1212.

[34] See Davidson (1972) 112, 125–126, 151 and (1992) 13–14, 30–31.

[35] Moraux (1942) 93, Sharples (1987) 1206.

[36] Throughout the Middle Ages beginning especially with Avicenna, the active intellect was construed both as the cause of the existence of informed material substances (the *dator formarum*), and as that which actualizes forms in the human material (potential) intellect, allowing man to cognize forms. The active intellect was thus a central theoretical notion of *both* cosmology and psychology. This view, as we just saw, goes back to Alexander. In the later medieval tradition, the active intellect was held to operate through emanation rather than through the divine power; and it was identified with the last emanated intelligence rather than with the Unmoved Mover. But these differences are of secondary importance here.

now seems well established.[37] The "divine body"—which means the sphere of the fixed stars, as well as, derivatively, the spheres of the planets[38]—is necessarily ensouled because "the best" cannot be soulless (*Mabādi'* §§ 7–8). Every heavenly sphere has its own soul, which is its nature and the cause of its eternal circular motion (§§ 16–19, 76–78, 96). This motion is brought about by an "impulse" or a "desire" (*ishtiqāq*) toward the Unmoved Mover (§§ 25, 29, 97, 100) that arises in the soul's intellect upon perceiving the Unmoved Mover (§ 79).[39] The Unmoved Mover is, thus, the cause of the daily circular motion of the sphere of the fixed stars, which in turn imparts to the spheres of the planets their diurnal westward motion (§ 55). By contrast, the planets' "opposite" motions, viz. their eastward "normal motions," are due to their intellects and belong to them essentially (*fī dhātihi*) (*Mabādi'* § 79; Bruns 1882, 1.25, 40.24–31). (The two kinds of motions, as we will see, are part of the providential scheme of the world.) The heavenly bodies and their circular motions being "prior" with respect to all other existing things and motions (§ 50; on the notion of "prior" here, see § 62), all motions in the sublunary world must go back to the heavenly ones (§ 50; see also § 94). Within the sublunary realm, the term "motion" now obviously refers not to displacement or locomotion alone, but encompasses also all kinds of qualitative change, including generation and corruption:[40] "the divine body and its circular motion are the first cause of the coming-to-be of existing things *and* of their natural motions" (§ 51). Specifically, owing to the diversity (*taghayyur*) and variety (*ikhtilāf*) of the sidereal eastward motions, the planets' "relationships" (*'iḍāfa*) to the sublunary material bodies—i.e., their distances, angles, and so forth—constantly change, and, as a consequence, the planets produce in these bodies multiple changes, viz. the multifarious forms of their generation and corruption (*Mabādi'* §§ 56, 127; Bruns 1882, 1.25, 41.12–19).[41] The heavenly bodies are cognizant of the generation and corruption they bring about in the sublunary world

[37] It was convincingly argued by Genequand (2001, 1–4) in his introduction to Alexander, *Mabādi*; see also (2001) 34 for Genequand's remarks on the Syriac version of the text. By juxtaposing statements taken from the *Mabādi* with convergent ones from works preserved in Greek, the present paper too lends indirect support to the thesis of authenticity.

[38] See Sharples (1983a) 156 (note to *De fato* 25). See also Simplicius' *In de caelo* (Heiberg 1894) 1.2–11, 116.12–15, 116.30–117.2, 117.15–19.

[39] The topic is discussed in detail in *Quaest.* 1.1 and 1.25.

[40] On the different kinds of motion, see *Phys.* 3.1, 201a10–14 and 5.2, 226a23–35.

[41] See Todd (1976) 225.30–34 on the planets' changing positions as causes of heating and cooling, causing and, thus, of the elements' changes into one another. See also Genequand (2001) 167 on *Mabādi'* § 141.

(§§ 85, 120), presumably by virtue of their souls.[42] Alexander's view, we note, is a generalization, indeed, a radicalization of Aristotle's account in *De gen. et corr.* 2.10: not only the Sun, but *all* the planets affect the sublunary world; and not only the Sun's annual motion on the ecliptic produces change down here, but the proper motions of *all* planets and the resulting changes of their positions with respect to the Earth share in determining generation and corruption.

This scheme vouchsafes providence (*Mabādi'* § 91): inasmuch as generation and corruption ultimately go back to the eternal unchanging Unmoved Mover which cognizes itself and in which the intellect, the intelligible, and the cognizing subject are one and the same (§§ 101, 110, 121, 123),[43] not only the obviously regular circular motions of the planets but also the changes of their mutual relationships follow an eternal order so that the process of generation and corruption in the sublunary world too "participates in [this] order and the eternity" (§ 56). Consequently, "it is not at all to be feared [as did, e.g., the author of the *De mundo*] that the world might perish" (§§ 57, 139–140). Indeed the cosmos forever "remains new and does not grow old":[44] superficial appearances to the contrary notwithstanding, inasmuch as generation and corruption are produced by the orderly motions of the planets, they too are orderly—they consist of transformations of contraries to one another rather than in corruption (§§ 138, 141). Alexander *identifies* sublunary *nature* with the effects of the motions of the heavenly bodies[45] and, therefore, takes it to be eternal and invariable. Alexander thus maintains that the Unmoved Mover produces "nature" through the intermediary of the heavenly bodies. This nature "is a *divine power*" (§ 132),[46] which "penetrates all parts

[42] See also *Mabādi'* § 3 and Genequand (2001) 145, Sharples (1983a) 26.

[43] See Genequand (2001) 161 on *Mabādi* § 101.

[44] Fragment VE 35 in Genequand (2001) 142. 21 (text), 143 (translation); see *Mabādi* § 143. See also *De fato* 25 (Bruns 1882, 195.24f.), where Alexander makes the point that the biological model in which the cause of an animal is an older animal is not transferable to the cosmos, since generation and corruption are co-eternal with the Unmoved Mover. The Stoics are clearly the target of this argument.

[45] Sharples (1983a) introduction, 26 and (1987) 1217.

[46] The Arabic translation uses both the term *quwwa ilāhiyya* (§ 131) and *quwwa ruḥāniyya* (§§ 128–129). The latter literally means "spiritual power," but I agree with Genequand that "spiritual" here is simply a synonym for "immaterial" or "divine" and that there is no reason to see here a Stoic "influence" as suggested by the late Shlomo Pines (1986b). See Genequand (2001) introduction, 18–19 (with n. 32), and 164f. on *Mabādi* § 128. See also Sharples (1988) 1188.

of the world and holds its parts together" (§§128; cf. §§129, 132) and which is thus "the cause of the unity and order of the world" (§128). This theory is a clear retort to Stoic physics: Alexander's "divine power" is the precise functional equivalent of the (also divine) *pneuma* in Stoic physics—in their respective schemes both produce "the unification of the whole" (Todd 1976, 223.6–14). The universe is, therefore, comparable to a well organized city, all of whose parts seek to be "connected" to the Unmoved Mover (§§129–130): "The continuity of things that come to be has a cause [viz. the Unmoved Mover], and it is on account of this that the universe is one and eternal, always organized in one and the same way."[47] This reassuring theory of providence, grounded in Alexander's notion of the divine Unmoved Mover as the cause of existence and persistence of sublunary things and of the entire cosmos, is the Aristotelian answer to alarming views such as those expressed in the *De mundo* (and by Stoicism). A central component of this cosmology is the idea that generation and corruption are caused by all the planets, a generalization of the Aristotelian view attributing sublunary generation and corruption to the motion of the Sun.

The orderliness in which a substance partakes is commensurate with its degree of perfection (*Mabādi'* §§134–135). Alexander does not spell out this idea here, but his view (to which we will come back) is that the heavenly spheres are the most perfect existing things and participate in the order established by the Unmoved Mover to the greatest extent; whereas the less perfect sublunary material substances do so (in various degrees) to a lesser extent (§§135–137). Note that in Alexander's view because providence is exerted through the eternal circular motions, it can extend only to entities that are eternal too, i.e., to species, and not to individuals.[48] Indeed, Alexander holds that a given configuration of the celestial bodies never occurs twice and that, consequently, two individuals will never exist twice.[49] Moreover, since the universe is eternal and so also is the providence as exerted by the heavenly bodies, it is not possible that the order of the world change: if, say, a planet ceased to exist,

[47] *De fato* 25 (Bruns 1882, 195.23–25) and commentary *ad loc.* in Sharples (1983a) 156, with the reference to the similar statement in Todd (1976) 223.9f.

[48] See, e.g., *Mabādi'* §§57, 80; Sharples (1982) 198, (1983a) introduction and 26, (1983b) 64 and (1994a).

[49] See the *testimonium* preserved in Averroes' *Epitome of De gen. et corr.* (Puig 1992, 60.3–61.10; Kurland 1958a, 126.52–62 and 1958b, 137–138) and the analysis in Rashed (2000) 234ff. See also Freudenthal (2000) 356–361.

the entire universe would collapse (§ 81), a statement that Alexander elaborates in greater detail in his *De providentia*, as we will now see.

In the *De providentia*, another of his treatises preserved in Arabic only, Alexander is not far from affirming that the general arrangement of our world—i.e., its natural order—is the best possible one.[50] The existing celestial arrangement is so minutely ordered, he says, that the slightest alteration therein would bring the sublunary world to naught:

> If the distance of the Sun to the Earth were different from what it is, or if its motion and displacement were not along the inclined ecliptic, or if it moved along the ecliptic but without following the rotation of the fixed stars, moving in its own [annual] motion alone, then we would be deprived of all things other [than the Sun] that exist for our sake in our abode and, furthermore, the generation of animals and plants would not be possible ... Indeed, if the Sun's distance to us were smaller than it is now, and not the one it actually is, then, owing to the proximity of the motion, the region around the Earth [i.e., its surface] would be heated beyond the temperate, and beyond what is proper for that region. And if, inversely, its distance were greater than it is, its heating would be less [than appropriate]. In both cases, there would be no generation of animal or plant species. Those who hold this opinion have sufficient evidence from the existence of regions on the Earth which are called "uninhabitable," [and which are so] on account of the overcoming of one or the other of these qualities [heat or cold].[51]
>
> (Thillet 2003, 11.4–18; Fazzo and Zonta 1999, 126)

Alexander devotes a page or two to elaborate this thesis apropos of the Sun, and then extends it to the Moon:

> Were [the Moon] closer [to us] than it is now, then it would prevent the constitution and the existence of the clouds, and the constitution and existence of waters [i.e., seas], because it would disperse and subtilize the rising exhalations ... Further, if its motion were not in an inclined plane but along an orbit parallel [to the equator], then it would afford neither the moderation of the cold nor the succor from the heat, which it provides at present in our abode ... Nor would it provide succor in coloring [variant reading: coming-to-be] and ripening of fruits, as it does now, being the foremost cause of these two processes ...[52]
>
> (Thillet 2003, 13.4–20; Fazzo and Zonta 1999, 132)

[50] Alexander's scheme implies that on the level of its cosmological order the world is the best possible one but not with respect to individuals, providence being limited to species. The crucial turning point in this respect is Avicenna, as has been persuasively shown notably in Rashed (2000) 236.

[51] This argument is echoed in Galen, *De usu part.* 3.10 (*in fine*).

[52] A similar argument in *De mundo*, 6, 399a20–30.

I quoted these passages at some length to highlight that Averroes' argument on the providential arrangement of the world presented above (p. 242) follows Alexander's line, whose text the Commentator paraphrases in part.[53]

Alexander concludes from these considerations that "the divine [= heavenly] bodies," through their "motions and the durability of the order they exhibit," are the "causes of the coming-to-be of things on Earth *and* the cause of their persistence" (Thillet 2003, 13.23–24; Fazzo and Zonta 1999, 134); the crucial importance of the heavenly bodies' *two* motions (daily and proper) is particularly emphasized in this context (see also *Quaest.* 1.25 [Bruns 1882, 40.34–41.4, 41.12–18]). "The good ordering of the things down here corresponds to the good ordering of the movements of the divine bodies" (Thillet 2003, 15.16–17; Fazzo and Zonta 1999, 138; similarly *De fato* 35 (Bruns 1882, 195.24–25); *Quaest.* 2.19), and both are due to God's Goodness (cf. *Mabādi'* §§ 11, 12, 25: see also Thillet 2003, 19.23; Fazzo and Zonta 1999, 150). On the strength of the assumption (of which more below) that the heavenly bodies act through a divine power issuing from them, Alexander holds that their effects on the sublunary world depend on the orderliness of their motions and on the harmonious proportionality of their distances to the affected material substances (Thillet 2003, 11.1–3; Fazzo and Zonta 1999, 126).

Let us now get down to the level of the concrete, physical workings of nature. How is divine Goodness exerted on the individual material substances? The Stoics had a straightforward answer to this question in terms of πνεῦμα, but Alexander vehemently rejected their materialist immanentist theory of matter (*De mixtione*). His alternative account is presented notably in the *De providentia*. The notion of "divine power," already briefly encountered in his *Mabādi'*, now moves to the center. Divine providence, Alexander states, is exerted through a "power" gushing forth (or issuing) from "the Sun, the Moon, and the other planets [lit. stars]" (Thillet 2003, 10.25–26; Fazzo and Zonta 1999, 124–126). This theory accounts both for the generation of forms in matter and the persistence of sublunary substances. The divine power ("which we also call

[53] That this is the case has been pointed out in Thillet (1979) vol. 4, introduction 65, 75–77; text 18–24 (unfortunately not recalled in Thillet 2003). Al-Kindī had quoted the very same passages in his *On the Proximate Efficient Cause of Generation and Corruption*; see Wiesner (1993) 54–56. Averroes also inserted a passage quoted verbatim from Alexander's *De providentia* into his *Epitome of the Metaphysics*; see Fazzo and Zonta (1999) 168. Rashed (2000) analyzes the reasons why Averroes follows Alexander, against Avicenna, on this issue.

'nature'"), Alexander says, "endows with existence and forms the substances in which it inheres, according to a certain proportion and a certain order" (Thillet 2003, 19.6–7; Fazzo and Zonta 1999, 150). As might be expected, the "certain proportion" and "certain order" depend on the motions of the planets: "the divine [= heavenly] bodies," Alexander specifies,

> through the intermediary of nature [= the divine power], exercise their providence on the coming-to-be of the things exisiting down here. But not only this: once the latter have come to be, [the heavenly bodies exercise their providence] also on their [continued] existence and their intactness.
> (Thillet 2003, 22.12–13; Fazzo and Zonta 1999, 160)

The coming-to-be of informed sublunary substances and their subsequent persistence are thus due to the divine power outpouring from the heavenly bodies.

This strong claim on behalf of the divine power naturally makes us curious about its exact *modus operandi*. For an answer, we turn to *Quaest.* 2.3.[54] Alexander's discussion in this intriguing short text, proceeds, as it seems to me, on three *presuppositions*:

(a) The heavenly, divine body produces in the adjacent (sublunary) matter a "power" (Thillet 2003, 47.28; Fazzo and Zonta 1999, 30), later referred to as "divine."
(b) This power is produced in the sublunary matter through the *motion* of the contiguous heavenly body (Thillet 2003, 47.28–30).
(c) This (divine) power is "the cause of the fact that some bodies are animate and other bodies inanimate, and that some animate [bodies] only possess a nutritive soul, other bodies a [soul] capable of sensation as well, and other bodies in addition to these possess also a rational [soul]" (Thillet 2003, 47.28, 48.12–22; Fazzo and Zonta 1999, 30), endowing human beings with reason and intellect (Thillet 2003, 48.20–22).

Assuming, then, that providence exists and that it is brought to bear through the divine power generated by the motion of the heavenly bodies, Alexander's specific problem is to determine on *what physical level* the divine power affects matter. He considers two possibilities.

[54] Text and Italian translation in Fazzo and Zonta (1999) 195–219, English translation in Sharples (1992) 93–98; see also Duhem (1965) 344–348, Moraux (1967).

A first possible view is that the divine power acts on the four elements, which are supposed to exist in themselves and possess a single essential quality, viz. heaviness or lightness (*Quaest.* 2.3 [Bruns 1882, 47.32, 49.2]).[55] What they do *not* possess (the discussion that follows makes this clear) are the four qualities: moistness, dryness, heat, or cold and, consequently, the capacity to combine into complex substances, specifically ensouled ones. Were the sublunary realm to consist only of the elements in their "primeval" state, i.e., if no divine power existed, then there would be no motion other than the rectilinear ones up and down toward the natural places; and there would be no forms, no life on Earth. In other words, the elements would then have no share in the divine and would therefore not possess—even potentially—form or soul (*Quaest.* 2.3 [Bruns 1882, 48.24–25]). But we know that the elements and the compounds *do* partake in the divine (for forms and life exist) and this is due to the (providential) divine power. On account of the divine power "sown in them," the elements are "better and more perfect" than they would otherwise have been (*Quaest.* 2.3 [Bruns 1882, 48.29, 49.4]). The perfection with which the divine power endows an element depends on the latter's proximity to the heavenly spheres (*Quaest.* 2.3 [Bruns 1882, 49.1, 2.6–9]) and it comes to the fore in its physical properties: fire, the purest, rarest, and also the most active element, is the most perfect one (*Quaest.* 2.3 Bruns 1882, 49.6; 50.22]).[56] The concomitance of proximity to the heavenly sphere with purity and rarity follows from the assumption that the elemental qualities are produced through the sphere's motion. Alexander's qualification of the divine power as "divine" is thus a natural corollary of its being produced through the motion of the heavenly (divine) sphere, a point Alexander explicitly affirms elsewhere.[57]

It is the share of divinity of which the elements partake that allows the generation of forms, specifically, of souls in compounds: the combinant

---

[55] The idea that heaviness and lightness are the only essential qualities of the elements is characteristic of Alexander; see the references in Sharples (1992) 94 n. 309.

[56] Contrary to what one might expect from this discussion, Alexander elsewhere says that the element fire is *less* hot than terrestrial fire down here; and he connects this with the fact that the former is rare whereas the latter inheres in dense matter (*Quaest.* 2.17 [Bruns 1882, 61.29–62.14]; Sharples 1990, 98–99). Perhaps his construal is an extrapolation from the premise that "a flame is less hot than red-hot iron" (*Quaest.* 2.17 [Bruns 1882, 62.1]): the rarer fire is, the less hot it is. This construal disconnects the heavenly substance (and consequently the "divine power") from heat.

[57] In *Mantissa* 23 (Bruns 1887b, 172.17f.), Alexander says that the divine power qualifies as "divine" because it arises in matter as a result of its proximity to the heavenly divine sphere. But see also n. 67 below.

elements or rather the compounds which they make up acquire a "psychic [motion]" (*Quaest.* 2.3 [Bruns 1882, 49.3]), i.e., the potentiality to have motions other than those depending on heaviness or lightness, beginning with the capacity to grow. Possibly Alexander thought that compounds are informed not only through the divine power already inhering in the constituting elements, but also by an *additional*, direct input of the divine power.[58] The perfection of the form that comes to be in a compound depends notably on the subtleness of the constituents. In a compound in which, say, the purer and more divine fire predominates, the soul brought forth will be more perfect than a soul brought forth in a body in which an earthy substance predominates—we will have an animal or even a human soul, rather than a vegetative one (*Quaest.* 2.3 [Bruns 1882, 49.4–15]).[59] The scale of being thus depends on the share of the divine power inhering in the components, and this in turn depends on their subtleness.[60] The resulting compounds are "more perfect and [even] animate" and they "share in a more divine nature and principle [than the elementary constituents]" (*Quaest.* 2.3 [Bruns 1882, 49.18, 22]).

According to this first view, then, the divine power acts on the four elements in proportion to their proximity to the sphere and, perhaps, on compounds too. It makes it possible for matter to be informed with the perfection—i.e., divinity—of the ensuing substances depending on the share in the divine of the composing elements. The divine power is produced in the elements through the motion of the heavenly bodies, providence's reliable instruments.

The second possible view situates the action of the divine power on a "lower" level of matter. It posits that what exists in itself, independently of the action of the divine power, is not the already constituted elements

[58] Moraux (1967, 164, 166) takes the divine power to affect the compounds, not the elements, a view shared by Sharples (1992) 96 n. 312. Although Alexander is not entirely clear, it seems to me that this interpretation does not do justice to the text (*Quaest.* 2.3 [Bruns 1882, 48.12–18, 49.20–23]), nor to Alexander's view that the divine power acts in proportion to the proximity to the heavenly body: a clod of matter becoming a man is not always higher up than one becoming an animal. In addition, as Sharples notes, it is also inconsistent with Alexander's position in his *De anima* (notably 7.8–13). See also n. 67 below.

[59] Following Aristotle and the medical tradition (see Tracy 1969), Alexander at times adds that the kind of soul emerging in a compound also depends on the good balance of its constituents, i.e., on the temperament of the living being, the best equilibrium giving rise to a rational soul (e.g., Thillet 2003, 18.19, 148, 20.19, 154; *Quaest.* 2.3, 50.18–26).

[60] Although Alexander does not say so, this is, of course, classic Aristotelian doctrine, fire, say, being ontologically "above" the other elements. See Gill (1989) 239.

but rather mere uninformed matter (ὕλη). On this view, the role of the divine power is to endow ὕλη with the four elementary forms: ὕλη "is given form and shaped by the power which comes to be [in] it from the divine bodies" (*Quaest.* 2.3 [Bruns 1882, 49.32]). Thus, fire comes to be: it is "given form" through the heat and dryness which the heavenly bodies, namely "the Sun, the Moon, and those other [wandering] stars," produce in the adjacent spherical layer of the ὕλη (*Quaest.* 2.3 [Bruns 1882, 49.30–50.3, 50.8]). The forms ensuing in the other layers of ὕλη depend on their remoteness from the heavenly bodies as well as on the "different relation of the [heavenly bodies] to things [down] here at different times, which results from their movement on a circle of this sort [i.e., an inclined circle]" (*Quaest.* 2.3 [Bruns 1882, 50.3–8]), a postulate already encountered more than once. Through the movements of the heavenly bodies, four layers of ὕλη each receive two pairs of qualities and thus each becomes one of the four sublunary elements (*Quaest.* 2.3 [Bruns 1882, 50.9–18, 23–24]).

Compounds can then be formed by virtue of the potentialities which the elements have acquired through the divine power: very roughly, and advisedly using anachronistic vocabulary, it can be said that the divine power endows matter with the capacity for self-organization. The perfection of each compound depends, as before, on its constituents: the greater the "share in the divine power" of the latter, the "more perfect" will the substance be.[61]

As is often the case, Alexander does not choose between the two accounts;[62] but this has no bearing for our purposes in the present context. Whether he held the divine power to affect matter on the level of the elements (and, perhaps, also on that of compounds) or on that of the ὕλη, the following corollary holds: according to Alexander, *matter has the capacity to be informed (and, notably, be ensouled) and, once informed or ensouled, to persist in its form; and it has this capacity by virtue of the divine power infused in it, a power which endows it with a share in the divine.*[63] A suitable clod of matter is informed into, say, a living being

[61] Recall that the perfection of a compound also depends on the good balance of the components; see n. 59 above.

[62] See Moraux (1967) 169 with n. 1.

[63] The forms that matter receives by virtue of the divine power and which thus underlie providence, it must be emphasized, are those of the species, excluding the traits of individual substances (see n. 50 above), a view which distinguishes Alexander from the Stoics and perhaps also the astrologers (see Sharples 1994b). In Alexander's view, the differences between individuals are attributable to matter (Thillet 2003, 23.10–13, 15).

by the sire's seed;[64] but this coming-to-be would not have been possible without the divine power inhering in that matter.[65] On Alexander's view, then, saying that a substance has come to be and persists "by nature" is tantamount to saying that it has in it "a divine thing" (*shai' 'ilāhī*) which is responsible for its form (Thillet 2003, 18.12; Fazzo and Zonta 1999, 148).[66] We saw that if the matter of a compound is subtle and balanced enough, then the form can be a soul, be it of a plant, or an animal or even a man (also Thillet 2003, 18.20–21; Fazzo and Zonta 1999, 148). This is the idea behind Alexander's statement that "it was through this [divine] power that providence made human being[s] rational creature[s]" (*Quaest.* 2.3 [Bruns 1882, 48.15–22]). Indeed, inasmuch as man received the greatest possible share of the divine thing, he is, among all sublunaries, "most divine" (Thillet 2003, 19.4; Fazzo and Zonta 1999, 150; *Mantissa* 23 [Bruns 1887b, 172.17–22]). This theory of matter, together with Alexander's characteristic physical doctrine according to which "the forms of the more complex natural bodies are built up out of those of the simpler bodies that go to make them up" (Sharples 1987, 1202) seems to underlie Alexander's well-known thesis according to which soul is the product of the combination (μῖξις) of the elements in a living body (Bruns 1887a, 24.21–23; cf. Sharples 1987, 1202–1203), "a result of the organization" (Hamelin 1953, 32) of a compound which emerges from the combination without yet being identical with it. (Famously Alexander compares the soul with a drug possessing a "specific (or 'occult') quality": both are a power that cannot be reduced to the powers of the ingredients.)[67]

See also Thillet (2003) 23.17, Fazzo and Zonta (1999) 162: "matter, owing to its innate specific fatigue, does not have the capacity to be entirely informed by the agent imprinting its form unto it."

[64] Sharples (1994a).

[65] "[Matter] strives toward that excellent nature, namely, the [divine] force issuing from the celestial body, by which it is informed, so that it partakes of *eternity and duration*," Alexander writes. This passage is not in the manuscripts of *De providentia*, but both Thillet and Zonta take it to belong to it: Thillet (2003) 133–134 (App. 1), Zonta in Fazzo and Zonta (1999) 164 n.

[66] Recall that "nature" and "divine power" are synonymous. See p. 252 f. above, and, e.g., Thillet 2003, 19.6; Fazzo and Zonta 1999, 150; *Mantissa* 12 (Bruns 1882, 172.19). See also Moraux (2001) 278 n. 69.

[67] See, e.g., Moraux (1942) 32–33, (1978) 296–299, and (2001) 356–359 (esp. n. 172); Thillet (1981). In his discussion of soul in *De anima*, Alexander does not mention the divine power, as noted by Moraux (1967) 169. Moraux's expectation that it be mentioned there derives from his view that the divine power informs matter on the level of compounds, not that of the elements (n. 58 above); see also Sharples (1992) 96 n. 312.

On Alexander's view, the causal chain ending with a soul emerging in a clod of matter has its beginning in the Unmoved Mover, which can thus be said to be the ultimate source of forms and of providence. Bearing in mind that Alexander identifies the Unmoved Mover with the active intellect, this amounts to saying, as some scholars have noted, that for Alexander the active intellect is the Giver of Forms.[68] Alexander thus created a synthesis of noetics, psychology, and cosmology in offering a largely Aristotelian alternative to the Stoic theory of providence. In this scheme, which posits a chain of causalities extending from God, the supreme Intellect thinking itself, down to material reality and the generation and the persistence of informed sublunary substances, the celestial bodies are the indispensable intermediaries which bridge the gap between pure form and matter and thereby maintain the sublunary world and all beings within it. By bridging the gap between the transcendent and the material realms, Alexander re-establishes on what he considers to be Aristotelian principles the unity of the entire cosmos: the divine power issuing from the heavenly bodies re-connects the world's two realms, the superlunary and the sublunary, which had been dissociated by Aristotle. This move—Alexander's alternative to Stoic immanentism[69]—largely hinges on the construal of the heavenly bodies as mediating between the two realms of reality. It indeed stands to reason that Alexander's "divine power" was qualified as divine not only on account of its origin in the divine heavenly bodies, but also because it was the functional equivalent of the Stoics' divine πνεῦμα.[70] (Some Stoics had themselves drawn on a notion of "divine force.")[71]

---

[68] E.g. Moraux (1942) 92–93, Thillet (1981) 20–22, Sharples (1987) 1208, Genequand (2001) introduction 22. See p. 250 above.

[69] See p. 255 above. This point is made on the basis of very different considerations in Rashed (1995), notably 344–350. Similarly, Moraux, followed by Sharples, sees Alexander's position as having resulted from the absorption of Stoic ideas into the globally Aristotelian framework—an instance of a "critical dialogue" between competing theoretical systems that produces emulation through opposition; see Moraux (1942) 162–164 and Sharples (1987) 1212 n. 137, 1216. Sharples (1992, 94 n. 307) has further suggested that Alexander's cosmology, especially his doctrine of the divine power, was "part of [his] response to Atticus' charge that Aristotle destroys the unity of the universe and denies providence by adopting different principles for the heavens and the sublunary world."

[70] Moraux (1967) 160 n. 2, has suggested that Alexander reacted to the immanentist notion of δύναμις as developed in pseudo-Aristotle's *De mundo*. See also Sharples (1992) 93–94 n. 307. Genequand (2001) introduction 18–19, has pointed out the important differences between Alexander's divine power and parallel ideas in Stoicism and in *De mundo*.

[71] See Cicero, *De natura deorum* (Pease 1955, 1.35–36, 1.39–40, 2,14, 2.55, 1.253n.).

Alexander's reunification of the sub- and superlunary realms is at the core of what I propose to call the *astrologization of the Aristotelian cosmos*. By this I wish to highlight four facets of an evolution beginning with Alexander and through which Peripatetic doctrine assimilated astrological motifs:

(a) Aristotle's original scheme in which the Sun and the Moon were the only celestial bodies having physical effects on the sublunary realm was generalized into one in which this role was ascribed to *all* planets.
(b) Aristotle was concerned only to show that generation and corruption, inasmuch as they are caused by the Sun's annual motion along the ecliptic, are cyclical, regular processes which, therefore, are not incompatible with the doctrine of the eternity of the world. Alexander by contrast argued that the heavenly bodies not only cause generation and corruption, but through their motions also *inform* and eventually ensoul sublunary matter.
(c) Whereas Aristotle merely hints at the role of the heavenly bodies in exerting σωτηρία (see p. 244 above), Alexander explicitly upheld their role in exerting providence on the sublunary world's species.
(d) Last, but certainly not least, Alexander embraced the distinctively astrological view that the affects of the planets upon sublunary substances vary with the direction of the planets' motion (retrograde or not) and with their "relationships" (*ʿiḍāfa*) to the sublunary material bodies (p. 251 above).

The "astrologization" thus consists, first, in a broadening of the *explanandum* of physical theory: it accounts not for the mere generation and corruption as a brute fact, but rather for forms and providence over species; and it consists, second, in extending its *explanans*, which maintains that *all* the planets produce effects in the sublunary world and that the nature of their effects changes with their motions and their relative positions. The use of the term "astrologization" should not be misunderstood to imply that Alexander accepted astrology: the fundamental astrological division of planets into beneficent and maleficent is of course totally absent from his work. Rather, the term "astrologization" is meant to convey that the doctrinal evolution which it denotes served to lower the

---

The notion of divine force also occurs in Cicero's *De divinatione*, where it is connected with divination; see Pines (1986a) 125 (117).

boundaries between Aristotelian physics and natural astrology. Note also that this *conceptual* characterization of the notion of "astrologization" is entirely unrelated to the historical question whether or not Alexander was acquainted with and indebted to astrology. For our purposes, the significant point is that Alexander's "astrologized" doctrine of the celestial influences on the lower world is a principled departure from Aristotle.

Alexander's position hinges entirely on the notion of the divine power issuing from the heavenly bodies. But how is that power produced, given especially that the latter are supposed to be constituted of impassible celestial matter?[72] Alexander attempts to tread a narrow path. On the one hand, he clearly associates the divine power with heat: the element getting the greatest share of the divine power is fire, which is, therefore, the hottest and subtlest as well as also the most perfect element. At first glance, this may appear to follow naturally from Alexander's assumption that the divine power is generated through the *motion* of the heavenly bodies. But Alexander seems to know better. The most noteworthy element in his account of the generation of the divine power (to which we will presently come) is what is *absent* from it: although Alexander alludes to the "contiguity" between the heavenly body and the layer of fire underneath it as a factor involved in the warming, he nowhere mentions the heavenly body's *friction* as a cause of heat, thus distancing himself from Aristotle's account, the deficiencies of which presumably did not escape him.[73] The quandary which he faced was to account for the generation of the divine power, which is associated with heat, through the motion of the celestial bodies but without invoking friction.

In the *De providentia* (Thillet 2003, 17.16; Fazzo and Zonta 1999, 144) Alexander adduces a model: just as a hot body warms neighboring bodies, so also the celestial spheres affect the adjacent sublunary matter. Alexander here implicitly posits a sort of "action at a distance" which leaves the medium unaffected, an idea we encounter also in his commentary on the *Meteorologica* (Hayduck 1899, 17.4–18.8; Moraux 2001, 279–281). A *testimonium* from Simplicius confirms that Alexander indeed upheld a theory according to which heat "overleaps" the layer of aether separating the Sun from the sublunary realm without affecting

---

[72] On Alexander's views of the fifth element, see e.g. Moraux (1963) coll. 1238–1239.

[73] Note that *De mundo* 2, 392b1 explains the heat as being due to "the rapidity of [the aether's] motion." Alexander, to be sure, does not deny that friction is a possible source of heat; see, e.g., Hayduck 1891, 147.24–25. See also n. 56 above.

it.[74] Alexander upheld a similar view with respect to certain other qualities, notably that of the lodestone.[75] The analogy between the warming by a hot body and the action of the heavenly bodies is revelatory: it clearly echoes the traditional conception that the heavenly bodies, especially the Sun, are fiery and as a result that they warm the sublunary world. But Alexander was of course aware that the heavenly bodies are not themselves hot and that Aristotle's attempts to "save" the traditional notion within his cosmology by invoking friction were not successful (see notes 56, 74). Still, on Aristotelian premises, the paradigmatic model of the transmission of qualities is an eminently hot body heating a neighboring body (Aristotle, *Meta.* 2.1, 993b24; Hayduck 1891, 147.1–19). To do justice to all these assumptions, I suggest, Alexander's reasoning was as follows. True, the heavenly bodies are not hot. By contrast, what these quality-less bodies *do* possess to the highest degree is existence, persistence, eternity, and divinity. Transmitting *these* from the supra- to the sublunary realm is precisely the role of the divine power; it is, as it were, the (more abstract or generalized) counterpart of heat, tailored to comply with the constraints of the Aristotelian five-element theory. The divine power, Alexander thus maintains, outpours or "emanates" (*ṣādirah*) from the celestial bodies and it reaches the sublunary matter owing to their contiguity (Thillet 2003, 18.7–11, 22.25; Fazzo and Zonta 1999, 146, 154); specifically, it reaches the "layer" of fire which is in direct contact with the celestial sphere. On receiving the divine power, this fiery layer communicates it downward to the other elements (Thillet 2003, 18.12–16; Fazzo

[74] Examining the question why the Sun "heats" differently at different times and why it heats more than the other heavenly bodies although, *ex hypothesi*, it does so by virtue of its motion, Simplicius writes:

> Alexander says that there are many qualities that things can acquire *via* intermediate substances, yet without the intermediate substances themselves acquiring the same [transmitted] quality; they rather pass it on to the things which do acquire it. Thus, for instance, he says, "dry bits of wood can be set afire by the Sun without the intervening substances [*scil.* the air] being set afire, and the quality [*scil.* numbness] is inflicted on men by the torpedo fish ... through the intervening [fisher's] nets, which are not themselves numb" (Heiberg 1894, 440.23–25).

I am grateful to Marwan Rashed for pointing out this passage to me, and to Alan Bowen for his help with translating it. On this passage, see also Moraux (2001) 280 n. 74.

[75] Alexander makes the point concerning the transmission of qualities without the medium being affected, and he adduces elsewhere too the example of the numbness transmitted *via* the fisher's net without the latter being itself numb; see Hayduck 1899, 18.8–28 and Moraux (2001) 279–282 (esp. n. 74 and the references). Alexander discusses the attraction exerted by the lodestone in similar words in *Quaestio* 2.23; see Sharples (1994) 122–123.

and Zonta 1999, 148).[76] Alexander affirms that fire is the sublunary body whose nature is most akin to that of the divine body: it is on account of this congruence (or affinity: *munāsaba*) of the agent and "the first patient" that the latter can "be affected" by the divine power emitted by the former (Thillet 2003, 18.13–16, Fazzo and Zonta 1999, 148), a thesis which again indicates a concern to do away with any reference to friction. Sublunary matter is all the more affected by the heavenly bodies since the latter, being eternal and divine, are not themselves affected reciprocally when they act (Thillet 2003, 17.23–18.2; Fazzo and Zonta 1999, 146).[77]

Alexander apparently struggled toward a notion of a divine power operating on the model of the transmission of heat, a view he presumably intended to replace Aristotle's "lame" idea that the Sun warms by virtue of some friction. Alexander's account is not as explicit as one might have wished, but this is apparently the farthest he could get in his attempt to ground the general postulate that the divine power pours forth from all the heavenly bodies, a postulate he made into the very cornerstone of his "astrologized" doctrine of providence. Once this postulate is accepted, the view that the heavenly bodies all affect the sublunary world receives solid theoretical grounding (contrary to Aristotle's corresponding, even weaker, affirmation).

3. *Averroes: An astrologized Aristotelian cosmos*

Averroes, we saw, followed Alexander in considering that the heavenly bodies, specifically the planets, vouchsafe God's providence in the sublunary world. Averroes' sentence "if we were to imagine the disappearance of a single movement or planet among them, then either no coming-to-be whatsoever, or that of some beings, would not be take place," quoted at the beginning of this paper (p. 243 above), is a paraphrase of Alexander's words (p. 254 above). But whereas the two commentators concurred on the nature and extension of providence, Averroes, as we shall see, went beyond Alexander in the role he ascribed to the heavenly bodies in bringing it about. In what follows, my purpose is thus to highlight where Averroes' astrologization of the Aristotelian cosmos builds upon and extends that of Alexander. It will naturally be impossible even to review the

[76] Similarly *De mundo* 6, 398b20–25.

[77] See *De mixtione* 11, 226.33 and 13, 229.6–8 on the absurdity of the (Stoic) assumption that something divine be acted upon.

developments that took place in the millennium separating the two great students of Aristotle and that culminated in Averroes' positions.

In the sentence following the one just quoted, Averroes writes: "It is clear that some existing things are specifically affected by a specific planet. This is why we find that those who have observed the stars in the past have divided existing things in accordance with them, and posited that existing thing A is of the nature of a star X and existing thing B of the nature of a star Y" (p. 243 above). What does Averroes have in mind in this statement, which would presumably have come as a surprise to Alexander?[78] He is nowhere very explicit, but some hints are available. In *De gen. an.* 2.2, 736a19–20, Aristotle remarks that semen being foam-like (ἀφρώδης), "the goddess who is supreme in matters of sexual intercourse was called after foam (viz. Aphrodite)" (Peck 1979, 165). Averroes, in his *Commentary*, renders this passage as follows:

> Having established this concerning the nature of semen [viz. that it is foam-like], it appeared to them [viz. the Ancients] that the planet Venus strongly affects [*yesh lo roshem ḥazaq*] sexual intercourse more than the other luminaries, so much so that they named it "the master of intercourse"; they also referred to it as "foam-like" (*qiṣpi, qiṣpit*).[79]

Aristotle's goddess Aphrodite has thus been replaced by the *planet* Venus. Whatever the exact historical development leading up to this, the significant fact is that Averroes has no qualms to assign a star X—in this case, Venus—to an "existing thing" A—in this case, sexual intercourse, a distinctively astrological thought. Similarly, when Averroes refers to "those who observed the stars in the past" and who "have divided the existents in accordance with them," he is likely to have in mind the association of the seven metals with the seven planets.[80]

[78] Maimonides similarly refers to "forces of the stars that are especially assigned to [individuals]" (see n. 13 above). His notion of heavenly influences on the sublunary world (through which astrology "comes in") was thus not far removed from that of Averroes (with whose writings Maimonides was unacquainted when he composed the *Guide*).

[79] Averroes, *Commentary on the Book of Animals*, Treatise 16 (= *De generatione animalium*): Hebrew translation by Jacob ben Makhir in MS Paris héb. 956, fol. 462b: 26–28. This work of Averroes is extant only in Hebrew translation. The passage is translated also in Shem-Ṭov Ibn Falaqera, *De'ot ha-pilosofim*, MS Leiden Or 4758 fol. 245va: 29–33. I am grateful to Resianne Fontaine for fruitful exchanges about this passage.

[80] Shem-Ṭov Ibn Falaqera, who, late in the 13th century, composed in Hebrew an encyclopedic account of the sciences which generally is very faithful to Averroes' spirit, explicitly invoked this association (although in another context): "Some say," he writes, that the colors of metals "are according to the planets [lit. stars] acting on them." For instance, the yellow color of the gold is related to the Sun, the white color of silver to the

Interesting is also the ambivalence that Averroes expresses in his commentary on Avicenna's medical poem *Arjūzah* (or: *Urjūzah*). Notwithstanding Avicenna's uncompromising rejection of astrology, the poem expresses an astrologically-inspired position concerning the influence of the Moon on the evolution of human illness (Michot 2006, 49*–50*). This elicits from Averroes comments on his attitude to astrology. He explains the astrological assumptions concerning beneficent and maleficent planets and concludes with the remark: "all this is at variance with what has been demonstrated in natural science, namely, that the actions of the planets are *all good*, and that the existing things [down] here all draw their existence from their motion."[81] Elsewhere, too, Averroes says that, inasmuch as astrology ascribes to planets effects now beneficent now maleficent (depending on their positions), it is invalidated by Aristotelian natural philosophy: the planets' local motion can produce either generation *or* corruption, but not both in alternation.[82] These statements are obviously in complete accord with Averroes' views of Providence described above, views to the effect that all celestial motions are providentially arranged so as to maintain order in the sublunary world. Averroes also remarks following al-Fārābī (see Druart 1978, 1979) that the "forms" which the astrologers identify in the heavens (viz. the constellations) are mere symbols, i.e., that they are purely conventional, "devoid of all truth"; and he consequently rejects the astrologers' contention that they exert influences on sublunary things having a "corresponding" form.[83] But elsewhere in the commentary on Avicenna's *Arjūzah*, Averroes is less clear. Thus, where he briefly mentions the Moon's effect on the crisis of an illness, Averroes first summarizes in some detail the standard medieval doctrine on the subject (going back to Galen),[84] and then goes on to

Moon, and so forth (*De'ot ha-pilosofim*, MS Parma 3156 [= De Rossi 164], fol. 122a). Ibn Falaqera borrowed this passage from the Ikhwān al-ṣafā' (Zonta 1996, 282). On the history of the association of the planets and the metals, which goes back to high antiquity, see Berthelot (1888) 73–85, Bouché-Leclerc (1899) 311–315, Halleux (1974) 151 ff., Crosland (1978) 79–81. Multhauf (1966, 38), in a felicitous turn of phrase, says that the differentiation of the metals owes more to astrology than to chemistry.

[81] MS Berlin 236 fol. 172a. This is a mid 13th-century Hebrew translation by Moses Ibn Tibbon.

[82] Averroes, *Middle Commentary on De generatione animalium*: see Alaoui (1995) 123.13–124:7; Kurland (1958a) 88.17–89.28 and (1958b) 101.

[83] *Long Commentary on Aristotle's Metaphysica*, in Bouyges (1973) 1688–1689; translation in Genequand (1984) 189. I am grateful to Ruth Glasner for having called my attention to this passage.

[84] For a concise summary, see Jacquart (1995) 204–206.

remark that "this is known through astrology [lit. the art of the stars]," which "classifies [the planets] into beneficent and maleficent" (MS Berlin 236 fol. 193b). He concludes his exposition by affirming his reserve with respect to this doctrine, which he does not, however, reject totally: this art is "feeble and *most of it* is false" (MS Berlin 236 fol. 194a). Elsewhere too he is circumspect, yet not totally dismissive:

> *If* what has been maintained by the observers of the effects of the planets [lit. stars] in ancient times is true, namely that some of them impart the heat and the dryness, some the heat and the moistness, some the cold and the moistness, and some the cold and the dryness, *then* the four qualities common to the celestial bodies and to the four elements are predicated of them equivocally or according to priority and posteriority.[85]

Thus, although (like all thinkers in the *falsafah* tradition) Averroes regarded judicial astrology as untrustworthy,[86] he nonetheless clearly echoes themes derived from the astrological tradition and his account of the sublunary world is astrologized much beyond that of Alexander.

The celestial influences on the sublunary world have a central role also in Averroes' view on the origin of forms in matter. We are here interested in his later position, the one he held after having relinquished the traditional idea of the active intellect as the Giver of Forms,[87] ascribing instead the coming-to-be of forms in matter to the formative powers of the celestial bodies. His point of departure is the Alexandrian view that whatever existence, persistence, and perfection is found in the sublunary world perforce goes back to the heavenly bodies:

> Existing things in motion are of two kinds: those whose motion is eternal and whose particular instances are neither generated nor corrupted; these are the most noble existing things among the things in motion and they exist by necessity. These are the heavenly bodies. The second kind of thing in motion is the one [of which particular instances] now exist and now are non-existent. These are the ones whose natures are susceptible to being more or less perfect. By contrast, it is in the nature of the first kind that [its instances] be of the utmost perfection. For this reason, this [first] kind of existing thing is *the cause of the existence of the second kind* … Existing things in motion of the first kind extend help to those of the second kind, exert providence over them, and educe them from non-

[85] *De substantia orbis*, in Hyman (1986) lines 166–169; translation on p. 95, modified; see also Glasner (2000) 319. Note that the phrase "observers of the *effects* of the stars" refers specifically to the astrologers, not to astronomers.

[86] See also Averroes, *Tahāfut al-Tahāfut* in Bouyges (1930) 510f., Van den Bergh (1954) 312.

[87] This change of mind has been identified in Davidson (1995).

> existence to existence, and bring them forth from a lifeless existence to a living existence and from a living existence to an ensouled existence, until they acquire the most perfect existence their nature is capable of receiving. And just as [the heavenly bodies] exert providence in that they move a deficient existence into a perfect existence, so also they exert providence by endowing [the sublunary things] with eternal and permanent persistence, as far as their nature allows.[88]

But Averroes ascribes not only the "existence" of the sublunary substances to the heavenly bodies, but also, more specifically, soul and life. Thus, he writes that "some of the compounded substances are ensouled owing to the heavenly bodies. This is why Aristotle says that a human being is produced by a human being and the Sun. The reason for his holding this view is that a human being is produced by a human being like himself and that because those [celestial] bodies are alive they can endow with life what is [down] here [in the sublunary world]. For only a body whose nature is to be ensouled can move matter to the animate [i.e., soul-] perfection."[89] Needless to say, this position is a far cry from those of either Aristotle or Alexander. Indeed, as I have argued elsewhere, in Averroes' later view, the celestial bodies assume a heavy load: they are the precise functional equivalent of the active intellect in Averroes' early view and are taken to produce whatever form, order, stability, and permanence there are in the world of generation and corruption.[90]

This brings us to our last question: How does Averroes account for the informing and vivifying powers of the planets? He seems to answer

---

[88] Averroes, *Commentary on Book of Animals*, Treatise 16 (= *De gen. an.*), Hebrew translation by Jacob ben Makhir in MS BNF héb. 956, fol. 225b: 1–11 (translation quoted from Freudenthal 2002, 120), roughly corresponding to Aristotle, *De gen. an.* 2.1, 731b24–33.

[89] Averroes, *Epitome of Aristotle's Metaphysica*: Quirós Rodríguez and Puig Montada [1919] (1998) 161 (§ 65), Van den Bergh (1924) 137.30 (translation quoted from Freudenthal 2002, 120–121). The passage (a late addition) is summarized in Davidson (1992) 241. See also the two redactions of another passage in the *Epitome of the Meta.*, translated in Davidson (1992) 239. In the *Long Commentary on the Metaphysica*, Averroes similarly writes that inasmuch as celestial bodies "are the principle [*mabdaʾ*] of animated and non animated things" they themselves necessarily are "animated" [*mutanāfisah*], and their principles "must be body and soul" (Bouyges 1973, 1534.8–10). In the same work, he similarly states that "it is the Sun and the other stars which are [the] principle of life for every being in Nature" (Bouyges, 1973, 1502.1–3). Averroes explains that if Aristotle invoked only the Sun in this context, this is simply owing to the fact that "it is the star whose action is most manifest in that respect" (viz. that of endowing life; see Bouyges (1973) 1502.7–8, translation quoted from Freudenthal (2002) 121).

[90] See Freudenthal (2002) for further details.

this question with reference to the "heats" (a term by which he means qualitatively different kinds of heat) of the planets: systematically using the term in the plural form, he holds that every planet has a different sort of heat. For instance:

> As for the "heats" generated by the "heats" of the stars, which produce each distinct species of animals—[each one of these "heats"] is potentially this or that species of animal. The power [*taqdīr*] present in each of the "heats" depends on the amount of the motions of the stars [i.e., their velocities] and their relative positions with respect to their closeness or distance. This power originates from the divine and intelligent work [*mihna*], which is analogous to the single form of the single primary art, to which other arts are subordinated.[91]

Whence derive these differentiated formative powers of the "heats"? In his *Commentary on Aristotle's De gen. an.*, Averroes, rephrasing 2.3, 736b30ff. (see n. 5 above), remarks that the heat in semen "is related" or "analogous" (*meyuḥas 'el*) to that of the Sun: both "give rise to and ensoul (*ma'amid … u-meḥayyeh*) bodies" (rather than destroy them, as ordinary fire does); indeed, without the concourse of the Sun and the spheres, the power in the semen's heat lacks the power to ensoul.[92] Specifically, in so-called spontaneous generation, "the power, which is like an instrument [or a substrate] carrying the formative power, is the heat engendered by the Sun and the planets, [when it cleaves to] the body suitable for this action, namely the aeriform body which resembles the body of semen that is engendered in sexually-reproducing animals."[93] This power, Averroes states, is the one Galen had called formative (*meṣayyer*).[94] The gist of the matter thus seems to be this: the semen of a male of each species, and even of each individual within a species, carries a distinct kind of heat, as the variety of the engendered individuals testifies; so also the heavenly bodies emit each a distinct kind of heat. (Averroes no doubt regarded the variety of living beings supposedly produced through "spontaneous generation" as confirming this view.)

---

[91] Bouyges (1973) 1501.16–1503.1, translation from Freudenthal (2002) 124.

[92] Oxford, Bodleian Library Opp. 1641 Qu (= Neubauer 1381 = Institute for Microfilmed Hebrew Manuscripts, Hebrew University, Jerusalem [= IMHM] # 22405), fol. 229a: 22–23. Similarly in Ibn Falaqera, *De'ot ha-pilosofim*, Leiden, MS Or 4758, fol. 246aa: 23.

[93] MS Oxford 1641 fol. 241b: 4–6, and fol. 255ab: 25–30, respectively. The additions in square brackets are words found in Ibn Falaqera's version, but not in Jacob ben Makhir's.

[94] E.g., MS Oxford 1641 fols. 229a: 29, 241b: 11 and fols. 246ab: 4, 255aa: 7–8, respectively. See, e.g., Galen, *De nat. fac.*, 1.5–6.

Averroes thus drew on the notion of a plurality of "heats" advisedly, considering it to be theoretically well-grounded. It seems, moreover, to have been entrenched in contemporary natural philosophy, for similar ideas can be found in al-Fārābī.[95] This topic calls for further research.

But how can these informative "heats" be generated by the quality-less heavenly bodies? So far as I see, Averroes comes closest to an answer in the *Epitome of De caelo*, where he explicitly raises the embarrassing question how the Sun and the other planets warm, although their substance is that of the "fifth body" (*Al-samāʿ wa-'l-ʿālam* 48.1). In answer, he writes: "The specific thing about the planets' [lit. stars'] and the Sun's heating is their light" (*Al-samā' wa-'i-'ālam* 49.3–4). How does this light connect to the heating? The question has more to it than meets the modern eye: contrary to post-17th-century physics, in the Aristotelian natural philosophy there is no causal connection whatsoever between heat ("fire") and light ("the activity of what is transparent *qua* transparent": *De an.* 2.7, 418b10; see also 419a9). Consequently, "light does not appertain to fire by virtue of its being fire."[96] To connect light and heat, Averroes argues:

> Light *qua* light warms the bodies down here by virtue of a divine power (*quwwa illāhiyya*) when it is reflected. This holds *a fortiori* when the rays (*al-ḫuṭūṭ al-shuʿāʿiyya*) fall on the warmed body in right angles, for then the reflection is greater. In fact, when the reflection is stronger, the heating is stronger, as we can see in the cases of the burning mirrors and the glass flasks which burn wool. *A fortiori* [this holds] when the reflecting body is polished.[97] (*Al-samāʿ wa-'l-ʿālam* 49.11–12)

Averroes' account entirely hinges on the notion of "divine power," which he in all likelihood borrowed from Alexander. However, the theoretical meaning he attributes to it is not the same as for Alexander. Whereas Alexander held that the divine power endowed matter with the capacity to be informed and ensouled, Averroes draws on it to connect light with heat. Does this mean that Averroes introduced the divine power merely as a sort of *ad hoc* theoretical gap-filler in order to account for the palpable fact that light, especially reflected light, warms? Is this an instance of *obscurum per obscurius* (which the qualification "divine" is

[95] Al-Fārābī invokes the "heats" and "lights" of the planets as informing causes; see Druart (1978, 1979 and 1981).

[96] In the *De substantia orbis*, Averroes writes that "the commentators ... were of the opinion that the capacity to produce heat through the reflection [of light] is not one of the accidents that are specific to fire; rather, it is one of the accidents common to the celestial bodies and to fire" (Hyman 1986, lines 156–158, translation 94 modified).

[97] Averroes, *Al-samāʿ wa-'l-ʿālam* 49.4–10; translation from Freudenthal (2002) 130.

intended to cover up)? This question must be left open at this stage.[98] What can be affirmed with confidence in any event is that in both Alexander's and Averroes' systems, the notion of "divine power" provides the theoretical foundation for the claim that the celestial bodies can influence natural processes in the sublunary realm: in both schemes, this concept bridges the gap created by Aristotle in separating the supra- from the sublunary world and it thus underlies whatever astrologization these schemes propound.[99]

4. *Conclusion. The astrologization of the cosmos: Aristotle, Alexander, and Averroes*

In this study, I have looked at three central stages in the evolution that I have dubbed "the astrologization of the Aristotelian cosmos"—Aristotle, Alexander, and Averroes, with some additional light coming from Maimonides. My purpose was not to describe the evolution stretching over a millennium and a half, but only to identify it. Aristotle posited a strict demarcation of the world into a sublunary and a superlunary realm. He recognized that the motion of the heavenly sphere, notably that of the Sun, affects the sublunary world, notably in creating the conditions for generation and corruption. This influence of the upper on the lower world could not be grounded in his theory of matter and Aristotle accounted for it *ad hoc* as due to the warming effect of motion. Alexander introduced the innovative idea that divine providence over the species of the sublunary world is exerted through the planets, thereby making a decisive step toward the astrologization of the Aristotelian cosmos. Alexander yet curtailed the heavenly bodies' causal role: while he

[98] It should be noticed that in later works, precisely when he replaced the account of forms in terms of the active intellect by the account in terms of the "heats" of the planets, Averroes does not seem to invoke again the notion of "divine power." The evolution of his views on this issue needs to be examined in the context of the evolution of his thought in general, and on the issue of the active intellect and the origin of sublunary forms in particular. Averroes' account(s) of the production of heat through light should also be compared with that offered by Simplicius in his Commentary on *De caelo* 2.7 (438:28–444:15 Heiberg). I am grateful to Marwan Rashed for this reference.

[99] The notion of divine force appears in many contexts in later Arabic philosophical literature, e.g., in natural philosophy Kraemer (1986, 176) thinks that this goes back to Alexander, and in accounts of dreams and divination (this may go back to the Stoics; see Pines 1986a).

departed from Aristotle in assigning a causal role to all planets and especially in invoking the latter's "relationships" to one another as a relevant factor, he did not take the planets to act specifically and directly on the sublunary substances ("existing thing A is of the nature of a star X and existing thing B of the nature of a star Y"). Alexander integrated the celestial influences into his system *via* the notion of "divine force," which he also took to produce the sublunary forms (including souls and life). Later authors continued to ascribe to the planets a role in the generation of sublunary ensouled substances (namely, by mixing the elements and thus "preparing" the mixture to receive forms), but the role of *informing* sublunary substances was ascribed to a separate entity (notably Avicenna's metaphysical notion of the Giver of Forms). When (in his later period) Averroes relinquished the active intellect and transferred to the planets also its function as a cause of existence, the latter were construed as the *givers of forms*, thus bringing the astrologization within the Aristotelian tradition to an its final stage.

Two further doctrines not discussed so far to which contemporary writers referred in this context should be now briefly mentioned. According to one standard view, man's psychic faculties follow upon his physical constitution, i.e., the temperament of his body. Thus, e.g., if "the hot" is dominant, the man is irascible.[100] Added to this was the assumption that the heavenly bodies, by virtue of their motions, mix the sublunary elements and that subsequently the resulting "blend" receives an appropriate form from the active intellect so that all substances, including man, ultimately owe their existence to the celestial bodies (see Davidson 1992). The two doctrines imply that the psychic faculties of a man *in fine* depend, at least partly, on the celestial bodies.[101] This conclusion grounded (and was in turn confirmed) by climatology, that is, a theory going back to Greek antiquity which derived the physical and psychic characteristics of peoples from the climate at their places of residence.[102] All these generally accepted ideas made it easy for astrologers to argue that man's character and, hence, his conduct, ultimately depend

---

[100] This theory, already present in Aristotle, was articulated notably in Galen's *Quod animi mores corporis temeramenta sequantur*. This treatise was widely diffused in its Arabic translation; see Biesterfeldt (1973). Alexander echoes this Galenic view in *De fato* 6, 170.20 ff.; see Thillet (1984) cii–ciii, Sharples (1983) 130. Maimonides also shared this view; see Freudenthal (2003).

[101] See Freudenthal (2004) and (2005) on al-Fārābī and Maimonides.

[102] See, e.g., Glacken (1967), Altmann [1944] (2005).

on the planets and can be predicted, at least with a great probability (see Freudenthal 2005). Maimonides certainly did not err in considering that astrology threatened to "come into" contemporary cosmology.

Just as astrology entered philosophy, so, too, natural philosophy penetrated astrology. As noted earlier, Ptolemy had already presented astrology in his *Tetrabiblos* as grounded in premises derived from natural philosophy.[103] This trend was subsequently pursued and amplified, especially in Arabic philosophy in which there was a notable rapprochement between philosophy and astrology in Neoplatonically-inspired quarters (especially in the works of al-Kindī, his student Abū Maʿashar al-Balkhī, and the Brethren of Purity).[104] These developments left their mark also on thinkers who dismissed astrology itself. Aristotle's bi-partition of the universe continued to be upheld on the level of principles, but it was in part undone on the level of natural philosophy.

The totality of medieval natural philosophy in the Aristotelian tradition posited the existence of celestial influences on the sublunary world.[105] As Maimonides put it: "there is a consensus of all philosophers to the effect that the governance of this lower world is perfected by means of the forces emanating to it from the [heavenly] sphere" (p. 244 above). This made dismissing astrology more tricky than it may seem: those philosophers who (grudgingly) conceded to compose a refutation of astrology[106] could not easily conduct their argumentation on the basis of an outright denial of any influences of the heavenly bodies on the sublunary world. Inasmuch as they did not question the very idea that the natural processes down here depend on the heavenly bodies, it was not possible to ground a rejection of astrology in physical theory.[107] Therefore, authors of refutations of astrology usually chose to argue on other

[103] See, e.g., Long (1982) 178–179, Riley (1987).

[104] On the introduction of astrology into the Arabic culture, see Ullmann (1972) esp. 271–277, Sezgin (1979) esp. 3–29, Saliba (1999), Fazzo and Wiesner (1993), Adamson (2002).

[105] For comprehensive studies, see North (1987), Grant (1994) 569–617, Freudenthal (2009).

[106] Philosophers usually viewed such a task as below their rank and mostly wrote refutations only as a response to a request from a third party; for them it was not a scientific issue.

[107] Maimonides however seems to have construed nature as being intrinsically indeterminist (Funkenstein 1977). He does not explicitly connect this to his rejection of astrology, but the two may have connected in his mind. See Freudenthal (2004) and (2005). The indeterminacy of nature is upheld on different grounds by Avicenna (Rashed 2000) as well.

fronts: they maintained that the precise effects of the heavenly influences are unknowable (too many factors are involved) and that astrological determinism can be undone by human free will.[108]

Acknowledgements

For helpful remarks on an early version of this paper I am grateful to Alan C. Bowen, Herbert A. Davidson, Emma Gannagé, Ruth Glasner, and Marwan Rashed. This is a substantially revised version of my paper "The Medieval Astrologization of The Aristotelian Cosmos: From Alexander of Aphrodisias to Averroes," *Mélanges de l'Université Saint-Joseph* 59 (2006) 29–68.

Bibliography

Adamson, P. (2002) "Abū Maʿashar, al-Kindī and the Philosophical Defense of Astrology," *Recherches de théologie et de philosophie médiévales* 69: 245–270.
Alaoui, J.E. (1995) *Talkhīṣ "Al-kaun wa-'l-fasād."* (Beirut).
Alexander of Aphrodisias, *De anima*. See Bruns (1887a), Fotinis (1980).
——, *De fato*. See Bruns (1882), Sharples (1983a), Thillet (1984).
——, *De mixitione*. See Todd (1976).
——, *De providentia*. See Thillet (1979) and (2003), Fazzo and Zonta (1999).
——, *In Aristotelis metaphysica commentaria*. See Hayduck (1891), Dooley (1989) and (1993), Dooley and Madigan (1992), Madigan (1993).
——, *In Aristotelis meteorologicorum libros commentaria*. See Hayduck (1899), Lewis (1996).
——, *Mabādī*. See Genequand (2001).
——, *Mantissa*. See Bruns (1887b), Sharples (1983b).
——, *Quaestiones*. See Bruns (1882), Sharples (1992) and (1994).
Altmann, A. (2005) "Judah Hallevi's Theory of Climates" [1944], *Aleph: Historical Studies in Science and Judaism* 5: 215–246.
Aristotle, *De generatione animalium*. See Peck 1979.
Averroes, *De substantia orbis*. See Hyman (1986).
——, *Epitome of Aristotle's De caelo*. See Rasāʾil Ibn Rushd (1947).
——, *Epitome of Aristotle's De generatione et corruptione*. See Rasāʾil Ibn Rushd (1947), Puig Montada (1992), Kurland (1958a) and (1958b).
——, *Epitome of Aristotle's Metaphysica*. See Quirós Rodríguez and Puig Montada [1919] (1998), Horten (1912), Van den Bergh (1924).

[108] See e.g. Druart (1978, 1979 and 1981) on al-Fārābī; Mehren (1884), Michot (2006) on Avicenna; Langermann (1991), Freudenthal (1993) on Maimonides.

——, *Middle Commentary on Aristotle's De generatione et corruptione*. See Alaoui (1995).
——, *Long Commentary on Aristotle's Metaphysica*. See Bouyges (1973), Genequand (1984).
——, *Tahāfut al-Tahāfut*. See Bouyges (1930), Van den Bergh (1954).
Avicenna, *Réfutation de l'astrologie*. See Michot (2006).
Barton, T.S. (1994) *Ancient Astrology* (London/New York).
Berthelot, M. (1888) *Collection des anciens alchimistes grecs. Introduction* (Paris) [= *Introduction à l'étude de la chimie des anciens et du moyen âge* (Paris, 1889)].
Biesterfeldt, H.H. (1973) "Galens Traktat 'Dass die Kräfte der Seele den Mischungen des Körpers folgen' in arabischer Übersetzung," *Abhandlungen für die Kunde des Morgenlandes*, XL (4).
Boll, F. (1922–1923) "Der Sternglaube in seiner historischen Entwicklung," in V. Stegemann (ed.), *Kleine Schriften zur Sternkunde des Altertums* (Leipzig 1950) 369–396.
Boll F.; Bezold, C.; and Gundel, W. [1917] (1931) *Sternglaube und Sterndeutung. Die Geschichte und das Wesen der Astrologie*, Heidelberg 4th edn (Leipzig).
Bouché-Leclerc, A. (1899) *L'astrologie grecque* (Paris).
Bouyges, M. (ed.) (1930) *Tahāfut al-Tahāfut* (Beirut).
——, (ed.) (1973) *Tafsīr Ma ba'd aṭabī'at* (= *Long Commentary on the Metaphysica*), 2nd edn (Beirut).
Bruns, I. (1882) *Quaestiones*, in Supplementum Aristotelicum 2.2 (Berlin).
——, (1887a) *De anima*, in Supplementum Aristotelicum 2.1 (Berlin).
——, (1887b) *De anima libri mantissa*, in Supplementum Aristotelicum 2.1 (Berlin).
Cicero, *De natura deorum*. See Pease (1955).
Crosland, M.P. (1978) *Historical Studies in the Language of Chemistry* (New York).
Cumont, F. (1909) *Les religions orientales dans le paganisme romain*, 2nd edn (Paris).
——, (1912) *Astrology and Religion among the Greek and Romans* (New York/London).
Davidson, H.A. (1972) "Alfarabi and Avicenna on the Active Intellect," *Viator* 3: 109–178.
——, (1992) *Alfarabi, Avicenna, and Averroes on Intellect: Their Cosmologies, Theories of the Active Intellect, and Theories of the Human Intellect* (New York/Oxford).
Dooley, W.E. (1989) *Alexander of Aphrodisias, On Aristotle, Metaphysics 1* (London).
——, (1993) *Alexander of Aphrodisias, On Aristotle, Metaphysics 5* (London).
Dooley, W.E. and Madigan, A. (1992) *Alexander of Aphrodisias, On Aristotle, Metaphysics 2 & 3* (London).
Druart, T.-A. (1978) "Astronomie et astrologie selon Farabi," *Bulletin de philosophie médiévale* 20: 43–47.
——, (1979) "Le second traité de Farabi sur la validité des affirmations basées sur la position des étoiles," *Bulletin de philosophie médiévale* 21: 47–51.

——, (1981) "Al Fārābī's Causation of the Heavenly Bodies," in Parviz Morewedge (ed.), *Islamic Philosophy and Mysticism* (Delmar, NY) 35–45.
Duhem, P. (1965) *Le système du monde*, vol. 2 (Paris).
Fazzo, S. (1988) "Alessandro d'Afrodisia e Tolomeo: aristotelismo e astrologia fra il II e il III secolo d.c.," *Rivista di storia della filosofia*, n.s. 4: 627–650.
——, (2003) "Alexandros d'Aphrodisias," in Richard Goulet (ed.), *Dictionnaire des philosophes antiques. Supplement* (Paris) 61–70.
Fazzo S. and Wiesner, H. (1993) "Alexander of Aphrodisias in the Kindī Circle and in al-Kindī's Cosmology," *Arabic Sciences and Philosophy* 3: 119–153.
Fazzo, S. and Zonta, M. (1999) *Alessandro di Afrodisia, La provvidenza. Questioni sulla provvidenza* (Milano).
Festugière, A.-J. [1949] (1983) *La révélation d'Hermès Trismégiste: vol. 2. Le dieu cosmique* (Paris).
——, [1950] (1986) *La révélation d'Hermès Trismégiste: vol. 1. L'astrologie et les sciences occultes* (Paris).
Forster, E.S. (1950) *De mundo* in W.D. Ross, *The Works of Aristotle Translated into English*, vol. 3 (Oxford).
Fotinis A.P. (1980) *The "De anima" of Alexander of Aphrodisias: A Translation and Commentary* (Washington, DC).
Freudenthal, G. (1986) "The Theory of the Opposites and an Ordered Universe: Physics and Metaphysics in Anaximander," *Phronesis* 31: 197–228.
——, (1993) "Maimonides' Stance on Astrology in Context: Cosmology, Physics, Medicine, and Providence" in F. Rosner and S.S. Kottek (edd.), *Moses Maimonides: Physician, Scientist, and Philosopher* (Northvale, NJ/London) 77–90.
——, (1995) *Aristotle's Theory of Material Substance: Form and Soul, Heat and Pneuma* (Oxford).
——, (2000) "Providence, Astrology, and Celestial Influences on the Sublunary World in Shem-Ṭov Ibn Falaquera's *De'ot ha-Filosofim*," in S. Harvey (ed.), *Medieval Hebrew Encyclopedias of Science and Philosophy* (Amsterdam) 335–370.
——, (2002) "The Medieval Astrologization of Aristotle's Biology: Averroes on the Role of the Celestial Bodies in the Generation of Animate Bodies," *Arabic Sciences and Philosophy* 12: 111–137.
——, (2003) "Four Implicit Quotations of Philosophical Sources in Maimonides' *Guide of the Perplexed*," *Zutot* 2: 114–125.
——, (2004) "La détermination partielle, biologique et climatologique, de la félicité humaine: Maïmonide *versus* al-Fārābī à propos des influences célestes," in T. Lévy and R. Rashed (edd.), *Maïmonide: philosophe et savant* (Louvain) 79–129.
——, (2005) "The Biological Limitations of Man's Intellectual Perfection According to Maimonides," in G. Tamer (ed.), *The Trias of Maimonides/Die Trias des Maimonides* (Berlin) 137–149.
——, (2009) "Cosmology: The Heavenly Bodies," in S. Nadler and T. Rudavsky (edd.) *The Cambridge History of Jewish Philosophy: From Antiquity through the Seventeenth Century* (Cambridge) 302–361.

Funkenstein, A. (1977) "Maimonides: Political Theory and Realistic Messianism," *Miscellanea Mediaevalia* 11: 81–103. (Reprinted in his *Perceptions of Jewish History* (Berkeley) 131–155.)

Furth, M. (1978) "Transtemporal Stability in Aristotelian Substances," *Journal of Philosophy* 75: 624–646.

——, (1988) *Substance, Form and Psyche: An Aristotelean Metaphysics* (Cambridge).

Genequand, C. (1984) *Ibn Rushd's Metaphysics: A Translation with Introduction of Ibn Rushd's Commentary on Aristotle's* Metaphysics, *Book Lām* (Leiden).

——, (2001) *Alexander of Aphrodisias: On the Cosmos* (Leiden).

Gill, M.L. (1989) *Aristotle on Substance: The Paradox of Unity* (Princeton).

Glacken, C.J. (1967) *Traces on the Rhodian Shore: Nature and Culture in Western Thought From Ancient Times to the End of the Eighteenth Century* (Berkley).

Glasner, R. (2000) "The Question of Celestial Matter in the Hebrew Encyclopedias" in S. Harvey (ed.) *Medieval Hebrew Encyclopedias of Science and Philosophy* (Amsterdam) 313–334.

Goulet, R. and Aouad, M. (1989) "Alexandros d'Aphrodisias," in Richard Goulet (ed.), *Dictionnaire des philosophes antiques*, vol. 1 (Paris) 125–139.

Grant, E. (1994) *Planets, Stars, & Orbs: The Medieval Cosmos, 1200–1687* (Cambridge).

Gressmann, H. (1925) *Die hellenistische Gestirnreligion* (Leipzig) (= *Beihefte zum Alten Orient*, Heft 5).

Gundel, W. and Gundel, H.G. (1966) *Astrologumena. Die astrologische Literatur in der Antike und ihre Geschichte* (Wiesbaden) (= *Sudhoffs Archiv*. Beiheft 6).

Guthrie, W.K.C. (1971) *Aristotle: On the Heavens* (London/Cambridge, MA).

Halleux, R. (1974) *Le problème des métaux dans la science antique* (Paris).

Hamelin, O. (1953) *La théorie de l'intellect d'après Aristote et ses commentateurs* (Paris).

Happ, H. (1971) *Hyle. Studien zum aristotelischen Materie-Begriff* (Berlin/New York).

Hayduck, M. (1891) *Alexandri Aphrodisiensis in Aristotelis Metaphysica commentaria* (Berlin).

——, (1899) *Alexandri Aphrodisiensis in Aristotelis Meteorologicorum libros commentaria* (Berlin).

Horten, M. (1912) *Die Metaphysik des Averroes*, nach dem Arabischen übersetzt und erläutert, *Abhandlungen zur Philosophie und ihrer Geschichte* 36 (Halle).

Hyman, A. (1986) *Averroes' De substantia orbis: Critical Edition of the Hebrew Text with English Translation and Commentary* (Cambridge, MA/Jerusalem).

Jacquart, D. (1995) "La scolastique médicale," in M.D. Grmek (ed.), *Histoire de la pensée médicale en Occident: vol. 1. Antiquité et moyen âge* (Paris) 175–210.

Kraemer, J.L. (1986) *Philosophy in the Renaissance of Islam* (Leiden).

Kurland, S. (1958a) *Averroes on Aristotle's "De generatione et corruptione": Middle Commentary and Epitome*, Hebrew text (Cambridge, MA).

——, (1958b) *Averroes on Aristotle's "De generatione et corruptione": Middle Commentary and Epitome* (Cambridge, MA).

Langermann, Y.T. (1991) "Maimonides' Repudiation of Astrology," *Maimonidean Studies* 2: 123–158.

Lawrence, M. (2006) "Hellenistic Astrology" in *The Internet Encyclopedia of Philosophy* (http://www.iep.utm. edu/a/astr-hel.htm). Seen January 20, 2008.
Lewis, E. (1996) *Alexander of Aphrodisias: On Aristotle Meteorology 4* (London).
Long, A.A. (1982) "Astrology: Arguments Pro and Contra" in J. Barnes, J. Brunschwig, M. Burnyeat, and M. Schofield (edd.), *Science and Speculation* (Cambridge) 165–192.
Longrigg, J. (1975) "Elementary Physics in the Lyceum and Stoa," *Isis* 66: 211–229.
Madigan, A. (1989) *Alexander of Aphrodisias: On Aristotle, Metaphysics 4* (London).
Maimonides, *The Guide of the Perplexed.* See Pines (1963) and Qafaḥ (1972).
Mancha, J.L. and Freudenthal, G. (2005) "Levi ben Gershom's Criticism of Ptolemy's Astronomy. Critical Editions of The Hebrew And Latin Versions and an Annotated English Translation of Chapter Forty-Three of the *Astronomy* (*Wars Of The Lord*, V.1.43)," *Aleph: Historical Studies in Science and Judaism* 5: 35–167.
Mehren, A.F.M. (1884) "Vues d'Avicenne sur l'astrologie," *Le Muséon* 3: 383–403.
Michot, Y. (2006) *Avicenna, Réfutation de l'astrologie. Édition et traduction du texte arabe, introduction, notes et lexique* (Beirut).
Moraux, P. (1942) *Alexandre d'Aphrodise, exégète de la noétique d'Aristote.* Bibliothèque de la Faculté de philosophie et lettres de l'Université de Liège 99 (Liège/Paris).
——, (1963) "Quinta essentia," in Pauly-Wissowa *Realencyclopädie*, XXIV(1) (Stuttgart) 1171–263.
——, (1967) "Alexander von Aphrodisias *Quaest.* 2.3," *Hermes* 95: 159–169.
——, (1970) *D'Aristote à Bessarion* (Laval).
——, (1978) "Le *De anima* dans la tradition grecque. Quelques aspects de l'interprétation du traité, de Théophraste à Thémistius" in G.E.R. Lloyd and G.E.L. Owen (edd.), *Aristotle on Mind and the Senses*, (Cambridge) 281–324.
——, (2001) *Der Aristotelismus bei den Griechen. Von Andronikos bis Alexander von Aphrodisias.* Dritter Band: *Alexander von Aphrodisias* (Berlin).
MS Berlin 236 (Foll. 1588) fols. 161a–221a. Staatsbibliothek, Berlin. Copy in Jerusalem, Institute for Microfilmed Hebrew Manuscripts, film no. 1816.
MS BNF héb. 956. Paris.
MS Leiden Or 4758. Bibliotheek der Rijksuniversiteit, Leiden.
MS Parma 3156. Biblioteca Palatina, Parma
Multhauf, R.P. (1966) *The Origins of Chemistry* (London).
North, J.D. (1987) "Medieval Concepts of Celestial Influence: A Survey," in P. Curry (ed.), *Astrology, Science and Society: Historical Essays* (Woodbridge/Wolfeboro) 5–17.
Oehler, K. (1969) "Ein Mensch zeugt einen Menschen. Über den Missbrauch der Sprachanalyse in der Aristotelesforschung," in K. Oehler, *Antike Philosophie und byzantinisches Mittelalter* (München) 95–145.
Pease, A.S. (1955) *M. Tulli Ciceronis de natura deorum*, 2 vols. (Cambridge, MA).
Peck, A.L. (1979) *Aristotle: Generation of Animals* (Cambridge, MA/London).
Pines, S. (1963) *Maimonides, Moses: The Guide of the Perplexed* (Chicago).

——, (1986a) "The Arabic Recension of *Parva naturalia*," in *The Collected Works of Shlomo Pines: vol. 2. Studies in Arabic Versions of Greek Texts and in Medieval Science* (Jerusalem/Leiden) 96–145.

——, (1986b) "The Spiritual Force Permeating the Cosmos According to a Passage in the Treatise on the Principles of the All ascribed to Alexander of Aphrodisias," in *The Collected Works of Shlomo Pines: vol. 2. Studies in Arabic Versions of Greek Texts and in Medieval Science* (Jeruslem/Leiden) 252–255.

Préaux, C. (1973) *La lune dans la pensée grecque* (Bruxelles).

Puig Montada, J. (1992) *Epitome del libro sobre la generación y la corrupción*, edición, traducción y commentario (Madrid).

Qafaḥ, Y. (1972) *Rabbenu Moshe ben Maimon, Moreh ha-nevukhim. Dalālah al-ḥa'īrīn*, 3 vols. (Jerusalem).

Quirós Rodríguez, C. and Puig Montada, J. [1919] (1998) *Compendio de la metafisica*, texto arabe con traduccíon y notas de Carlos Quirós Rodríguez (Madrid), reprinted with an introduction by J. Puig Montada (Cordoba).

Rasa'il Ibn Rushd (1947) *Averroes; Epitome of Aristotle's* De caelo (Hyderabad).

Rashed, M. (1995) "Alexandre d'Aphrodise et la 'magna quaestio'," *Les études classiques* 63: 295–351.

——, (2000) "Théodicée et approximation: Avicenne," *Arabic Sciences and Philosophy* 10: 223–257.

——, (2002) "La Préservation (*soteria*), objet des *Parva naturalia* et ruse de la nature," *Revue de philosophie ancienne* 20: 35–59.

Riley, M. (1987) "Theoretical and Practical Astrology: Ptolemy and His Colleagues," *Transactions of the American Philological Association* 117: 235–256.

Saliba, G. (1999) "Astronomy and Astrology in Medieval Arabic Thought," in R. Rashed and J. Biard (edd.), *Les doctrines de la science de l'antiquité à l'âge classique* (Louvain) 131–164.

Schroeder, F.M. (1981) "The Analogy of the Active Intellect to Light in the 'De anima' of Alexander of Aphrodisias," *Hermes* 109: 215–225.

Sezgin, F. (1979) *Geschichte des arabischen Schrifttums: vol. 6. Astrologie, Meteorologie und Verwandtes* (Leiden).

Sharples, R.W. (1982) "Alexander of Aphrodisias on Divine Providence: Two Problems," *Classical Quarterly* 32: 198–211.

——, (1983a) *Alexander of Aphrodisias: On Fate* (London).

——, (1983b) *De anima libri mantissa XXII–XXV*, in Sharples (1983a) 94–115.

——, (1983c) "The Unmoved Mover and the Motion of the Heavens in Alexander of Aphrodisias," *Apeiron* 17: 62–66.

——, (1987) "Alexander of Aphrodisias: Scholasticism and Innovation," in W. Haasse (ed.), *Aufstieg und Niedergang der römischen Welt*, Teil II: *Principat: vol. 36. Philosophie, Wissenschaften, Technik*, 2. Teilband: *Philosophie* (Berlin/New York) 1176–243.

——, (1990) "The School of Alexander?" in R.R.K. Sorabji (ed.), *Aristotle Transformed: The Ancient Commentators and Their Influence* (London/Ithaca, NY) 83–111.

——, (1992) *Alexander of Aphrodisias, Quaestiones 1.1–2.15* (London).

——, (1994a) "On Body, Soul and Generation in Alexander of Aphrodisias," *Apeiron* 27: 163–170.
——, (1994b) *Alexander of Aphrodisias, Quaestiones 2.16–3.15* (London).
——, (2001) "Schriften und Problemkomplexe zur Ethik" and "Bibliographie," in Moraux (2001) 513–618 and 619–650.
Solmsen, F. (1960) *Aristotle's System of the Physical World* (Ithaca, NY).
Stern, J. (1997) "The Fall and Rise of Myth in Ritual: Maimonides *versus* Nahmanides on the *Huqqim*: Astrology, and the War against Idolatry," *The Journal of Jewish Thought and Philosophy* 6: 185–263.
Tester, J. (1987) *A History of Western Astrology* (Woodbridge, Suffolk).
Thillet, P. (1979) *Alexandre d'Aphrodise. "Traité de la providence,"* Thèse de Doctorat d'État, 5 vols. (Sorbonne, Paris).
——, (1981) "Matérialisme et théorie de l'âme et de l'intellect chez Alexandre d'Aphrodise," *Revue philosophique* 106: 5–24.
——, (1984) *Alexandre d'Aphrodise. Traité du destin* (Paris).
——, (2003) *Alexandre d'Aphrodise. Traité de la providence* (Lagrasse).
Todd, R.B. (1976) *Alexander of Aphrodisias on Stoic Physics* (Leiden).
Toomer, G.J. (1985) "Galen on the Astronomers and Astrologers," *Archive for the History of Exact Science* 32: 193–206.
Tracy, T.J. (1969) *Physiological Theory and the Doctrine of the Mean in Plato and Aristotle* (The Hague/Paris).
Ullmann, M. (1972) *Die Natur- und Geheimwissenschaften im Islam* (= *Handbuch der Orientalistik*, erste Abt., Ergänzungsband VI, 2. Abschnitt) (Leiden).
Van den Bergh, S. (1924) *Die Epitome der Metaphysik des Averroes*. Übersetzt und mit einer Einleitung und Erläuterungen versehen (Leiden).
——, (1954) *Averroes' "Tahafut al-Tahafut" (The Incoherence of the Incoherence): Translated from the Arabic with Introduction and Notes* (Oxford/London).
Wiesner, H.S. (1993) *The Cosmology of al-Kindi*, Ph.D. Dissertation, Harvard University (Cambridge, MA).
Zonta, M. (1996) "Mineralogy, Botany and Zoology in Hebrew Medieval Encyclopedias," *Arabic Sciences and Philosophy* 6: 263–315.

ARISTOTLE'S *DE CAELO*: A BIBLIOGRAPHY[1]

A. *Medieval and early modern manuscripts*[2]

Family a

Parisinus graecus 1853 (E) *ca* 10th century, *De caelo* on folios 69–106 v°.
Vaticanus gr. 253 (L after Bekker 1870) 13th century.
Vaticanus Ottobonianus gr. 188, 15th century, based on *Vaticanus gr.* 253 and the exemplar used by William of Moerbeke in his Latin translation (Γ).

Family b

Vindobonensis phil. gr. 100 (J) 9th century (second half), *De caelo* on folios 56–86.
Marcianus gr. 214 (H^a in Bekker 1870) 12th century.
Marcianus gr. 200 (Q in Bekker 1870) 1457, copied from J by Jean Rhosos.
Marcianus gr. 210 (F^m in Mioni 1958, U in Longo 1961) 13th century, combines *Marcianus gr.* 214 and *Vaticanus gr.* 253.
Marcianus gr. 211 (E^b in Bekker 1870, V in Moraux 1954 and in Longo 1961) 13th century, derived from *Marcianus gr.* 210.
Marcianus gr. 212 (G^a in Bekker 1870) 15th century, derived from *Marcianus gr.* 214 with variants from *Marcianus gr.* 211.
Marcianus gr. 206 (f in Bekker 1870) 1467, copied at Rome by Charitonymus Hermonymus of Lacedemonia, based on *Marcianus gr.* 211.

[1] The editors are very grateful to Andrea Falcon for his valuable suggestions and additions to this bibliography, and to Veronica Shi for her editorial work. We also thank Gerhard Endress for his kindness in supplying us with the information for the entries on the Arabic tradition.

[2] Following Moraux (1965): cf. Moraux (1954).

Laurentianus 87,17 (j in Longo 1961) 14th century.
Vindobonensis phil. gr. 64 1457, copied under the direction of Jean Rhosos.
Laurentianus 87,11, attributed to Jean Rhosos.
Riccardianus gr. 14 (O^b in Bekker 1870) 15th–16th century.
Vaticanus gr. 1027 (H after Bekker 1870) 12th century.
  Same subgroup as *Vaticanus gr.* 1027
    Vaticanus Urbinas gr. 37 (M after Bekker 1870) 14th century.
    Parisinus Coislinianus 169 (m in Longo 1961) 14th century.
    Laurentianus 87,7 (F after Bekker 1870) 14th century.
    Vaticanus gr. 499 (k in Bekker 1870, f in Longo 1961) 13th–14th century, related to *Laurentianus* 87,7.
    Parisinus gr. 2595 (e in Longo 1961).

Unconnected subgroups

Ambrosianus gr. 403, G 61 sup. (G in Longo 1961) 13th century.
  Direct descendants
    Vindobonensis phil. gr. 2 1496, copied in the Abruzzes by Robert Maioranos.
    Vindobonensis phil. gr. 27, early 16th century.
    Parisinus suppl. gr. 642 (s in Moraux 1954, P in Longo 1961) 14th century.
    Parisinus gr. 2032 (i in Bekker 1870, p in Longo 1961) 14th century.
    Parisinus Coislinianus 166 (o in Bekker 1870, O in Moraux 1954) end of 14th century.
    Parisinus gr. 1860 (l in Bekker 1870) 15th century.
    Vaticanus gr. 249, 15th century.
    Laurentianus 81,1 (S in Bekker 1870 and in Fobes 1913, Q in Longo 1961) end of 13th century, copied by Jean Panaretes.
    Vaticanus Palatinus 161 (P in Moraux 1954) 1442, copied by Jean Scutariotas from *Laurentianus* 81, 1.
    Ambrosianus gr. 512, M 46 sup. (A in Moraux 1954, W in Longo 1961) 13th century.
    Vaticanus Urbinas gr. 38 (Q^a in Bekker 1870) 15th century, copied by Michel Apostolios and modeled on *Ambrosianus gr.* 512, M 46 sup.

Other manuscripts prior to 1600

Alexandrinus Bibl. Patriarc. 87 (26. 606. 76) 16th century, with commentary by Theodore Metochites in the margins.
Ambrosianus gr. 725, R 119 sup. 16th century.
Ambrosianus gr. 1026. G 115 inf. 16th century.
Berolinensis gr. In fol. 67^2, 13th–14th century.
Bononiensis Bibl. Univ. 2302 (*olim* 216) 15th–16th century, copied at Padua by Zacharias Kallierges the Cretan.
Cantabrigiensis Bibl. Univ. 190 (Dd. IV 16), copied at Florence in 1441.
Cantabrigiensis Bibl. Univ. Add. 1732, beginning of 14th century.
Laurentianus 87,20 (v in Bekker 1870, A in Longo 1961) 14th century.

Lugduno-Batavus Vossianus gr. Q 3, 13th century.
Lugduno-Batavus Vossianus gr. F 63, end of 15th century.
Matritensis Bibl. Nat. 4563 (*olim* N 26) 1470, copied by Constantin Lascaris.
Monacensis gr. 200, 15th–16th century.
Oxoniensis Bodl. Bywater 4[1], 1418, copied on Crete by Jean Chionopoulos.
Oxoniensis Coll. Corp. Christi 104, 15th century, copied by Jean Thessalos Scutariotas, first three books only.
Parisinus gr. 1861 (c in Bekker 1870) 15th century, copied by George Gregoropoulos.
Parisinus gr. 1866, 15th century, excerpts only.
Parisinus gr. 2033, 15th century, copied by Michel Apostolios.
Parisinus gr. 2063 (Z in Longo 1961) 14th century, stops at A 3.269b33.
Parisinus gr. 2595[2], 13th–14th century.
Scorialensis Σ III 2 (101 Revilla) 16th century.
Scorialensis T II 13 (152 Revilla) 15th century, copied by Antonio Damilas of Milan.
Scorialensis T II 21 (160 Revilla) 16th century, first three books only.
Vaticanus Reginae Suecorum gr. 123, 16th century.
Vaticanus gr. 252, 16th century.
Vaticanus gr. 254, 14th century, with Simplicius' commentary.
Vaticanus gr. 2201.
Vindobonensis phil. gr. 75, 1445, copied by Demetrios Kykandyles.

B. *Modern critical editions of the Aristotle's* De caelo

Allan, D.J. (1955) *Aristotelis de caelo libri quattuor*, 3rd edn 1961 (Oxford).
Bekker, I. (1870) *Aristotelis opera* (Berlin).
Longo, O. (1961) *Aristotele. De caelo* (Florence).
Moraux, P. (1965) *Aristote. Du ciel* (Paris).
Prantl, C. (1881) *De coelo, et de generatione et corruptione* (Leipzig).

C. *Commentaries and translations*

C1. *Ancient to early modern commentaries in Greek and Latin*[3]

4th century

Themistius, *In libros Aristotelis* De caelo *paraphrasis, hebraice et latine. Sēfer ha-shāmayim we-ha-ʿōlām le-Arisṭō ʿim pērūsh Tāmisṭiʾūs* (ed. Landauer 1902).[4]

[3] Following Edward Grant in *Planets, Stars, and Orbs: The Medieval Cosmos, 1200–1687* (Cambridge, 1994). The original dates of publication and/or modern editions are listed where available.

[4] Hebrew translation by Zeraḥyah ben Yiṣḥāq ben Sheʾaltiʾēl, made in 1284, from

6th century

Simplicius of Cilicia, *In Aristotelis de caelo commentaria* (ed. Heiberg 1894).

9th century

Pseudo-Avicenna, *Liber celi et mundi* (ed. Gutman 2003).

12th century

Gerard of Cremona, *Aristoteles, Liber de celo et mundo* (ed. I. Opelt 1971). Translation from the Arabic of ibn al-Biṭrīq (9th century) into Latin.

13th century

Albertus Magnus, *De coelo et mundo* (ed. Hossfeld 1971).
Alvaro of Toledo, *De substantia orbis* (ed. Alonso Alonso 1941).
Gunsalvi of Burgos (ed. Renaud 1973).
John of Jandun, *Quaestiones in Aristotelis de caelo.*
Michael Scot, *Averrois Cordubensis commentum magnum super libro de caelo et mundo Aristotelis* (ed. Arnzen 2003).
Petrus de Alvernia, *Quaestiones supra librum de caelo et mundo* (ed. Galle 2003).
Thomas Aquinas, *In libros Aristotelis de caelo et mundo expositio* (Commissio Leonina 1886, ed. Spiazzi 1952).
Thomas de Bungeye, *Quaestio in Aristotelis de caelo et mundo* (ed. Parker 1968).
William of Moerbeke, *Simplicii in Aristotelis de caelo* (ed. Bossier, Vande Veire, and Guldentops 2004).

14th century

Albertus de Saxonia, *Quaestiones in libros Aristotelis de caelo et mundo* (ed. Patar 2008).
Ioannis Buridanus, *Expositio et quaestiones in Aristotelis de caelo* (ed. Moody 1942, ed. Patar 1996).
Nicole Oresme, *Le livre du ciel et du monde* (ed. Menut and Denomy 1968).
Paul of Venice, *Liber celi et mundi.*

15th century

Johannes de Magistris, *De caelo et mundo.*
Thomas Bricot, *Quaestio in Aristotelis de caelo et mundo* (1486).

---

the lost Arabic version of Yaḥyā ibn ʿAdī, died 974), and the Latin translation from the Hebrew by Moses Alatino (16th century).

16th century

Quaestiones in de caelo
Conimbricenses, or Coimbra Jesuits (1592, reprinted 1598).
Galileo Galilei (ed. Wallace 1977).
Johannes Major (1526).

17th century

Quaestiones in de caelo
Pedro Hurtado de Mendoza (1615).
Bartholomew Amicus (1626).
Raphael Aversa (1627, vol. 2).
Roderigo de Arriaga (1632).
Bartholomaeus Mastrius and Bonaventura Bellutus (1640).
Franciscus de Oviedo (1649).
Thomas Compton-Carleton (1649).
Franciscus Bonae Spei (1652).
Melchior Cornaeus (1657).
Sigismundus Sebellonus (vol. 2, 1663).
George de Rhodes (1671).
Illuminatus Oddus (1672).
Johannes Poncius (1672).

C2. *Medieval translations and commentaries in Arabic and Hebrew*

9th century

Yaḥyā (Yūḥannā) ibn al-Biṭrīq, *Arisṭūṭālīs, Kitāb al-Samā' wa-l-ʿālam [Book on the Heaven and the World]* (ed. Badawī 1961). Translation from the Syriac.[5]
Ḥunayn ibn Isḥāq (?), [*Sixteen Questions on Aristotle's Book on the Heaven and the World*] (ed. Gutman 2003), Latin version.[6]

[5] A defective printing from one ms., presenting for book 1 the revised version of Mihrān (v. infra); critical edition in preparation by Gerhard Endress, *Aristoteles Semitico Latinus* (Leiden 2010). Variants against the Greek ms. tradition have been discussed, on the basis of the notes in Endress 1966, by Tarán (1974).

[6] Sixteen *quaestiones* on all parts of *De caelo*. The authorship of Ḥunayn, which can be deduced from a note in Ibn al-Nadīm (died 980), *al-Fihrist*, 250.30 (Flügel) is uncertain. The medieval tradition attributes the text to Avicenna. The Hebrew version, while made from a Latin version, has numerous revisions and additions in comparison to the extant Latin text; see Glasner (1996). While there are certain parallels with the commentary by Themistius, this is not a paraphrase of the latter; see Glasner (1996) *pace* Alonso Alonso (1951).

10th / 11th century

Abū l-Farağ ʿAbdallāh ibn al-Ṭayyib, *Arisṭūṭālis, [Kitāb al-Samā' wa-l-ʿālam].* Arabic translation of the *De caelo*.[7]

Abū Sahl al-Masīḥī, ʿĪsā ibn Yaḥyā, *Talḫīṣ kitāb al-Samā' wa-l-ʿālam li-Arisṭū [Commentary—Paraphrase of De caelo].* Ms. Leiden, Biblioteek der Rijksuniversiteit, Acad. 44, no. 4.[8]

Abū ʿAlī ibn Sīnā (Avicenna), *Kitāb al-Šifā' [The Book of the Cure]*, Part II, *al-Ṭabīʿīyyāt [Physica]*, vol. 2: *al-Samā' wa-l-ʿālam [On the Subject Matter of De caelo]*, (ed. Qāsim 1968).[9]

Themistius, *In libros Aristotelis de caelo paraphrasis, hebraice et latine. Sēfer ha-šāmayim we-ha-ʿōlām le-Arisṭō ʿim pērūš Tāmisṭi'ūs* (ed. Landauer 1902).[10]

12th century

Ibn al-Ṣalāḥ (Abū l-Futūḥ Aḥmad ibn Muḥammad ibn al-Sarī), *Qaul fi Bayān al-ḫaṭa' al-ʿāriḍ fī maʿnā maḫkūr fī l-maqāla al-ḫāliḫa min kitāb al-Samā' wa-l-ʿālam [Explanation of an Error Occurring on a Matter Mentioned in Book 3 of the Book On the Heaven and the Worlds]* (ed. Türker 1964).[11]

Averroes (Abū l-Walīd Muḥammad ibn Aḥmad ibn Rushd), *Kitāb al-Samā' wa-l-ʿālam [Epitome of De caelo]*, see Rasā'il ibn-Rushd 1946, repr. ed. Jéhamy 1994 (ed. Eichner 2010).

——, *Talḫīṣ [kitāb] al-Samā' wa-l-ʿālam [Middle Commentary on De caelo]. Talḫīṣ al-Samā' wal-ʿālam* (ed. Jamāl-al-Dīn al-ʿAlawī [Jamal-Eddine Alaoui] 1984).

——, *Šarh (Tafsīr) kitāb al-Samā' wa-l-ālam. [Commentary on Aristotle's Book On the Heaven and the Universe].* Facsimile of ms. 11821, National Library, Tunis (ed. Endress 1994), Latin version by Michael Scotus (ed. Arnzen 2003).

——, *Ma'amar be-ʿEṣem ha-Galgal [De substantia orbis].* Critical edition of the Hebrew text of lost Arabic original with English translation and commentary (ed. Hyman 1986), Latin version by Alvaro de Toledo (ed. Alonso Alonso 1941).

---

[7] Large fragments preserved in ms Paris, Bibliothèque nationale de France, arabe 2281. The translator, a physician of Baghdad who died in 1043, was the author of a number of commentaries of Aristotelian works. The lemmata of Aristotle's text are accompanied by a concise literal commentary in the tradition of the Alexandrian *lexis*, preserving also the division into lecture courses, *pragmata*, Arabic *taʿlīm*.

[8] Abū Sahl, a Christian physician from Khurasan who died in 1010, was a teacher of Avicenna in medicine.

[9] Translated by Gunsalvi of Burgos *ca.* 1280. The subject matter of *De caelo* 3 is treated in the context of the following part of the *Šifā', al-Kawn wa-l-fasād* ("*De generatione et corruptione*"), also edited by Qāsim.

[10] See n. 4 above.

[11] A critique of *De caelo* 3.8 306b3–8.

Mihrān ibn Manṣūr al-Masīḥī, *Kitāb al-Samā' wa-l-'ālam* (ed. Badawi 1961). Revision of Ibn al-Biṭrīq's translation of book 1, dated 1158.[12]

C3. *Modern editions and translations of pre-modern commentaries*

Alaoui, J.-E. (Ǧamāl-al-Dīn al-'Alawī) (1984) *Averroes, Talḫīṣ [kitāb] al-Samā' wa-l-'ālam (Middle Commentary on De caelo). Talḫīṣ al-Samā' wal-'ālam*, Manšūrāt Ǧāmi'at Sīdī Muḥammad ben 'Abdallāh, Kulliyyat al-ādāb wa-l-'ulūm al-insaniyya bi-Fās 1 (Fes).

Alonso Alonso, M. (1941) *Commentario al "De substantia orbis" de Averroes (Aristotelismo e Averroismo) por Alvaro de Toledo* (Madrid).

Arnzen, R. (2003) *Averrois Cordubensis commentum magnum super libro De celo et mundo Aristotelis, ex recognitione Francis James Carmody* (Leuven). Latin translation by Michael Scotus.

Aujac, G.; Brunet, J.-P.; and Nadal, R. (1979) *Autolycos de Pitane. La sphère en mouvement, Levers et couchers héliaques, Testimonia* (Paris).

Badawī, 'Abd-al-Raḥmān (1961) *Arisṭūṭālīs, al-Samā'wa-l-Āṯār al-'ulwiyya* (Cairo). A revision of Ibn al-Biṭrīq' translation of book 1 by Mihrān ibn Manṣūr al-Masīḥī, dated 1158.

Bossier, F.; Vande Veire, C.; and Guldentops, G. (2004) ed. *Simplicius. Commentaire sur le traité du ciel d'Aristote. Traduction de Guillaume de Moerbeke*, vol. 1 (Leuven).

Bowen, A.C. (2003–2008) "Simplicius' Commentary on Aristotle, *De caelo* 2.10–12: An Annotated Translation," *SCIAMVS* 4: 23–58, *SCIAMVS* 9: 25–131.

Capra, L. (1993) *Torquato Tasso. Note di Torquato Tasso a "De Caelo" di Aristotele* (Rome).

Commissio Leonina, (1886) *S. Thomae de Aquino in libros Aristotelis de caelo et mundo expositio*. (Rome/Paris).

Eichner, H. (2010) *Averroes: Kitāb al-Samā' wa-l-'ālam (Epitome of De caelo), Averrois Opera*, ser. arab. 23 (Paderborn).

Endress, G. (1994) *Šarh (Tafsīr) kitāb al-Samā' wa-l-ālam. Commentary on Aristotle's Book* On the Heaven and the Universe, Facsimile ed., reproduced from ms. 11821, National Library, Tunis. Publications of the Institute for the History of Arabic-Islamic Science, Series C: Facsimile editions, vol. 57 (Frankfurt am Main).

Galle, G. (2003) *Petrus de Averni: Questions on Aristotle's De Caelo* (Leuven).

Gutman, O. (2003) *Avicenna, Liber celi et mundi: A Critical Edition with Introduction* (Leiden).

Hankinson, R.J. (2006) *Simplicius: On Aristotle's "On The Heavens 1.10–12"* (Ithaca).

Heiberg, I.L. (1894) *Simplicii in Aristotelis de caelo commentaria*, Commentaria in Aristotelem Graeca 7 (Berlin).

Hossfeld, P. (1971) *Alberti Magni opera omnia edenda curavit Institutum Alberti Magni Coloniense*, v. 1, *De caelo*. (Cologne).

[12] Mihrān was a Christian physician at the Artuqid court of Diyābakr.

Hyman, A. (1986) *Averroes'* De substantia orbis: *Critical Edition of the Hebrew Text with English Translation and Commentary [Ma'mar b-'eṣem ha-galgal le-Ibn Rushd b-targūm 'ibrī]*, Corpus Philosophorum Medii Aevi, *Opera Averrois*, Medieval Academy Books 96 (Cambridge MA/Jerusalem).
Jéhamy, G. (1994) *Rasā'il Ibn Rušd al-falsafiyya, 2: Risālat [!] al-Samā' wa-l-'ālam* (Beirut).
Landauer, S. (1902) *Themistius. In libros Aristotelis de caelo paraphrasis*, Commentaria in Aristotelem Graeca 5 (Berlin).
Menut, A.D. and Denomy, A.J. (1968) *Le Livre du ciel et du monde* (Madison).
Moody, E.A. (1942) *Iohannis Buridani quaestiones super libris quattuor de caelo et mundo* (Cambridge).
Mueller, I. (2004) *Simplicius: On Aristotle's "On The Heavens 2.1–9"* (London).
——, (2005) *Simplicius: On Aristotle's "On The Heavens 2.10–14"* (London).
Parker, B.S. (1968) *Thomas de Bungeye's Commentary on the First Book of Aristotle's* De caelo. PhD Thesis, Tulane University (New Orleans, LA).
Patar, B. (1996) *Ioannis Buridani expositio et quaestiones in Aristotelis de caelo* (Louvain).
——, (2008) *Alberti de Saxonia quaestiones in Aristotelis de caelo* (Louvain).
Renaud, M. (1973) "Le "De celo et mundo" d'Avicenne," *Bulletin de philosophie médiévale* 15: 92–120.
Qāsim, Maḥmūd (1968) *Abū 'Alī Ibn Sīnā (Avicenna), Kitāb al-Šifā' [The Book of the Cure]*, Part II, *al-Ṭabī'iyyāt [Physica], vol. 2: al-Samā' wa-l-'ālam [On the Subject Matter of* De caelo*]* (Cairo).
Raimondo, S. (1952) *Thomas Aquinas. In Aristotelis libros de caelo et mundo, de generatione et corruptione, meterologicorum expositio* (Turin).
Rasa'il ibn Rušd (1946) *Averroes: Epitome of Aristotle's* De caelo (Hyderabad).
Spiazzi, R. M (1952) *S. Thomae de Aquino in libros Aristotelis de caelo et mundo expositio* (Rome/Paris).
Türker, M. (1964) "İbnü 'ṣ-Ṣalaḥ'ın *De Coelo* ve onun şerhleri hakkındaki tenkitleri," *Araştırma* 2: 1–79.
Wildberg, C. (1987) *Philoponus: Against Aristotle, on the Eternity of the World*, (London).
——, (1991) *Simplicius: Against Philoponus on the Eternity of the World* (London).

C4. *Modern editions, translations, and commentaries*

Bekker, I. (1870) *Aristotelis opera* (Berlin).
Dalimier, C. and Pellegrin, P. (2004) *Aristote. Traité du ciel* (Paris).
Elders, L. (1996) *Aristotle's Cosmology: A Commentary on the* De caelo (Assen).
Gigon, O. (1950) *Vom Himmel, Von der Seele, Von der Dichtkunst* (Zurich).
Guthrie, W.K.C. (1939) *On The Heavens*, Loeb Classical Library (Cambridge).
Jori, A. (1999) *Il cielo* (Milan).
Leggatt, S. (1995) *Aristotle: On The Heavens I and II, with an Introduction, Translation and Commentary* (Warminster).
Longo, O. (1961) *Aristotele. De caelo* (Florence).

Moraux, P. (1965) *Aristote. Du ciel* (Paris).
Prantl, C. (1857) *Aristoteles' Vier Bücher über das Himmelsgebäude und Zwei Bücher über Entstehen und Vergehen* (Leipzig).
——, (1881) *De coelo, et de generatione et corruptione* (Leipzig).
Stocks, J.L. (1902) *De Caelo*, in W.D. Ross (ed.) *Works of Aristotle* (Oxford) vol. 2. Repr. in R. McKeon (ed.) *The Basic Works of Aristotle* (New York, 1941) 395–466, and in J. Barnes (ed.) *The Complete Works of Aristotle: The Revised Oxford Translation* (Princeton, 1981) 447–511.
Tricot, J. (1949) *Aristote. Traité du ciel. Traduction et notes*, (Paris).

D. *Modern monographs, collected studies, and articles on Aristotle's* De caelo

Algra, K. (1995) *Concepts of Space in Greek Thought* (Leiden).
Allan, D.J. (1950) "Mediaeval Versions of Aristotle, *De caelo*, and of the Commentary of Simplicius," *Mediaeval and Renaissance Studies* 2: 82–120.
Alonso Alonso, M. (1951) "Ḥunayn traducido al latín por Ibn Dāwūd y Domingo Gundisalvo," *Al-Andalus* 16: 37–47.
Ashley, B.M. (1958) "Aristotle's Sluggish Earth: The Problematics of the 'De caelo'," *Albertus Magnus Lyceum* 18. Repr. in *The New Scholasticism* 32 (1958): 1–32 and 202–234.
Barnes, J. (1981) *The Complete Works of Aristotle: The Revised Oxford Translation* (Princeton).
Bernadete, J. (1957) "Aristotle's Argument from Time," *Journal of Philosophy* 54: 151–152.
Berti, E. (2002) "La causalità del Motore immobile secondo Aristotele," *Gregorianum* 83: 637–654.
Blair, P.S.F. (1988) *Aristotle* De caelo *1.9 279a18–b3: Who Rules the Cosmos?* MA Thesis, University of Texas at Austin.
Bödéus, R. (1990) "La prétendue intuition de Dieu dans le 'De coelo' d'Aristote," *Phronesis* 35: 245–257.
Bodnár, I.M. (1997) "Movers and Elemental Motions in Aristotle," *Oxford Studies in Ancient Philosophy* 16: 81–117.
——, (2005) "Aristotle's Rewinding Spheres: Three Options and Their Difficulties," *Apeiron* 38: 257–275.
Bodnár, I. and Pellegrin, P. (2006) "Aristotle's Physics and Cosmology," in M.L. Gill and P. Pellegrin (edd.) *Blackwell Companions to Philosophy: A Companion to Ancient Philosophy* (Oxford) 270–291.
Bogen, J. and McGuire, J.E. (1987) "Aristotle's Great Clock: Necessity, Possibility, and The Motion of The Cosmos In 'De Caelo I:12'," *Philosophy Research Archives* 12: 387–448.
Bolton, R. (2009) "Two Standards for Inquiry in Aristotle's *De caelo*," in A.C. Bowen and C. Wildberg (edd.) *New Perspectives on Aristotle's* De caelo (Leiden) 51–82.
Bolzan, J.E. (1973) "Aristoteles, 'De caelo,' 310B11–14," *Journal of the History of Philosophy* 11: 443–451.

——, (1981) "Democrito, Aristoteles y la flotacion de los cuerpos," *Dialogos* 16: 109–114.

Bos, A.P. (1988) "'*Mantei*' in Aristotle, 'De caelo' II 1," *Apeiron* 21: 29–54.

Bossier, F. (1975) *Filologisch-historische navorsingen over die middeleeuwse en humanistische Latijnse vertalingen van den commentaren van Simplicius*, 3 vols, Doctoral dissertation (Leuven).

——, (1987) "Traductions latines et influences du commentaire *In de caelo* en Occident (XIII'–XIV' siecles)" in I. Hadot (ed.) *Simplicius, sa vie, son œuvre, sa survie* (Berlin) 289–325.

Bourgey, L. (1975) "Observation and Experiment in Analogical Explanation," in J. Barnes, M. Schofield, and R. Sorabji (edd.) *Articles on Aristotle: 1. Science* (London) 175–182.

Bowen, A.C. and Wildberg, C. (2009) edd. *New Perspectives on Aristotle's* De caelo (Leiden).

Brandis, C.A. (1836) *Scholia in Aristotelem* (Berlin).

Broadie, S. (1982) *Passage and Possibility: A Study of Aristotle's Modal Concepts* (Oxford).

——, (2009) "The Possibilities of Being and Not-Being in *De caelo* 1.11–12," in A.C. Bowen and C. Wildberg (edd.) *New Perspectives on Aristotle's* De caelo (Leiden) 29–50.

Burnyeat, M.F. (2004) "Introduction: Aristotle on the Foundations of Sublunary Physics," in F.A.J. de Haas, and J. Mansfeld (edd.) *Aristotle's* On Generation and Corruption *I*, Symposium Aristotelicum (Oxford) 7–24.

Byrne, C. (2002) "Aristotle on Physical Necessity and the Limits of Teleological Explanation," *Apeiron* 35: 19–46.

Capra, L. (1993) *Torquato Tasso. Note di Torquato Tasso a 'De Caelo' di Aristotele* (Rome).

Carmody, F.J. (1982) *Innovations in Averroes'* De caelo (Berkeley).

Charlton, W. (1983) "Prime Matter—a Rejoinder," *Phronesis* 28: 197–211.

Chroust, A.-H. (1975) "Aristotle *De Caelo* 279a18–35 (or 279b3), a 'Fragment' of the Lost Aristotelian *On Philosophy*," *The Thomist* 39: 332–340.

——, (1977) "Some Observations on Aristotle's Doctrine of the Uncreatedness and Indestructibility of the Universe," *Rivista critica di storia della filosofia* 32: 123–143.

Claghorn, G.S. (1954) *Aristotle's Criticism of Plato's Cosmology* (London).

Cohen, S. "Aristotle on Elemental Motion," *Phronesis* 39: 150–159.

Crowley, T.J. (2005) "On the Use of '*Stoicheion*' in the Sense of 'Element'," *Oxford Studies in Ancient Philosophy* 29: 367–394.

Denyer, N. and Makin, S. (2000) "Aristotle on Modality: II—Nicholas Denyer, Never Will and Cannot," *Aristotelian Society: Supplementary Volume* 163–178.

Dicks, D.R. (1970) *Greek Astronomy from Thales to Aristotle* (London).

Drabkin, I.E. (1935) "Aristotle, *De caelo* 2.6, 288a, 22," *The Classical Weekly* 29: 93–96.

Dreyer, J.L.E. (1906) *History of the Planetary Systems from Thales to Kepler* (Cambridge).

Duhem, P.M.M. (1954–1973) *Le système du monde: histoire des doctrines cosmologiques de Platon à Copernic*. 10 vols. (Paris).
Easterling, H.J. (1961) "Homocentric Spheres in 'De Caelo'," *Phronesis* 6: 138–153.
——, (1964) "*Quinta Natura*," *Museum Helveticum* 21: 73–85.
Effe, B. (1970) *Studien zur Kosmologie und Theologie der Aristotelischen Schrift "Über die Philosophie"* (Munich).
Endress, G. (1966) *Die arabischen Übersetzungen von Aristoteles' Schrift De caelo* (Frankfurt).
——, (1995) "Averroes *De Cælo*: Ibn Rushd's Cosmology in His Commentaries on Aristotle's *On the Heavens*," *Arabic Sciences and Philosophy* 5: 1–41.
——, (2001) "Alexander of Aphrodisias on the First Cause: Aristotle's First Mover in an Arabic Treatise Attributed to Alexander of Aphrodisias," in C. D'Ancona and G. Serra (edd.), *Aristotele e Alessandro di Afrodisia nella tradizione araba* (Padova), 19–74.
Falcon, A. (2001) *Corpi e movimenti: il De caelo di Aristotele e la sua fortuna nel mondo antico* (Napoli).
——, (2005) *Aristotle and the Science of Nature: Unity without Uniformity* (Cambridge).
Federspiel, M. (1995) "Notes sur le traité aristotelicien *Du ciel*," *Revue des Études Anciennes* 97: 505–516.
Fiorentino, F. (1974) "Osservazioni sul commento di S. Tommaso al "De caelo" di Aristotele," *Sapienza* 27: 429–440.
Fobes, F. (1913) "A Preliminary Study of Certain Mss of Aristotle's *Meteorology*," *Classical Review* 27: 249–252.
Freeland, C. (1994) "Aristotle on Perception, Appetition, and Self-Motion," in M.L. Gill and J.G. Lennox (edd.) *Self-Motion from Aristotle to Newton* (Princeton) 35–64.
Freudenthal, G. (2009) "The Astrologization of the Aristotelian Cosmos: Celestial Influences on the Sublunar World in Aristotle, Alexander of Aphrodisias, and Averroes," in A.C. Bowen and C. Wildberg (edd.) *New Perspectives on Aristotle's* De caelo (Leiden) 239–281.
Furley, D. (1994) "Self-Movers," in M.L. Gill and J.G. Lennox (edd.) *Self-Motion from Aristotle to Newton* (Princeton) 3–14.
Gigon, O. (1950) *Vom Himmel, Von der Seele, Von der Dichtkunst* (Zurich).
——, (1952) "Aristoteles-Studien 1," *Museum Helveticum* 19: 113–136.
Gill, M.L. (2009) "The Theory of the Elements in *De caelo* 3 and 4," in A.C. Bowen and C. Wildberg (edd.) *New Perspectives on Aristotle's* De caelo (Leiden) 139–161.
Gill, M.L. and Lennox, J.G. (edd.) (1994) *Self-Motion from Aristotle to Newton* (Princeton).
Glasner, R. (1996) "The Hebrew Version of *De Celo et Mundo* attributed to ibn Sīnā," *Arabic Sciences and Philosophy* 6: 89–112.
Gould, J.B. (1993) "Aristotle on Time and Possibility in 'De Caelo' I.12," *Philosophical Inquiry* 15: 59–74.
Granada, M.A. (2004) "Aristotle, Copernicus, Bruno: Centrality, the Principle

of Movement and the Extension of the Universe," *Studies in History and Philosophy of Science* 35: 91–114.
Grant, E. (1994) "Appendix II: The Anatomy of Medieval Cosmology—The Significance of the 'Catalog of Questions' in Appendix I," in E. Grant, *Planets, Stars, and Orbs: The Medieval Cosmos, 1200–1687* (Cambridge) 742–775.
Hadot, I. (1987) "Recherches sur les fragments du commentaire du Simplicius sur la *Métaphysique* d'Aristote," in I. Hadot (ed.) *Simplicius, sa vie, son oeuvre, sa survie* (Berlin/New York) 225–245.
Hahm, D.E. (1982) "The Fifth Element in Aristotle's *De Philosophia*: A Critical Reexamination," *Journal of Hellenic Studies* 102: 60–74. Repr. in J.P. Anton and A. Preus (edd.) *Essays in Ancient Greek Philosophy* 2 (Albany, 1983) 404–428.
Hall, J.J. (1971) " 'Planets' in Simplicius *De caelo* 471.1 ff.," *The Journal of Hellenic Studies* 91: 138–139.
Hankinson, R.J. (2009) "Natural, Unnatural and Supernatural Motions: Contrariety and the Argument for the Elements in *De caelo* 1.2–4," in A.C. Bowen and C. Wildberg (edd.) *New Perspectives on Aristotle's* De caelo (Leiden) 83–118.
Hanson, N.R. (1963) "On Counting Aristotle's Spheres," *Scientia* 98: 223–232.
——, (1974) *Constellations and Conjectures* (Dordrecht).
Happ, H. (1965) "Kosmologie und Metaphysik bei Aristoteles," in K. Flasch (ed.) *Parusia: Festschrift für Johannes Hirschberger* (Frankfurt/Main) 155–187.
Heath, T.L. (1932) *Greek Astronomy* (New York).
Hintikka, J. (1957) "Necessity, Universality, and Time in Aristotle," *Ajatus* 20: 65–90.
——, (1973) *Time and Necessity: Studies in Aristotle's Theory of Modality* (Oxford).
Hoffman, P. (1987) "Simplicius' Polemics: Some Aspects of Simplicius' Polemical Writings against John Philoponus: From Invective to a Reaffirmation of the Transcendency of the Heavens," in I. Hadot (ed.) *Simplicius, sa vie, son oeuvre, sa survie* (Berlin/New York) 57–83.
Hon, G. and Goldstein, B.R. (2008) "What Keeps the Earth in Its Place? The Concept of Stability in Plato and Aristotle," *Centaurus* 50: 305–323.
Hugonnard-Roche, H. (2003) "Aristote le Stagirite, *De caelo*. Tradition syriaque et arabe," in R. Goulet (ed.) *Dictionnaire des philosophes antique: Supplément* (Paris) 283–294.
Johansen, T. (2009) "From Plato's *Timaeus* to Aristotle's *De caelo*: The Case of the Missing World Soul," in A.C. Bowen and C. Wildberg (edd.) *New Perspectives on Aristotle's* De caelo (Leiden) 9–28.
Judson, L. (1994) "Heavenly Motion and the Unmoved Mover," in M.L. Gill and J.G. Lennox (edd.) *Self-Motion from Aristotle to Newton* (Princeton) 155–171.
——, (1991) ed. *Aristotle's Physics: A Collection of Essays* (Oxford).
——, "Eternity and Necessity in *De caelo* 1.12 (A Discussion of Waterlow, 1982)" *Oxford Studies in Ancient Philosophy* 1: 217–255.
Kahn, C.H. (1985) "The Place of the Prime Mover in Aristotle's Teleology," in A. Gotthelf (ed.) *Aristotle on Nature and Living Things* (Pittsburg) 183–205.
Kalan, V. (2005) "Phenomenological Interpretations of Aristotle's Cosmology," *Phainomena* 14: 97–111.

Kerferd, G.B. (1991) "Aristotle's Treatment of the Doctrine of Parmenides," in H. Blumenthal and H. Robinson (edd.) *Aristotle and the Later Tradition* (Oxford) 1–7.

King, H.R. (1956) "Aristotle without *Prima Materia*," *Journal of the History of Ideas* 17: 370–339.

Kosman, A. (1994) "Aristotle's Prime Mover," in M.L. Gill and J.G. Lennox (edd.) *Self Motion from Aristotle to Newton* (Princeton) 135–153.

Kouremenos, T. (2003) "Infinite Bodies and the Prime Mover in Aristotle's *Phys.* 8. 10," *Philologus* 147: 44–55.

Kronz, F.M. (1989) "Aristotle, the Direction Problem, and the Structure of the Sublunar Realm," *The Modern Schoolman* 67: 247–257.

Lang, H.S. (1998) *The Order of Nature in Aristotle's Physics: Place and the Elements* (Cambridge).

——, (2005) "Aristotle's Science of Nature," in S.V. Astrid (ed.) *Geschichte der Mathematik und der Naturwissenschaften in der Antike: 3. Physik/Mechanik* (Stuttgart) 77–92.

Lennox, J.G. (2009) "De caelo 2.2 and Its Debt to *De incessu animalium*," in A.C. Bowen and C. Wildberg (edd.) *New Perspectives on Aristotle's* De caelo (Leiden) 187–214.

Leunissen, M. (2009) "Why Stars Have No Feet: Explanation and Teleology in Aristotle's Cosmology," in A.C. Bowen and C. Wildberg (edd.) *New Perspectives on Aristotle's* De caelo (Leiden) 215–237.

Liske, M.T. (1995) "In welcher Weise hängen Modalbegriffe und Zeitbegriffe bei Aristoteles zusammen?," *Zeitschrift für Philosophische Forschung* 49: 351–377.

Lloyd, G.E.R. (1996) "Heavenly Abberations," in Lloyd, G.E.R. (ed.) *Aristotelian Explorations* (Cambridge) 160–183.

Longrigg, J. (1970) "Aristotle, *De caelo*: Review of P. Moraux," *The Classical Review* 20: 171–174.

Machamer, P.K. (1978) "Aristotle on Natural Place and Natural Motion," *Isis* 69: 377–387.

Martin, T.-H. (1871) *Mémoire sur les hypothèses astronomiques d'Eudoxe, de Callippe, d'Aristote et de leur école* (Paris).

Martineau, E. (1979) "*Aion* chez Aristote ⟨⟨De caelo⟩⟩ I, 9. Théologie cosmique ou cosmo-théologie?," *Revue de Metaphysique et de Morale* 84: 32–69.

——, (1980) "Response à M. Denis O'Brien," *Revue de Metaphysique et de Morale* 85: 519–528.

Matthen, M. (2001) "Holistic Presuppositions of Aristotle's Cosmology," *Oxford Studies in Ancient Philosophy* 20: 171–199.

——, (2009) "Why Does Earth Move to the Center? An Examination of Some Explanatory Strategies in Aristotle's Cosmology," in A.C. Bowen and C. Wildberg (edd.) *New Essays on Aristotle's* De caelo (Leiden) 119–138.

Matthen, M.P. and Hankinson, R.J. (1993) "Aristotle's Universe: Its Form and Matter," *Synthese* 96: 417–435.

McCue, J.F. (1962) "Scientific Procedure In Aristotle's 'De Caelo'," *Traditio* 18: 1–24.

McKeon, R. (1941) *The Basic Works of Aristotle* (New York).

Merlan, P. (1966) "Two Theological Problems In Aristotle's *Met Λ*: 6–9 and *De Caelo* A1: 9," *Apeiron* 1: 3–13.

Mignucci, M. (1990) "Aristotle's *De Caelo* I,12 and His Notion of Possibility," in D. Devereux and P. Pellegrin (edd.) *Biologie, logique et métaphysique chez Aristote* (Paris) 321–334.

Moraux, P. (1949) "Einige Bemerkungen über den Aufbau von Aristoteles' Schrift *de caelo*," *Museum Helveticum* 6: 157–165.

——, (1951) "Recherches sur le *De caelo* d'Aristote. Objet et structure de l'ouvrage," *Revue Thomiste* 170–196.

——, (1954) "Notes sur la tradition indirecte du 'De caelo' d'Aristote," *Hermes* 82: 145–182.

——, (1961) "La méthode d'Aristote dans l'étude du ciel: *De caelo* I.1–II.12," in S. Mansion (ed.) *Aristote et les problèmes de méthode* (Paris) 173–194.

——, (1963) "Quinta Essentia," in Pauly-Wissowa, *Realencyclopädie der classischen Alterumswissenschaft* 24: 1171–263.

Moraux, P. and Wiesner, J. (2001) *Der Aristotelismus bei den Griechen. Von Andronikos bis Alexander von Aphrodisias* (Berlin).

Morison, B. (2002) *On Location: Aristotle's Concept of Place* (Oxford).

Moss, J.D. (1991) "Ludovico Carbone's Commentary on Aristotle's *De caelo*," in D.O. Dahlstrom (ed.) *Nature and Scientific Method. Essays in Honor of William A. Wallace* (Washington) 169–192.

O'Brien, D. (1980) "Aristote et l' 'Aion'. Enquête sur une critique recente," *Revue de Metaphysique et de Morale* 85: 94–108.

——, (1997) "Aristotle's Theory of Movement" in J.J. Cleary and W. Wians (edd.) *Aristotle's Theory of Movement*, Proceedings of the Boston Area Colloquium in Ancient Philosophy 11 (Lanham) 47–86.

O'Keefe, T. and Thorsrud, H. (2003) "Aristotle's 'Cosmic Nose' Argument for the Uniqueness of the World," *Apeiron* 36: 311–326.

Owen, G.E.L. (1961) "*Tithenai ta phainomena*," in S. Mansion (ed.) *Aristote et les problèmes de méthode* (Paris) 83–103.

——, (1970) "Aristotle: Method, Physics, and Cosmology," in C.C. Gillispie (ed.) *Dictionary of Scientific Biography*, vol. 1 (New York) 250–258.

Palmer, J. (2000) "Aristotle on the Ancient Theologians," *Apeiron* 33: 181–205.

Pellegrin, P. (2009) "The Argument for the Sphericity of the Universe in Aristotle's *De caelo*: Astronomy and Physics," in A.C. Bowen and C. Wildberg (edd.) *New Perspectives on Aristotle's* De caelo (Leiden) 163–185.

Praechter, K. (1924) "Simpl. in Aristot. *de caelo* p. 370,29 ff. H.," *Hermes* 59: 118–119.

Quandt, K. (1983) "'Ai gàr ton enantíon apodeixeis aporíai perì ton enantíon eisin': Philosophical Program and Expository Practice in Aristotle," *Classical Antiquity* 2: 279–298.

——, (1984) "On the Sense and Authenticity of *De caelo* 280A27–34," *Classical Philology* 79: 187–198.

Repellini, F.F. (1980) "*De caelo* of Aristotle as a Dialectic Reply to Plato's *Timaeus*," *Rivista critica di storia della filosofia* 35: 99–126.

van Rijen, J. (1989) *Aspects of Aristotle's Logic of Modalities* (Dordrecht).

Rini, A.A. (2003) "When Time Is of the Essence: *Prior Analytics* I.15 and *De caelo* I.12," *Logique et Analyse* 46: 419–440.

Robinson, H.M. (1974) "Prime Matter in Aristotle," *Phronesis* 19: 168–188.

Roochnik, D. (2006) "Aristotle's Commonsensical Cosmology," in N. Ranasinghe (ed.) *Logos and Eros: Essays Honoring Stanley Rosen* (South Bend) 134–146.

Rosen, E. (1985) "The Dissolution of the Solid Celestial Spheres," *Journal of the History of Ideas* 46: 13–31.

Sandbach, F.H. (1982) "Plutarch and Aristotle," *Illinois Classical Sudies* 7: 207–232.

Sauvé Meyer, S. (1994) "Self-Movement and External Causation," in M.L. Gill and J.G. Lennox (edd.) *Self-Motion from Aristotle to Newton* (Princeton) 65–80.

Savoie, D. (2003) "Problèmes de datation d'une occultation observée par Aristote," *Revue d'histoire des sciences* 56: 493–504.

Seeck, G.A. (1964) *Über die Elemente in der Kosmologie des Aristoteles: Untersuchungen zu "De Generatione et Corruptione" und "De Caelo"* (Munich).

Sharples, R.W. (1976) "Responsibility and the Possibility of More than One Course of Action: A Note on Aristotle *De caelo* II, 12," *Bulletin of the Institute of Classical Studies of the University of London* 23: 69–72.

Sinnige, T.G. (1973) "Cosmic Religion in Aristotle," *Greek, Roman and Byzantine Studies* 14: 15–34.

Sisko, J. (2002) Review of D. Sedley (ed.) *Oxford Studies in Ancient Philosophy* 20 (2001), *Notre Dame Philosophical Reviews* 2002.03.01

Solmsen, F. (1952) "Aristotle De *caelo* A9, 278a28 ff.," *Classical Philology* 47: 94–95.

——, (1958) "Aristotle and Prime Matter: A Reply to Hugh R. King," *Journal of the History of Ideas* 19: 243–252.

——, (1958b) "Aristotle and Presocratic Cosmogony," *Harvard Studies in Classical Philology* 63: 265–282.

——, (1960) *Aristotle's System of the Physical World: A Comparison with His Predecessors* (Ithaca).

Stephenson, F.R. (2000) "A Lunar Occultation of Mars Observed by Aristotle," *Journal for the History of Astronomy* 31: 342–344.

Stocks, J.L. (1902) "De Caelo," in W.D. Ross (ed.) *Works of Aristotle*, vol. 2 (Oxford).

Tarán, L. (1974) "Review of P. Moraux. 1965. *Aristote. Du ciel*. Paris," *Gnomon* 46: 121–142. = L. Tarán, *Collected Papers* 1962–1999 (Leiden 2001), 372–403.

Thorp, J. (1982) "The Luminousness of The Quintessence," *Phoenix* 36: 104–123.

Tigner, S.S. (1974) "Empedocles' Twirled Ladle and the Vortex-Supported Earth," *Isis* 65: 433–447.

Toulmin, S.E. and Goodfield, J. (1963) *The Fabric of the Heavens* (London).

Ugaglia, M. (2004) *Modelli idrostatici di moto da Aristotele a Galileo* (Rome).

Verdenius, W.J. (1969) "Critical and Exegetical Notes on *De caelo*," in I. Düring (ed.) *Naturphilosophie bei Aristoteles und Theophrast* (Heidelberg) 268–284.

Vuillemin, J. (1979) "L'argument dominateur," *Revue de Metaphysique et de Morale* 84: 225–257.

Wallace, A. (1977) *Galileo's Early Notebooks: The Physical Questions. A Translation from the Latin, with Historical and Paleographical Commentary* (Notre Dame).

Waterlow, S.: see Broadie, S.

Wieland, W. (1960) "Die Ewigkeit der Welt," in D. Henrich, W. Schulz and K.-H. Volkmann-Schluck (edd.) *Die Gegenwart der Griechen in neuerer Denken: Festschrift für H.G. Gadamer* (Tübingen).

Wildberg, C. (1988) *John Philoponus' Criticism of Aristotle's Theory of Aether*, Peripatoi 16, (Berlin/New York).

Williams, C.J.F. (1966) "Aristotle and Corruptibility: A Discussion of Aristotle *De Caelo* I, XII (Part I)," *Religious Studies* 1: 95–108.

Wolfson, H.A. (1958) "The Plurality of Immovable Movers in Aristotle and Averroës," *Harvard Studies in Classical Philology* 63: 233–253.

Zahlfleisch, J. (1897) "Die Polemik des Simplicius gegen Alexander und andere in dem Commentar des ersteren zu der Aristotelischen Schrift *de Caelo*," *Archiv für Geschichte der Philosophie* 10: 191–227.

INDEX OF PASSAGES CITED

INDEX OF SUBJECTS

INDEX OF MODERN AUTHORS

INDEX OF ANCIENT AND MEDIEVAL AUTHORS

PHILOSOPHIA ANTIQUA

A SERIES OF STUDIES ON ANCIENT PHILOSOPHY

EDITED BY

K.A. ALGRA, F.A.J. DE HAAS, J. MANSFELD
C.J. ROWE, D.T. RUNIA, Ch. WILDBERG

Recent volumes in the series

13. Nicolaus Damascenus. *On the Philosophy of Aristotle.* Fragments of the First Five Books, Translated from the Syriac with an Introduction and Commentary by H. J. Drossaart Lulofs. Reprint of the 1st (1965) ed. 1969. ISBN 90 04 01725 9
14. Edelstein, L. *Plato's Seventh Letter.* 1966. ISBN 90 04 01726 7
17. Gould, J. B. *The Philosophy of Chrysippus.* Reprint 1971. ISBN 90 04 01729 1
18. Boeft, J. den. *Calcidius on Fate.* His Doctrine and Sources. 1970. ISBN 90 04 01730 5
20. Bertier, J. *Mnésithée et Dieuchès.* 1972. ISBN 90 04 03468 4
21. Timaios Lokros. *Über die Natur des Kosmos und der Seele.* Kommentiert von M. Baltes. 1972. ISBN 90 04 03344 0
23. Iamblichus Chalcidensis. *In Platonis dialogos commentariorum fragmenta.* Edited with Translation and Commentary by J. M. Dillon. 1973. ISBN 90 04 03578 8
24. Timaeus Locrus. *De natura mundi et animae.* Überlieferung, Testimonia, Text und Übersetzung von W. Marg. Editio maior. 1972. ISBN 90 04 03505 2
26. Gersh, S. E. *Κίνησις ἀκίνητος.* A Study of Spiritual Motion in the Philosophy of Proclus. 1973. ISBN 90 04 03784 5
27. O'Meara, D. *Structures hiérarchiques dans la pensée de Plotin.* Étude historique et interprétative. 1975. ISBN 90 04 04372 1
28. Todd, R. B. *Alexander of Aphrodisias on the Stoic Physics.* A Study of the *De Mixtione* with Preliminary Essays, Text, Translation and Commentary. 1976. ISBN 90 04 04402 7
29. Scheffel, W. *Aspekte der platonischen Kosmologie.* Untersuchungen zum Dialog 'Timaios'. 1976. ISBN 90 04 04509 0
31. Edlow, R.B. *Galen on Language and Ambiguity.* An English Translation of Galen's *De Captionibus* (On Fallacies), With Introduction, Text and Commentary. 1977. ISBN 90 04 04869 3
34. Epiktet. *Vom Kynismus.* Herausgegeben und übersetzt mit einem Kommentar von M. Billerbeck. 1978. ISBN 90 04 05770 6
35. Baltes, M. *Die Weltentstehung des platonischen Timaios nach den antiken Interpreten.* Teil 2. Proklos. 1979. ISBN 90 04 05799 4
39. Tarán, L. *Speusippus of Athens.* A Critical Study with a Collection of the Related Texts and Commentary. 1982. ISBN 90 04 06505 9
41. O'Brien, D. *Theories of Weight in the Ancient World.* Four Essays on Democritus, Plato and Aristotle. A Study in the Development of Ideas 2. Plato: Weight and Sensation. The Two Theories of the 'Timaeus'.1984.ISBN 90 04 06934 8
45. Aujoulat, N. *Le Néo-Platonisme Alexandrin: Hiéroclès d'Alexandrie.* Filiations intellectuelles et spirituelles d'un néo-platonicien du Ve siècle. 1986. ISBN 90 04 07510 0
46. Kal, V. *On Intuition and Discursive Reason in Aristotle.* 1988. ISBN 90 04 08308 1
48. Evangeliou, Ch. *Aristotle's Categories and Porphyry.* 1988. ISBN 90 04 08538 6

49. Bussanich, J. *The One and Its Relation to Intellect in Plotinus.* A Commentary on Selected Texts. 1988. ISBN 90 04 08996 9
50. Simplicius. *Commentaire sur les Catégories.* Traduction commentée sous la direction de I. Hadot. I: Introduction, première partie (p. 1-9, 3 Kalb fleisch). Tra duc tion de Ph. Hoffmann (avec la collaboration d'I. et P. Hadot). Commentaire et notes à la traduction par I. Hadot avec des appendices de P. Hadot et J.-P. Mahé. 1990. ISBN 90 04 09015 0
51. Simplicius. *Commentaire sur les Catégories.* Traduction commentée sous la direction de I. Hadot. III: Préambule aux Catégories. Commentaire au premier chapitre des Catégories (p. 21-40, 13 Kalbfleisch). Traduction de Ph. Hoffmann (avec la collaboration d'I. Hadot, P. Hadot et C. Luna). Commentaire et notes à la traduction par C. Luna. 1990. ISBN 90 04 09016 9
52. Magee, J. *Boethius on Signification and Mind.* 1989. ISBN 90 04 09096 7
54. Fortenbaugh, W. W., et al. (eds.) *Theophrastes of Eresos.* Sources for His Life, Writings, Thought and Influence. 1992. ISBN 90 04 09440 7 *set*
55. Shankman, A. *Aristotle's* De insomniis. A Commentary. ISBN 90 04 09476 8
56. Mansfeld, J. *Heresiography in Context.* Hippolytos' *Elenchos* as a Source for Greek Philosophy. 1992. ISBN 90 04 09616 7
57. O'Brien, D. *Théodicée plotinienne, théodicée gnostique.* 1993. ISBN 90 04 09618 3
58. Baxter, T. M. S. *The Cratylus.* Plato's Critique of Naming. 1992. ISBN 90 04 09597 7
59. Dorandi, T. (Hrsg.) *Theodor Gomperz. Eine Auswahl herkulanischer kleiner Schriften (1864-1909).* 1993. ISBN 90 04 09819 4
60. Filodemo. *Storia dei filosofi. La stoà da Zenone a Panezio* (PHerc.1018). Edizione, traduzione e commento a cura di T. Dorandi. 1994. ISBN 90 04 09963 8
61. Mansfeld, J. *Prolegomena.* Questions to be Settled Before the Study of an Author, or a Text. 1994. ISBN 90 04 10084 9
62. Flannery, s.j., K.L. *Ways into the Logic of Alexander of Aphrodisias.* 1995. ISBN 90 04 09998 0
63. Lakmann, M.-L. *Der Platoniker Tauros in der Darstellung des Aulus Gellius.* 1995. ISBN 90 04 10096 2
64. Sharples, R.W. *Theophrastus of Eresus.* Sources for his Life, Writings, Thought and Influence. Commentary Volume 5. Sources on Biology (Human Physiology, Living Creatures, Botany: Texts 328-435). 1995. ISBN 90 04 10174 8
65. Algra, K. *Concepts of Space in Greek Thought.* 1995. ISBN 90 04 10172 1
66. Simplicius. *Commentaire sur le manuel d'Épictète.* Introduction et édition critique de texte grec par Ilsetraut Hadot. 1995. ISBN 90 04 09772 4
67. Cleary, J.J. *Aristotle and Mathematics.* Aporetic Method in Cosmology and Metaphysics. 1995. ISBN 90 04 10159 4
68. Tieleman, T. *Galen and Chrysippus on the Soul.* Argument and Refutation in the *De Placitis* Books II-III. 1996. ISBN 90 04 10520 4
69. Haas, F.A.J. de. *John Philoponus' New Definition of Prime Matter.* Aspects of its Background in Neoplatonism and the Ancient Commentary Tradition. 1997. ISBN 90 04 10446 1
71. Andia, Y. de. *Henosis.* L'Union à Dieu chez Denys l'Aréopagite. 1996. ISBN 90 04 10656 1
72. Algra, K.A., Horst, P.W. van der, and Runia, D.T. (eds.) *Polyhistor.* Studies in the History and Historiography of Ancient Philosophy. Presented to Jaap Mansfeld on his Sixtieth Birthday. 1996. ISBN 90 04 10417 8
73. Mansfeld, J. and Runia, D.T. *Aëtiana.* The Method and Intellectual Context of a Doxographer. Volume 1: The Sources. 1997. ISBN 90 04 10580 8
74. Slomkowski, P. *Aristotle's* Topics. 1997. ISBN 90 04 10757 6
75. Barnes, J. *Logic and the Imperial Stoa.* 1997. ISBN 90 04 10828 9
76. Inwood, B. and Mansfeld, J. (eds.) *Assent and Argument.* Studies in Cicero's *Academic Books.* Proceedings of the 7th Symposium Hellenisticum (Utrecht, August 21-25, 1995). 1997. ISBN 90 04 10914 5

77. Magee, J. (ed., tr. & comm.) *Anicii Manlii Severini Boethii* De divisione liber. Critical Edition, Translation, Prolegomena, and Commentary. 1998. ISBN 90 04 10873 4
78. Olympiodorus. *Commentary on Plato's* Gorgias. Translated with Full Notes by R. Jackson, K. Lycos & H. Tarrant. Introduction by H. Tarrant. 1998. ISBN 90 04 10972 2
79. Sharples, R.W. *Theophrastus of Eresus.* Sources for his Life, Writings, Thought and Influence. Commentary Volume 3.1. Sources on Physics (Texts 137-223). With Contributions on the Arabic Material by Dimitri Gutas. 1998. ISBN 90 04 11130 1
80. Mansfeld, J. *Prolegomena Mathematica.* From Apollonius of Perga to Late Neoplatonism. With an Appendix on Pappus and the History of Platonism. 1998. ISBN 90 04 11267 7
81. Huby, P. *Theophrastus of Eresus.* Sources for His Life, Writings, Thought and Influence. Commentary Volume 4. Psychology (Texts 254-327). With Contributions on the Arabic Material by D. Gutas. 1999. ISBN 90 04 11317 7
82. Boter, G. *The* Encheiridion *of Epictetus and Its Three Christian Adaptations.* Transmission and Critical Editions. 1999. ISBN 90 04 11358 4
83. Stone, M.E. and Shirinian, M.E. *Pseudo-Zeno. Anonymous Philosophical Treatise.* Translated with the Collaboration of J. Mansfeld and D.T. Runia. 2000. ISBN 90 04 11524 2
84. Bäck, A.T. *Aristotle's Theory of Predication.* 2000. ISBN 90 04 11719 9
85. Riel, G. Van. *Pleasure and the Good Life.* Plato, Aristotle, and the Neoplatonists. 2000. ISBN 90 04 11797 0
86. Baltussen, H. *Theophrastus against the Presocratics and Plato.* Peripatetic Dialectic in the *De sensibus*. 2000/ ISBN 90 04 11720 2
87. Speca, A. *Hypothetical Syllogistic and Stoic Logic.* 2001. ISBN 90 04 12073 4
88. Luna, C. *Trois Études sur la Tradition des Commentaires Anciens à la Métaphysique d'Aristote.* 2001. ISBN 90 04 120074 2
89. Frede, D. & A. Laks (eds.) *Traditions of Theology.* Studies in Hellenistic Theology, its Background and Aftermath. 2001. ISBN 90 04 12264 8
90. Berg, R.M. van den. *Proclus' Hymns.* Essays, Translations, Commentary. 2001. ISBN 90 04 12236 2
91. Rijk, L.M. de. *Aristotle – Semantics and Ontology.* 2 volumes. Volume I. General Introduction. The Works on Logic. 2002. ISBN 90 04 12324 5 Volume II. The Metaphysics. Semantics in Aristotle's Strategy of Argument. 2002. ISBN 90 04 12467 5
92. Finamore, J.F. & J.M. Dillon. *Iamblichus* De Anima. Text, Translation, and Com men tary. 2002 ISBN 90 04 12510 8
93. Fortenbaugh, W.W., R.W. Sharples, & M.G. Sollenberger. *Theophrastus of Eresus.* On Sweat, on Dizziness and on Fatigue. 2003. ISBN 90 04 12890 5
94. Tieleman, T. *Chrysippus'* On affections. Reconstruction and Interpretation. 2003. ISBN 90 04 12998 7
95. Görler, W. *Kleine Schriften zur hellenistisch-römischen Philosophie.* Herausgegeben von C. Catrein. 2004. ISBN 90 04 13736 X
96. Polito, R. *The Sceptical Road.* Aenesidemus' Appropriation of Heraclitus. 2004. ISBN 90 04 13742 4
97. Fortenbaugh, W.W. *Theophrastus of Eresus.* Sources for his Life, Writings, Thought and Influence. Commentary Volume 8. Sources on Rhetoric and Poetics (Texts 666-713). 2005. ISBN 90 04 14247 9
98. Perkams, M. & R.M. Piccione (Hrsg.) *Proklos. Methode, Seelenlehre, Metaphysik.* Akten der Konferenz in Jena am 18.-20. September 2003. 2006. ISBN 90 04 15084 6
99. Schäfer, C. *Philosophy of Dionysius the Areopagite.* An Introduction to the Structure and the Content of the Treatise *On the Divine Names.* 2006. ISBN 90 04 15094 3

100. Delcomminette, S. *Le* Philèbe *de Platon*. Introduction à l'Agathologie Platonicienne. 2006. ISBN 90 04 15026 9
101. Fortenbaugh, W.W. *Aristotle's Practical Side*. On his Psychology, Ethics, Politics and Rhetoric. 2006. ISBN-10: 90 04 15164 8, ISBN-13: 978 90 04 15164 2
102. Brancacci, A. & P.-M. Morel (eds.). *Democritus: Science, The Arts, and the Care of the Soul*. Proceedings of the International Colloquium on Democritus (Paris, 18-20 September 2003). 2007. ISBN-10: 90 04 15160 5, ISBN-13: 978 90 04 15160 4
103. Huby, P. *Theophrastus of Eresus*. Sources for his Life, Writings, Thought and Influence. Commentary Volume 2. Logic. With Contributions on the Arabic Material by Dimitri Gutas. 2007. ISBN-10: 90 04 15298 9, ISBN-13: 978 90 04 15298 4
104. Sider, D. & C.W. Brunschön (eds.). *Theophrastus of Eresus*. On Weather Signs. 2007. ISBN-10: 90 04 15593 7, ISBN-13: 978 90 04 15593 0
105. Mirhady, D.C. (ed.). *Influences on Peripatetic Rhetoric*. Essays in Honor of William W. Fortenbaugh. 2007. ISBN 978 90 04 15668 5
106. Bobonich, C. & P. Destrée (eds.). Akrasia in *Greek Philosophy*. From Socrates to Plotinus. 2007. ISBN 978 90 04 15670 8
107. D'Ancona, C. (ed.). *The Libraries of the Neoplatonists*. Proceedings of the Meeting of the European Science Foundation Network "Late Antiquity and Arabic Thought. Patterns in the Constitution of European Culture" held in Strasbourg, March 12-14, 2004 under the impulsion of the Scientific Committee of the meeting, composed by Matthias Baltes†, Michel Cacouros, Cristina D'Ancona, Tiziano Dorandi, Gerhard Endreß, Philippe Hoffmann, Henri Hugonnard Roche. 2007. ISBN 978 90 04 15641 8
108. Bonelli, M. *Timée le Sophiste: Lexique Platonicien*. 2007. ISBN 978 90 04 15887 0
109. Deslauriers, M. *Aristotle on Definition*. 2007. ISBN 978 90 04 15669 2
110. Bloch, D. *Aristotle on Memory and Recollection*. Text, Translation, Interpretation, and Reception in Western Scholasticism. 2007. ISBN 978 90 04 16046 0
111. Roskam, G. *'Live unnoticed'*. On the Vicissitudes of an Epicurean Doctrine. 2007. ISBN 978 90 04 16171 9
112. Berg, R.M. van den. *Proclus' Commentary on the Cratylus in Context*. Ancient Theories of Language and Naming. 2008. ISBN 978 90 04 16379 9
113. Eliasson, E. *The Notion of* That Which Depends on Us *in Plotinus and Its Background*. 2008. ISBN 978 90 04 16614 1
114. Mansfeld, J. and Runia, D.T. *Aëtiana*. The Method and Intellectual Context of a Doxographer. Volume II: The Compendium. 2 Parts. 2009. ISBN 978 90 04 17206 7 *set*
115. Chiaradonna, R. & F. Trabattoni (eds.). *Physics and Philosphy of Nature in Greek Neoplatonism*. Proceedings of the European Science Foundation Exploratory Workshop (Il Ciocco, Castelvecchio Pascoli, June 22–24, 2006). 2009. ISBN 978 90 04 17380 4
116. Calzolari, V. et J. Barnes (réd.). *L'oeuvre de David l'Invincible et la transmission de la pensée grecque dans la tradition arménienne et syriaque*. 2010. ISBN 978 90 04 16047 7
117. Bowen, A.C. and C. Wildberg (eds.). *New Perspectives on Aristotle's* De caelo. 2010. ISBN 978 90 04 17376 7
118. Mansfeld, J. and Runia, D.T. *Aëtiana*. The Method and Intellectual Context of a Doxographer. Volume III: Studies in the Doxographical Traditions of Ancient Philosophy. 2010. ISBN 978 90 04 18041 3

www.ingramcontent.com/pod-product-compliance
Lightning Source LLC
LaVergne TN
LVHW010543300926
849162LV00002B/222